Toward a Global Science

RACE, GENDER, AND SCIENCE
Anne Fausto-Sterling, *General Editor*

TOWARD A
GLOBAL SCIENCE

Mining Civilizational Knowledge

Susantha Goonatilake

INDIANA UNIVERSITY PRESS BLOOMINGTON & INDIANAPOLIS

This book is a publication of

Indiana University Press
601 North Morton Street
Bloomington, Indiana 47404-3797 USA

www.indiana.edu/~iupress

Telephone orders 800-842-6796
Fax orders 812-855-7931
Orders by e-mail iuporder@indiana.edu

The paper used in this publication meets the minimum require-
ments of American National Standard for Information Sciences
—Permanence of Paper for Printed Library Materials, ANSI
Z39.48-1984.

Manufactured in the United States of America

Library of Congress Cataloging-in-Publication Data

Goonatilake, Susantha.
    Toward a global science : mining civilizational knowledge /
Susantha Goonatilake.
        p.   cm. — (Race, gender, and science)
    Includes bibliographical references and index.
    ISBN 0-253-33388-1 (cl : alk. paper). — ISBN 0-253-21182-4
(pa : alk. paper)
    1. Science—Asia—History—15th century. 2. Science—Eu-
rope—History—20th century. 3. Asia—Intellectual life—History—15th century. 4. Europe—Intellectual life—History—20th
century. 5. Europeans—Cultural assimilation—Asia—History—
20th century.   I. Title.   II. Series.
Q127.A65G66   1998
509.5—dc21   98-43205

1 2 3 4 5 03 02 01 00 99 98

Contents

SECTION 3
MORE IMAGINATIVE EXPLORATIONS

Preface

During the last few centuries, since the European Renaissance, the scientific enterprise has emerged as perhaps the major intellectual movement in the world. In very many ways, it is the most powerful. Initially, the new thrust of science was fed by many sources: the earlier Greek tradition, the Arabs and, as recent research is unearthing, other civilizations further afield, such as those of South Asia and East Asia. But since the sixteenth century, this has been largely a European exercise, with names like Galileo, Copernicus, Kepler, Brahe, Newton, Leibniz, Linnaeus, Darwin, Wallace, Hertz, Maxwell, Einstein, Schrödinger, and Bohr as key figures.

The demographics of the scientific practitioners of today (and tomorrow) are changing this imbalance. Already, in developing countries like India and China there are more people with scientific training than there are in most developed countries. Scientists in large developing countries are, for the moment, less productive and creative than their counterparts in developed countries, for a variety of institutional and other sociological reasons. But these gaps are being overcome with globalization and instant communication—witness the emergence of India as a superpower of software development in less than ten years of concentrated effort. Just as importantly, laboratories and universities in developed countries are being increasingly staffed by migrants from Asia, while the majority of graduate students in science and technology at prestigious American institutions are Asian. The key names in future discoveries would increasingly reflect this multiple ethnic mix. As the scientific enterprise takes hold across the globalized world, there is no doubt that due purely to demographics, future Nobel Prize winners, if not future Einsteins and Darwins, are going to emerge outside the European roster. Already the yearly count of Nobel Prize winners reflects this shift, non-European winners being largely Asians working in the US.

But in these activities, Asian workers are operating no differently from their European counterparts. If they have just emigrated from Asia, they lead somewhat schizophrenic lives, their thought processes still tied in large part to their original culture. If they are the children of migrants, then for all purposes they have been socialized within a European milieu, and their internal culture is very little different from their counterparts of European origin.

This book is dedicated to bringing a different type of non-European dimension to science than these demographically generated ones. It is driven by a strong conviction that the pre-Renaissance acquisition of aspects of Asian knowledge—for example, South Asian algebra and arithmetic, Chinese printing, gunpowder, and compass—did not exhaust the contribution that non-European civilizations can give to science. In fact, there are elements of valid knowledge, still lying in what I call civilizational stores, that can be grafted onto the contemporary scientific enterprise.

The legitimacy of this task has been made easier recently by the recognition that similar stores also exist among the simplest of social groupings, such as those of forest dwellers. The quest for the ethnomedical and ethnobiological knowledge of such peoples is part of the current scientific enterprise. But here I extend the quest to much more sophisticated realms accessible only through the accumulation of knowledge through formal collection, recording, and debate—namely usable knowledge in civilizations.

For reasons of cultural familiarity, I will take South Asia as an example of a civilization to be tapped. Similar exercises can undoubtedly be done for the cases of West Asia and East Asia, among other candidates. In dealing with South Asia, I draw upon the historical tradition of the region roughly between the river basins of the Indus (whence the names "Hindu" and "Indian") and Ganges, including the Himalayas to the north and Sri Lanka to the south. The area covers cultural elements that historically arose from both endogenic transactions and interactions with regions outside, and correspond to those state entities known today as Pakistan, India, Nepal, Bangladesh, and Sri Lanka. I also include geographic elements that lie outside these boundaries but are part of this broad culture, including countries or regions like Tibet, Burma, and Southeast Asia. I often use the word "Indian" in reference to these cultural entities. While this is not valid politically or geographically today, it is a convenient shorthand, in the broad original sense of lands more or less bounded by the Indus. For the record, I myself am not a citizen of the geographic and political entity known today as India.

Several belief and knowledge systems now straddle this area, but I concentrate only on the system which arose primarily out of South Asia. An important element in the history of science is that of Islam, which constitutes a vibrant cultural factor in the region; however, I deal with Islamic contributions only in terms of the past and as transmitters from South Asia. I do not deal with their current potential, which I believe could be very significant.

The book has three broad sections. First, a brief introduction states the epistemological assumptions of the book and explores the knotty problems of attempting to recover scientific elements from past non-Western traditions. I especially distinguish the secular efforts I advocate in the book from attempts such as those which falsely claim that all knowledge systems are equivalent, and from purely religion-driven attempts like some aspects of the "Islamic Science" movement.

In the three chapters of section 1, I set the framework for the mining exercise by viewing science without Eurocentric blinkers. Proposing a model of the civilizational construction of science, I examine how the modern scientific enterprise was built by transfers from other civilizations, apart from Europe's own contributions. I also show that from a comparative perspective, much of the given historiography of science has to be rethought, especially key events and their social impacts.

Section 2, on "mining" for contemporary uses, includes an introductory chapter on the uses of noncivilizational, non-Western knowledge —that is, indigenous knowledge—in the modern knowledge enterprise. I then describe how in the last decade, South Asian civilizational input has either become incorporated or shows promise and potential in three specific areas of the scientific enterprise.

In section 3, I make further, more imaginative explorations, touching on cases where South Asian ideas have been used or have the potential to be used in three areas. I explore how South Asian input can be useful in navigating the philosophical and ethical problems raised by two technologies that will dominate the future, namely biotechnology and information technology. I also show how a fruitful marriage of one aspect of this technology—virtual reality—with South Asian philosophy can enliven both the technology and the philosophy. I also examine how various South Asian positions can be used to feed key discussions on science with underpinnings in philosophy.

And, in a final chapter, I look at the role of South Asian metaphors for "mining" and attempt to evaluate what the future holds for mining civilizational knowledge.

As an exercise in dispelling the ghosts of Eurocentrism, throughout the book I give examples of "parallels and antecedents" to illustrate that several European contributions, hitherto assumed unique, had surfaced either previously or contemporaneously in South Asia. This is primarily an activity in removing the pervasive Eurocentric bias that guides today's dominant perspective on science.

In this book, I forage in areas outside my formally trained fields of engineering and sociology. But in many of these efforts, I have tried to keep to fields that I had formally explored elsewhere for other professional ends. Thus, I have explored the prospects for biotechnology and information technology for several UN agencies. I have also written on evolutionary theory and interacted professionally with those now attempting to build general models of evolution. These different streams have fed my writing here. In addition, I also bring in more imaginative and speculative material, carried over from my general reading and other interests. I attempt to use this material with the same rigor that I bring to my more formal interests.

Toward a Global Science

ONE

INTRODUCTION

Prescribing programs for science can lead to misperceptions of one's position. It is perhaps a hazardous task. So this introductory chapter is both a disclaimer of positions that may be attributed to me as well as a statement of what I see as possibilities in expanding scientific knowledge.

First, to state my position clearer, I hope I am allowed a biographical note. Before I started a second career in the social sciences, I had practiced in the physical sciences. Trained as an engineer in my own country (in a system cloned from the British) as well in Britain and Germany, I have designed, constructed, operated, and repaired or helped perform projects in electrical engineering and the cement industry. In the case of a power outage in my Third World country of Sri Lanka, it would be my duty—with consumers screaming curses—to get the system going again or, if a factory burned down, to get it functioning in the shortest time. So I am fully wedded to science-based technologies (and their constituent sciences) that work. If something does not work, I should know why, so that I can deliver a satisfactory product or service. And if something goes wrong due to my ignorance, I will be held accountable, ultimately in a court of law: if I were a civil engineer and designed a building that collapsed and killed people, or if I were a medical practitioner whose patient died due to my malpractice, I would be again held accountable.

As this aside indicates, modern science's applications work, and their efficacy is held to be valid in the realm not only of the "laws" of nature, but also of human law. Modern science and many of its applied fields, like engineering and medicine, are based on testable laws and are expected to be governed rigorously. But as a student of the sociology of science and technology, I also know that sciences came into their present form through a set of particular social pressures.

Yet science, whatever its social, political, psychological, or philosophical roots, is ultimately "that which works." To simplify a bit, it is a black box into which one puts a question—and out comes a reasonable and testable answer. Whatever the trajectory of the present delivery system of science, this is the ultimate test. Consequently, all knowledge systems cannot be equivalent. I do not believe that a forest dweller's knowledge of laws of motion comes anywhere near that of a Galileo, Copernicus, Kepler, or Newton. To hold such a position is, at the most, romantic ignorance. But I do believe that the forest dweller has a deep knowledge of the forest, its flora and fauna. I believe, with many anthropologists who have studied the problem, that the intellectualism that drives the forest dweller is in an ultimate sense not different from that of a Newton. He or she happens to have faced a different set of problems, a different set of historical givens, and come to different positions. And I also believe that if Newton were to be magically transported to the forest, he would learn much systematic, testable knowledge of plants from the local groups, and he would have little to contribute on the subject in return. On the other hand, if we were to transport a Linnaeus to the forest, there would indeed arise some enlivened talk on categorizing plants.

But if we shift these hypothetical meetings to civilizational entities, where formal systems for knowledge gathering and transmission exist, there would be deeper discussions. If Leibniz, Newton's contemporary and rival, were to be transported to twelfth-century China, he would find much in common with his concepts of monads and point-moments, as Joseph Needham has documented. Similarly, if Leibniz were to be transported to fifth-century Sri Lanka, where atomic theories of time were elaborated in great detail, he would again have invigorating discussions on the nature of time. Or, returning to more modern times, Mach, the philosopher who influenced Einstein the most, would recognize South Asian positions on epistemology—as indeed he discovered on his own. Cognitive scientists, evolutionists, or computer specialists would find familiar ideas in the South Asian scene, as some have indeed already found out.

In a recent public statement, US scientists warn of a creeping relativism in dealing with sciences whereby all knowledge systems are held to be equivalent; the statement also complains of a parallel return to obscurantism through multicultural courses and postmodernism.[1] Being a cosmopolitan person, speaking several languages and having lived, studied, and worked in different cultural milieus, I am fully wedded to multiculturalism. But there are shades of differences in the use of the word "multicultural," especially when applied to the sciences. Let me digress.

I have visited, say, Athens, Angkor Wat, and the Museum of Modern Art in New York. I found the Greek sculpture boring in its naturalness, while works of art at the other two locales struck deep responsive chords. I found especially congenial some of the West African-influenced pieces by a Picasso or a Braque at MOMA. Later, in West Africa itself, I saw some of the types of masks that influenced Cubism. But this does not mean that given a choice of mathematics, I would select a West African system (which writers like Zaslavasky have documented) instead of the Euclidean system to do my scientific problems. In another setting, if a sophisticated non-Euclidean geometry were to emerge from these West African ideas and, if it were consonant with the scientific task at hand, I would choose it. So multiculturalism in science and multiculturalism in the humanities are two different things. In the case of sciences, everyday reality is always the reference point: it should work. Art does not have to "work"; it does not solve problems about nature, except in an indirect, allusory sort of way.

Further, my reading of postmodernism is not just as a system (or more appropriately a nonsystem) of thought. To me, postmodernism is a reflection of the zeitgeist of today's Western intellectual life, which is seemingly exhausted. The initial thrust of the Enlightenment has lost both its vigor and its intellectual certainty. Current problems about the limits of tinkering with the environment have laid to rest some of the simplistic readings of Bacon's writings on science as the torturer of nature. Current epistemological problems in several fields—including the seeming fountainhead of them all, physics—have questioned the Cartesian dichotomy of subject and object. And the project of mathematizing the "true knowledge" of science and completing the rational project at a full foundational level has also collapsed, because mathematics, after Gödel, has lost its earlier assumed certainty.

So the Enlightenment and modern project is floundering, not because of postmodernist relativism, but on the terms set by the project itself. As for postmodernism, I find the writings insightful, but at times the authors are boring and trivial, seeming to waste hundreds of pages to prove some rather simple points. If more serious stuff in that philosophical genre is needed, required reading would be, among others, Nagarjuna (first century), Aryadeva (first century) and Bhartuhari (eighth century)—three South Asians. Compared to some of the trivialities of postmodernism, here lies real meat in the same broad genre.

So we have to read a different message into these postmodern times of the Western psyche: the modern agenda has run out of steam, and postmodernism is both a symptom and a reflection of this exhaustion. It is not an agenda to replace modernism, which would be a contradiction in terms. The new agenda, instead, has to come from the remnants of the earlier certainties that still have validity and the new certainties that can come from other cultures, including other civilizational spheres.

In addition, on the threshold of the twenty-first century, we are faced with a set of new and different challenges in the scientific sphere. Science is not what it seemed at the beginning of its journey in Newton's time. Since then—taking into consideration the number of scientists working and the number of scientific journals or other fora for the dissemination of discoveries—science has been growing exponentially, doubling its population of practitioners and output every few years.[2] If science proceeds at this pace, scientists would soon outnumber the rest of the human population, and the weight of all the scientific papers ever published would be greater than that of the earth.[3] The result is that in terms of present practice, certain quantitative growth limits will be reached in science within the next generation or so. These quantitative changes will require qualitative shifts in the nature of science. Twenty-first-century science is not going to be seventeenth-century science.

Part of the solution to this quantitative problem is to let the machine-computer sphere take over some the work now done in the human sphere. This is part of a trend in human problem-solving, which has always sought shortcuts, replacing once cumbersome procedure with simpler, more powerful ones. Thus there were changes in scientific productivity from the Greco-Roman system of calculation, to the place-value arithmetic of the Renaissance, through logarithms, through the slide rule and to the computer. All the calculations that Newton and others of his time took years to do by hand are today done at the touch of a button by my teenaged nephew's astronomy program: almost instantaneously, he can see the conjunctions and outcomes of the planetary system for eons to come, from any vantage point in space or time —by means of a software that in a developed country today costs the same as a meal in a restaurant. Or to take another example, at the click of a button, a whole set of individual equations and problems—say, the floor design of a multi-storied building, or expected results from a particle accelerator—can be produced. In a similar fashion, the use of automation in laboratories is extending experimentation to the machine sphere; Galileo is no longer dropping objects from the tower of Pisa. Library research is also changing, with a tremendous increase in productivity because of the availability of extensive databases.

In the eighteenth century, Diderot and the Encyclopedists had the then farfetched dream of recording all knowledge.[4] Speaking of his encyclopedia, Dennis Diderot wished:

> to collect all the knowledge scattered over the face of the earth, to present its general outlines and structure to the men with whom we live, and to transmit this to those who will come after us, so that the work of the past centuries may be useful to the following centuries, that our children, by becoming more educated, may at the same time become more virtuous and happier, and that we may not die without having deserved well of the human race.[5]

Today, the technology potentially exists that would allow almost total access to the entire formal written information system. Thus the dreams of Diderot, according to information specialists, may now be on the verge of realization, as far as Western written cultural knowledge is concerned.[6] So, this is one other instance where the Enlightenment Project is running out of its agenda.

Because of this knowledge explosion, the means of accessing and comprehending a mass of data have become central to the scientific enterprise. This automation has already been extended to search for theory. In fact, a decade or so ago, a program interestingly called "Bacon" after the Scientific Revolution's theorist of knowledge, "rediscovered" several laws in physics, such as Ohm's Law and Snell's Law.[7] Current experimental techniques like those involved in the Human Genome Project and the Hubble Telescope are spewing out so much data that automated processing at a higher level, using artificial intelligence techniques that mimic some aspects of human problem-solving, have become essential.[8] Such programs see patterns in the data, analyze them, and come to conclusions.

Serious research is also being conducted on what have been called "knowbots" and "discovery machines," which will swim in a sea of data and come out with discoveries.[9] But such artifacts do not exist in a vacuum. They are distillations, in hardware and software, of theories and debates on the cognitive process. Here discussions of psychology, the mind, perception, knowledge, and so on become of indirect—yet sometimes central—concern to the scientific enterprise.

This book does not take the position, for instance, that the ancient Egyptians knew of modern batteries, a charge leveled in the US scientists' statement (cited above) as an example of the untruths some multicultural courses are teaching. Or for that matter, I do not hold that ancient Sri Lankans knew how to fly simply because the Sanskrit poetic epic *Ramayana* mentions that its villain Ravana, the king of Lanka, used an alleged flying contraption. In this book, I have rested on formally accepted fact in the essentially Western realm of discourse. Also, although science has many incongruities in its growth, as Feyerabend has pointed out, I do not take his position that "anything goes" in science and that there is no method.[10] There are *methods*, and different sciences use different approaches.

In that sense, all systems of knowledge are not equivalent. There are also hierarchies of explanations possible for many phenomena. Some explanations are more parsimonious than others, explaining many phenomena with a smaller number of variables; others predict better; and in still others, the results of scientific enquiry in to a phenomenon can be replicated. There are also phenomena that cannot, by definition, be repeated, like the Big Bang or the beginning of life on earth. One can only simulate such events.

There is also an incommensurableness that extends to different disciplinary approaches. Thus a physicist would use Newton's equations for a calculation of a planetary orbit, and then he could well run the system in reverse, as time in the Newtonian world is reversible. At the same time, that physicist, in designing an efficient heat engine, would work with the field of thermodynamics, which has an arrow of time. The research program of molecular biologists is not the same as that of systems theorists, or for that matter that of biological evolutionists, or botanical taxonomists. Their views of their fields, definitions of problems, and approaches to potential solutions are all different.

So a variety of epistemological and even ontological positions (such as appear in quantum physics) exist in today's sciences. Hence the choices for potential interventions in the enlarged sciences of the twenty-first century are much wider than those normally understood as scientific methodology. As the shift to increasing automation of knowledge gathering must necessarily gather momentum, so these other deeper dimensions in science become important. With automatic laboratories, knowbots, and discovery machines, the seventeenth-century Scientific Revolution's agenda can be partly left to machines. This situation begs the need to broaden scientific horizons: the new sciences must incorporate into the emerging mixture results in ontology, epistemology, logic—results already arrived at in disciplines outside the boundaries of traditional Western scientific endeavor. Thus the enlarged science project of the twenty-first century must incorporate other cultural elements.

These cultural elements include those values that consciously or unconsciously seep into the scientific project. The decision to emphasize the Manhattan Project and the intense interest in particle physics was such a value decision, fueled by the Second World War and the Cold War. Now other values prevail, and so the big particle accelerator is dropped, blocking, for the time being, any new discoveries to be revealed through its massive particle-smashing power. Questions revolving around values also dog such subjects as biotechnology and biomedicine.

Often, values in science are incorporated unconsciously. Can we incorporate them consciously? Yes, if our lives or other important matters are at stake. Thus the perception of an endangered environment has stopped particular scientific and technological developments. And the values of Green movements are reflected in the globally binding Rio Declaration. These restraining factors have already explicitly guided human knowledge.

But how far can we consciously bring in elements from outside traditional Western science? Can one have a Hindu science, a Buddhist science, a Christian science, a Marxist science, or for that matter an Islamic science? With the recognition of the social basis of science, especially its capitalist underpinnings, there were attempts to develop a socialist science in the former Soviet Union and at times in China. Some of these efforts led to the crudities of Lysenkoism and the anti-Intellectualism of the Cultural Revolution. These failures prove that completely totalizing changes are no longer possible in science. Science is like the biological tree of evolution that already has its own rigidities. Changing it completely is an impossible task, as impossible as starting a new biological system, replacing the 4,000 million years old existing one. The time for total revision has passed; the existing system has enough entropic and other rigidities. The existing science may be capitalistic, Eurocentric, patriarchal, and/or class-based. But to grow a new one wholesale is no longer possible. One can only graft elements to the existing tree, and such grafts only take if there is some compatibility. One could make piecemeal adjustments, however, and some of these could have a major impact if they are grounded in deep epistemological or ontological changes. But all such changes have to be governed within the paradigmatic and other needs of a given discipline. So there are limits to, say, an Islamic or any other religion-based science.

More importantly in this porous world, fundamentalist projects based only on a priori assumptions are doomed. Witness the failed attempts in the Soviet Union and in China. The outside has a tendency to come streaming in, upsetting the most antiseptic of cultural enclaves. Thus, a Khomeini could use a Sony tape recorder as a tool for disseminating his message (as he did); but behind every Sony and its tape there is a world of technical and scientific culture providing the scientists and technicians among Khomeini's flock with a different set of cultural messages. Khomeini's message becomes "only" a surface resting on a deeper structure of a given scientific culture.

In this sense, unlike in the remote past, searches for absolute fundamentalist sovereignty in scientific epistemology are doomed. Today, one cannot without contradictions "build socialism in one country" or a regime of pure Islam. Eastern Europe, China, and Cambodia all have in this sense imploded from their earlier searches for purity, because of the dynamics of the globalized system. The enemy is no longer across the border, it is within, it is part of oneself; we are now exposed to more messages than any individual or country can contain.

I import concepts and results from several South Asian belief systems into discussions in the following chapters, including Jain mathematics, Hindu-Buddhist psychologies, and ontologies from different traditions. The South Asian belief systems all can be described as philosophical systems with observational elements, in comparison to the revelatory Judeo-Christian religious traditions. I use such imported elements only if they help the problems at hand. The rest of the religious baggage is not imported. These elements are only made to fit the problems at hand, rather than the opposite—trying to fit the problems into a priori assumptions brought from a belief system.

I do not want to replace the Western Doctor of Philosophy with the witch doctor. I believe, however, that there are elements that the witch doctor—or in our more sophisticated cases, "civilizational knowledge carrier"—can contribute to the knowledge base of doctors in the different sciences. Some of these contributions may yield completely fresh and sophisticated approaches, as in the case of civilizational knowledge. Thus there may be strict limits attainable through a solely Mullah-driven science, but on the other hand there would be many elements in the great Islamic scientific traditions that could still be drawn upon. Very probably there are important nuggets that did not get translated to Latin from Arabic in the manuscripts in Cordoba and elsewhere, concepts still worth examining. Similarly, in the other great civilizational areas like East Asia, and lesser ones like the pre-Columbian Americas, there are mines of knowledge yet to be adequately explored.

If it sounds like I am accepting the "totalizing" hegemony of modern science, I am. I want to enlarge it if possible, not destroy it. I want to reach beyond the Enlightenment and the modern projects and some of their Eurocentric limitations. But the modern sciences, when taken individually, are not monolithic, ontologically and epistemologically totalizing projects. There are too many differences and even contradictions in the approaches of the different disciplines as to methodology, epistemology, and at times ontology. So science as a totalizing project is totalizing only to the extent that it is an organized skeptical attempt to gather valid knowledge. With that pursuit I am perfectly comfortable. I want only to increase the skepticism, to make it more valid, and to enlarge the catchment area. In this endeavor, I have chosen as illustrations of possibilities those topics that appear to me to have been successfully mined in areas in which I have some modicum of intellectual familiarity. I could have in the process left out some interesting areas of promise. I have also, on the other hand, let the imagination roam and at times made speculative suggestions. But to me these are but floundering beginnings in a longer journey.

SECTION **1**

VIEWING SCIENCE WITHOUT EUROCENTRIC BLINKERS

TWO

THE TRAJECTORIES OF
CIVILIZATIONAL KNOWLEDGE

But there is neither East nor West
Border, nor Breed nor Birth

—RUDYARD KIPLING,
"The Ballad of East and West"

The social processes that sociologists have documented which influence the development of science provide a particular perspective on the evolution of knowledge. According to such a view, a knowledge tree evolves buffeted by social forces in the environment as well as within the scientific community itself. In other words, scientists take facts—including those observed in the laboratory and other observations—and combine them with concepts to open a cognitive window to the physical world.

These concepts could arise from past traditions of a discipline, from other disciplines or other sources within the culture, from other cultures, or from entirely new constructs made by individual researchers. The different shoots of the knowledge tree that inch upward from this social construction are also—to mix the metaphors a bit—particular windows to the outside world. Such a social evolutionary perspective provides for a more enlarged, civilizational view of the social nature of science than do other formulations. Here I will broadly summarize this larger view of civilizational knowledge, which I have sketched elsewhere.[1]

If one imagines the "whole" of physical reality to be a blackboard, then the scrawls a socially constructed knowledge trace makes are only one possible exploration on the surface of this physical reality. Given different social factors, both within the knowledge community as well as in the environment outside, one could have different scrawls. Thus, the knowledge tree that science has delivered to us since the eighteenth century becomes but one particular tree in a possible forest. Given different "starting points" and historical runs—that is, different problems, communities of knowledge-makers, usable conceptual elements, and histories—one would conceivably have different sets of knowledge trees. They could emerge, one should note, with component disciplines all possessing the same sort of rigor (as to the nature of evidence and validity of theories) that governs the contemporary scientific enterprise. Thus, if each of the old regional civilizations (Europe, West Asia, South Asia, or East Asia) could have had different "historical runs" of knowledge—especially after an institutional establishment of science similar to those what arose in seventeenth- and eighteenth-century Europe—then one would have different knowledge trees corresponding to each region.

The form, content, and areas of reality explored would be different, and so would the ensuing body of scientific knowledge. In fact, the different contemporary national profiles of science, which sociologists have documented, could be beginnings of different branchings-out, which if allowed to grow autonomously, would grow into different knowledge trees. This broad view of the possibility of multiple knowledge trees allows for an enlarging of the terrain of physical reality that we can explore, increasing the ensuing findings and the accompanying growth of concepts.

There are some further consequences of an evolutionary social epistemology of civilizational knowledge. A particular twig, a particular social scrawl on reality's blackboard, is but a particular window to the world delivering, as it were, a worm's-eye view from a particular discipline at a particular moment of development. Put another way, as a discipline propels itself forward, it moves through a tunnel bounded by its epistemological boundaries and guarded jealously by its members.[2]

The scientist working at the frontier of his or her discipline crawls up through the historical tunnel of the discipline, a tunnel built by the predecessors, and glimpses only what is immediately ahead, and neither what is behind in the tunnel or outside it. This highly structured view is literally a tunnel vision of the world outside, from the discipline's present point of view. Such a view gives meaning from a particular discipline, but is in fact one of many possible views—although members of the discipline may be unaware of the views from the other tunnels. It is therefore both a source of knowledge (derived from what is viable within the discipline) as well as a source of ignorance—ignorance of what is outside its boundaries.

Knowledge of the physical world, because it is a product of particular historical trajectories, must therefore by necessity always be incomplete. There are by definition large areas of the "blackboard of reality" virtually bare of knowledge scrawls. But given the impossibility of ever truly envisioning the total blackboard of reality, is there a means of enlarging the existing stock of our (albeit incomplete) knowledge?

ENLARGING THE SOCIAL-KNOWLEDGE FIELD

One can think of the different strands of knowledge within a knowledge tree as conceptual turnings taken by particular branches as they attempt to develop views on the world they are sampling. The turn each one takes depends on the particular environment of the moment—the concepts being used and the different part of reality being explored. The process is also like changing lenses to get different perspectives. How can one use these different concepts, these lenses, to get a larger, more inclusive view?

Imagine going from the present to the past down the pathways traced by a given knowledge tree. It is in many ways like what astronomers do when they look at the sky and see the distant past unfolding as a set of physical changes, where the universe gets less and less complex as one heads toward the beginning of the Big Bang, when all particles and energy were undifferentiated. One can do the same with a knowledge tree, going back to rediscover the chain of intellectual events that gave rise to the particular set of branches (or windows to the external world) that constitute any discipline's sampling of reality at any given moment. And as one goes back and comes to succeeding turning points, one sees the corresponding changes in the conceptual structure. What was taken in, what was lost, is now seen. And yet what was lost could have significance for another knowledge tree at another, future time. This has been the case during the resurrection of some past debates in the contemporary West. This broad approach could be invaluable as a general technique, in mining other traditions for knowledge.

The present set of sciences was built essentially within the historical milieu of the West's last few centuries. Scientists during this period selectively chose facts and concepts to fit into a prevailing situation, leaving out other facts, areas of enquiry, and conceptual elements. This process was the result of explicit as well as implicit social acts. But although the Western tree of the last few centuries is dominant today, there have been other explorations elsewhere, in other cultures and other civilizations that are today marginalized. Now, the knowledge these marginalized groups possess is not insignificant.

The local, informal knowledge held by small groups and scattered around the world once fed regional centers of knowledge accumulation, when after the neolithic transition, city cultures emerged in different parts of the world. These regional civilizations, through the play of particular local social dynamics, developed their own broader knowledge terrains, demarcating areas of enquiry, modes of analysis, and "discipline" arrays. The hegemonizing tendencies that are the legacy of rapid developments since the eighteenth century have continued to dominate institutional science, muting and in many ways stilling these local traditions.

The new ways of knowing the world were built up initially in Europe, assembled from cultural elements within Europe itself as well as from those transmitted from outside. Since then, the number of external cultural contributors to this stock of knowledge has diminished. But extra-European knowledge still exists in regional civilizations, exemplified by the South Asian cases taken up in this book.

The contents of these local stores have empirical, ideological, factual, and false aspects. Some are patently spurious and unusable in the modern scientific enterprise. On the other hand, there are many areas that could be very useful.

"STARTING POINTS" OF SCIENCE

One of the greatest theoretical physicists of this century, Schrödinger, noted that science depends on the theories and data of past researchers, an "outcome of selections formerly made." These selections were the results of certain trains of thought working on the mass of experimental data *then* at hand. And so, if we go back through an indefinite series of stages in scientific advance, "we shall finally come to the first conscious attempt of primitive man to understand and form a logical, mental picture of events, observed in the world around him."[3] The "origin of science" is, according to Schrödinger, "without any doubt in the very anthromorphic necessity of man's struggle for life."[4]

Conventional histories of the scientific enterprise begin with Greece as just such a "Schrödinger starting point" for science. This is seen, for example, in the seemingly encyclopedic work of J.D. Bernal nearly two generations ago. Further, debates on present knotty problems in science, especially those crossing the domain of philosophy, sometimes go back to this seeming seed-bed of modern science for inspiration and the reexamination of core problems and assumptions.

Yet the "anthromorphic necessity" to know physical reality had never been limited to the early Greeks. In fact, a vast set of proto–starting points for potential scientific trajectories (per Schrödinger, above) has now been laid bare by anthropological research in many parts of the world, discussed in the last chapter. Further, the broad Greek heritage itself has been pushed backwards in time and sideways in space and shown to have been socially constructed from contributions of diverse elements outside of Greece, especially from Africa and Asia. This unraveling of a tight historical lineage of knowledge that was, up to a generation ago, the received wisdom in the West lays bare not only vast areas of already explored knowledge terrains in the non-Western world, but also a pool of potential additional contributions to existing—or new—knowledge branches.

Recent research indicates that the view of Greece as a particularly unique civilizational starting point was established only in the nineteenth century. Other historical research (see my discussions of the South Asian region) also indicates that some contributions to the Western knowledge base and previously considered unique to the Greeks were, in fact, known to other contemporary civilizations at about the same time, or even earlier. We now know that considerable traffic of ideas existed between the different regions in the ancient world.

One of these recent revisionist ideas, posited recently by Martin Bernal, is that there are many African roots to Greek civilization.[5] His views, transferred to the history of science, alter considerably the well-known, tight Marxist-Eurocentric trajectory for science developed by his father, J.D. Bernal.[6] Martin Bernal argues that the concept of Greece as the principal source of Western knowledge and identity was socially constructed in nineteenth-century Europe. His argument highlights how the ancient Greeks themselves—including Herodotus, the "father" of Greek history—acknowledged their strong debt to Egypt. Bernal goes on from there to argue that aspects of this Egyptian heritage of Europe were transmitted through the centuries, without a break, from the ancient Greeks to the Romans, into the Renaissance and to the Rosicrucians of the seventeenth century, to the Freemasons of the eighteenth century. His picture offers a point of departure from the ethnocentricity of the purely Greek model.

He argues persuasively that the Greek model was a nineteenth-century invention deeply implicated in the rise of European racism and imperialism. In his words, for "eighteenth-century romantics and racists, it was simply intolerable for Greece, which was seen not merely as the epitome of Europe but also as its pure childhood, to have been the result of the mixture of native Europeans and colonizing Africans and Semites."[7] With the onset of slavery and the emergence of colonial empires, the idea that dark-skinned African people were the true originators of Greek civilization was now incompatible with creativity and civilization.

By pushing back the primacy of Greek civilization to older cultural roots in Egypt and to earlier civilizations in the region, like Sumer, Martin Bernal clears a path toward recognizing how ideas from these older civilizations helped Greece. Other comparative historical research already done on various contemporary scientific traditions indicates that the primacy of Greece has to be considerably rethought. The studies that I later relate in detail regarding South Asian worlds indicate many parallels and antecedents with the Greek tradition. (A similar exercise could undoubtedly be done comparing the Greek tradition with, say, the East Asian.)

Thus the unique nature of the Greek heritage is problematized, and the data on parallels and antecedents to the Greek era also problematize the sociological givens of the rise of science.

This situation calls for a fresh, long-range historical perspective on science. What I call a "civilizational perspective" will help provide a metahistorical view of the nature of science. In our perspective, the present Western science is a civilizational science. In this light, we can see what social and other elements went (and go) into the establishment of Western science. This knowledge will help universalize the processes of hunting for scientific knowledge and give hints and perspectives on my own objective—mining other civilizations for science.

I will begin by looking at what social scientists have been saying recently about the sociological factors underlying the growth of science. From there, it is easy to extrapolate to the larger view, extending these results beyond the Western domain to a civilizational model of the historical geopolitics of science.

THE SOCIOLOGICAL EXPLANATIONS OF "WESTERN" SCIENCE

The rise of Western science has to be connected not to some inherent superiority, but to major shifts in the socioeconomic and cultural systems of modern Europe. Such shifts can result in strong repercussions within the universally human impulse to create science.

Studies in this long-range macro-perspective have attempted to relate the emergence of mathematized science in the sixteenth and seventeenth centuries to the Renaissance and mercantile capitalism;[8] Newtonian optics and ballistics to the social needs of the early mercantilist period and its interests in navigation;[9] and the emergence of quantum theory in post–World War I Germany to social and intellectual conditions in the Weimar Republic.[10] These macro-cognitive shifts corresponding to macro-social changes are more difficult to prove than the social factors at the level of the nation or the small group. Yet they are suggestive. The emergence of European economic hegemony thus accompanies the emergence of Western science and provides the framework for it.

While the cultural impact of a large economic transformation affecting a whole continent is difficult to demonstrate, evidence for this economic/cultural model does come from twentieth-century examples. If science has been influenced in our time by social and economic forces, one could reasonably argue that similar pressures helped to *create* Western science and direct it in particular ways. I will begin by giving examples of the impacts of socioeconomic factors in different countries, which gave rise to different national profiles of science.

National Differences in Science

According to one empirical study, science and technology practices of the US, Japan, Britain, France, Sweden, and Germany vary, depending upon such factors as the country's history, funding sources, research and development allocations, and coordinatory mechanisms.[11] Country- and discipline-specific studies indicate how these different factors operate in practice. Steven Yearley, who studied the growth of science and scientific institutions in Ireland, has shown how science in Ireland was shaped by a variety of specific social factors. The principal social factors included the peripheral position of Ireland vis-à-vis the rest of Europe as well as British colonial influences on Irish culture generally. Some of these influences parallel the growth of science in other small, peripheral countries, and these forces continue to shape Irish science.[12]

Between 1900 and 1930, science developed vigorously in France due to strong government support, a strong lobby for science, and improvements in institutions. Yet France was later overtaken by others because of ideological factors, epistemological differences, and institutional conflicts. In early twentieth-century France, "supply push" forces from within the scientific community gradually helped science, whereas in other countries a "demand pull" of economic and political factors influenced the development of science.[13]

The field of high-temperature superconductivity provides another illustration of the different contemporary responses of national scientific and technological systems to important breakthroughs. With the advent of high-temperature superconductivity, the Japanese developed a national strategy within national institutions. The US, on the other hand, did not have a strategy at the national level. The US emphasis is on individuality, and so hundreds of strategies exist, each having different objectives and impeding a concerted US response.[14]

National differences also show in publications behavior. A study of Nordic articles in cardiovascular research indicated a strong correlation between the regional location and specialization of the citing, as well as the journals cited. For instance, an article published in a US journal was cited by authors of articles in other US journals more often than those authors cited research published outside the US. Because of the larger US scientific community, this has resulted in higher citation frequencies for US publications in cardiovascular research compared to work published in Nordic journals—resulting in a problem of cognitive access.[15] This also implies that a paper in a setting more peripheral than a Nordic one, a Third World journal, would not reach the scientific mainstream even if it contained significant findings.

This marginalization in developing countries is shown by studies comparing the publishing practices of those who write for local journals and those who write for Western journals. Studies of two developing-country contexts—Korea and Philippines—indicate that those who publish locally tend to cite national sources more than those who publish internationally—who tend to cite international sources.[16]

The differences between developed and developing countries are not limited to cognitive access. A survey conducted in 78 developing countries, combined with interviews and a bibliometric study of about one-fourth of the nearly eight hundred scientists in the sample, revealed strong dependency links. That is, developing country nationals were heavily dependent on outside sources for their training, the building of their institutions, and the funding of their research. Researchers in developing countries chose their research topics on cues from colleagues in developed countries, used the same kinds of equipment, and relied on the same international scientific literature.[17]

Other detailed case studies demonstrate the impact of national social and political forces on science, including the biometry/Mendelism controversy;[18] the spontaneous generation/biology controversy;[19] statistics;[20] nineteenth-century cerebral anatomy;[21] and Lysenkoism, the last an example from a socialist country.[22] It is clear that social factors at the level of the nation permeate the practice of science deeply.

Micro Studies in Science

But let us now take the sociological argument further, in the direction of micro-examples. Such studies are impossible for any period earlier than this century; the evidence is simply lost. But they permit us to argue a civilizational, geopolitical case by showing how the construction of science is so deeply related to material circumstances. Detailed case studies done over the last decade and a half have discussed in considerable detail how the construction of science takes place within the small groups that actually "do" science. These small groups include members of laboratories, networks of practitioners or theorists constituting "invisible colleges," and key gatekeepers in a discipline, such as editors and referees. Sufficient case studies now exist to describe in detail the social and contingent nature of the knowledge so produced.

Thus, research on scientists working on lasers and gravitational waves has indicated how particular interpretations of science have been socially mediated.[23] Similar results have been obtained through other studies, such as those on radar motor research,[24] quantum mechanics,[25] and the weak neutral current.[26] Pinch has suggested that "scientific theories themselves are multi-dimensional and that what constitutes a theory in science is a variable and will mean different things to different groups of scientists."[27] And a study of a sub-atomic phenomenon, the so-called J phenomenon, has confirmed that formal scientific rationality is at least partially constructed from social commitments.[28] Among other key studies, mention should be made of Latour and Woolgar,[29] who explored the social construction of scientific facts within the laboratory, as well as the seminal work of Fleck nearly fifty years ago.[30] This research reveals that the subculture of a scientific discipline appears to be "far more than the setting for scientific research; it is the research itself."[31]

The social influences extend also to mathematics. In 1948 Dirk Struik attempted to place mathematics in its social setting, by way of a Marxian approach. The first book-length monograph of a sociology of mathematics, by Sal Restivo, argues that there are different styles and modes of doing mathematics, depending on the context. Thus problems are posed—and solutions reached—according to various influences within and outside of the discipline. Consequently, branches of mathematics develop because they are determined by social—not abstract, "scientific"—pressures.[32]

The social conditioning extends also to journals and publishing. Despite pressures toward uniformity brought about by the handful of publications consortia that control scholarly publications, each scholarly journal develops a particular style of its own because of social factors. This is influenced by such factors as research area, frequency of the journal, and the market size; the role of professional associations; aesthetic demands of editors, authors and the scholarly community; and each journal's unique history.[33]

Again, these micro-studies support my argument that the preconditions of scientific culture exist universally, but social, economic and other circumstances limit and condition the sites where scientific activity will flourish. In the ancient world there were many possible sites. Since the emergence of Western hegemony, there have been fewer such places and the geopolitics of science have been limited by wealth, access to resources, and other factors. Yet we do not want to conceive the social and political too narrowly. To do so would be to miss the assistance offered to a geopolitical civilizational analysis by feminist commentators on science.

The Feminist Critique of Science: Gendered Science

Another examination of the social conditioning of science comes from feminist thinkers who have argued that gender dominates science—through the predominance of male scientists and masculine metaphors.[34] Feminist critiques have dissected the intellectual history of science and shown it to be patriarchal in origin, with close ties to capitalism. The present science enterprise, according to feminist discourse, raises questions relating to the received heritage of Plato's view of knowledge, Francis Bacon's empiricism, and Frederick Taylor's division of labor. Also according to this view, the exploitation of nature and the stigmatization of motherhood is ingrained in modern science.[35]

Science's patriarchal bias, according to some feminist interpretations, has essentially given rise to a sociobiological explanation for inequalities in gender, class, and race. This patriarchal world view also affects science in that it gives primacy more to controlling nature, than to understanding it. For example, according to one argument, when science searches for "master" molecules, it presupposes a hierarchical model of nature rather than a more interactive one. Patriarchal science rests heavily on linear thinking, quantification, and reductionism, in preference to many other ways of obtaining and organizing information.[36]

Evelyn Keller has also shown that images of knowledge within Western intellectual and scientific history are often based on sexuality.[37] These biases also exist in such contemporary interests as artificial intelligence, or AI. Historical studies in language, analyses of texts, and empirical data show that the disciplines covered by logical-mathematical intelligence are usually gendered. Jansen has thus shown that AI—its models and disciplinary assumptions—seem masculinized. The result is a "phantom objectivity" that ignores alternative constructions of reality.[38]

Several writers have explored the empirical indications of a gender bias in a larger epistemological vein. Among the most thought-provoking are Sandra Harding and Donna Haraway.[39] In her writings on environmental factors, Vandana Shiva has combined feminist and South Asian perspectives.[40] All have described how knowledge is deeply influenced by its gender relations and how existing knowledge is therefore only partial. Feminist epistemology has again highlighted the connection between knowledge and politics and the fact that the social position of the producer of knowledge affects the knowledge outcome.[41]

The influences documented by feminists thus include an additional set of social factors that shape the trajectories of science. Other factors—starting points and subsequent historical influences, including those at the level of the small group and nation—guide its eventual trajectory. Science thus manifests itself as a social field consisting of struggles, forces, and relationships. Those relationships determine important problems, acceptable methods, and correct knowledge, governing the practices and results of science.[42]

Geopolitical entities thus jostle with each other in the manner of plate tectonics in geology, or like icebergs in the polar regions. Science is only the tip of the iceberg, the jostling underneath being largely invisible. Society, in its historical unfolding, becomes the deep structure of science, and science emerges as a partial reflection of the dynamics of this deep structure. And if the dynamics include racism, then one has a *Racial Economy of Science,*[43] the title of a recent collection on the theme of racial bias in the construction of science. Interestingly, this volume is edited by Sandra Harding, a leading feminist theorist. Clearly there are many points at which feminist and civilizational critiques of science intersect.

But how are the scientific results arising from these social dynamics actually built up? And what are the social sources of scientific theories and explanations?

SCIENCE'S COGNITIVE ELEMENTS

Theories and explanations are themselves socially constructed. Especially at paradigmatic and other breaks in the cognitive structure, cognitive fields are created from arrangements of new conceptual elements or rearrangements of old ones. This process could be likened to a young child's using his building blocks playfully to make a structure that pleases him and his cognitive orientation at the moment; in the case of science, the building blocks would be concepts. Fresh arrangements of concepts occur when the existing conceptual structure does not match the reality presented by new facts. At that point, new concepts arise in the imaginations of the theoreticians and come from a variety of sources. They could be dredged from past intellectual traditions.

For example, the corpuscular hypothesis of atomism—whose roots in the Western tradition go back to Democritus—was revived by Boyle, displacing the existing Aristotelian view that was opposed to atomism.[44] Galileo, in a similar manner, moved away from the Aristotelian tradition to that of Plato, replacing "the descriptive science of Aristotle [with] the structural science of Plato."[45]

Astronomy provides another such example, in Laplace's view of the solar system's evolution. Laplace believed that a large dust cloud under gravitational attraction collapsed, giving rise to the solar system. His theory was later superseded by others, including the notion that the planets were formed by a passing star pulling out the sun's matter. Yet in recent decades, the central perspective of Laplace has been revived to indicate that the formation of the planetary system was indeed the result of the gravitational collapse of a dust cloud.

Other examples of similar revivals include Boltzmann's atomistic-kinetic theory.[46] Similarly, Heisenberg's indeterminacy principle had been first demonstrated by Grimaldi in the seventeenth century,[47] but was soon forgotten because it had no apparent scientific relevance at the time.[48] Einstein revived Newton's particle view of the nature of light after the wave theory's period of dominance.[49]

A less abstract knowledge package was revived recently by the Israelis. Literary evidence suggests that about 2,000 years ago, the Negev desert in the Middle East was the site of a thriving agricultural civilization. The remains of old farms could still be seen. So when Israel was created, there was an attempt in the fifties to revive this system. The first attempt, based on twentieth-century perceptions of what the irrigation system should be, was not successful. The old system was then studied archaeologically and on the basis of historical documents: the old system was more sophisticated than was first assumed, incorporating complicated hydrological principles. These insights, obtained from the past, were then incorporated in another attempt at regeneration, which was successful. The system was even extended beyond its original confines.[50]

Other disciplines, or a general culture itself, could be further sources of fresh conceptual constructs. As an example, one could note that biological evolution influenced several other disciplines, mapping in them its different aspects.[51] Biology in turn was deeply influenced by its own intellectual environment. Generally, the nineteenth-century "debate on evolution could not be considered in isolation from the theological, philosophical, literary, social, political and economic debates in the same period."[52] The ideas and cultural language of Darwin's cultural circle were crucial to his work; his methodology and metaphysics, in particular, were deeply influenced by the views of the Scottish realist school of philosophy.[53] The ideas of Adam Smith, Malthus, and theologian William Paley also helped shape Darwinism.[54]

Feuer has given many examples of a priori concepts introduced into different branches of science.[55] Mendeleev's discovery of the periodic law of chemical elements is an example, Mendeleev being guided by mystical values. A more recent example is the theoretical physicist Gellmann's injunction that "anything which is possible is compulsory."

Gerald Holton has documented a set of a priori elements or "themes" which inform theory.[56] Themes usually occur in pairs of opposites, such as complexity-simplicity, reductionism-holism, and discontinuity-continuity. Holton demonstrated that not more than fifty couples or triads of themes have been historically sufficient to negotiate the great variety of discoveries in science.[57]

Whether concepts flow in from the past, from other disciplines, or from other cultures, their core elements often appear in the form of metaphors. The knowledge of the world is forged through metaphors, which could be considered the basic form of symbolism. They are "the pregnant mother"[58] to science and to all innovative thought.[59] Creative personnel in different areas, whether they be scholars, poets, or scientists, are therefore all metaphorists. Daniel Rothbart has argued that concept-formation using metaphors is an essential part of scientific reasoning and that the formation of concepts in science is largely a metaphoric process. Generally speaking, metaphors implicitly transfer features from one semantic field to another, entirely different one.[60] Julie Thompson Klein summarized the broad role of a metaphor as helping to explain and interpret what cannot be understood through existing approaches. Metaphors integrate diverse elements or suggest hypothesized answers, probe into a problem, or give insight into categories observed in another system. They can juxtapose the familiar with the unfamiliar and reveal similarities and differences between a new problem and an existing one.[61] Clearly, the extradisciplinary impact of metaphors could influence modern science in many ways, helping its social construction as science moves along, buffeted by various social forces.

GENERAL APPROACHES TO "MINING"

How could the project of enlarging the world's scientific knowledge base, which I have alluded to, be generally done? In fact, can we assemble some "how-to" tips on mining procedures?

At the present, there is no possibility of starting fresh trajectories of science from new Schrödinger starting points in the many thousands of small groups around the world. Neither is it possible to continue the trajectories of slow accumulation in regional civilizations that occurred up to the last few centuries, until the latter were overwhelmed by the exponential growth of new scientific knowledge from Europe. Neither, for that matter, are fresh nongendered knowledge trajectories, *ab-novo*, possible.

Today a vast body of workable, accumulated scientific knowledge exists in the world. It may be gendered, it may have left out vast areas of existing knowledge, but it cannot, even if this were desirable, be ignored or wished away. The situation is exactly parallel to another evolutionary trajectory of information, namely that of biology and the existing system of living things. The possibility of "restarting" the biological evolutionary system with a new starting point to create a new biological system is no longer possible. The present globe is filled with the results of a four-thousand-million-year history of a particular biological evolution and its attendant outcomes. However much we understand the biological process and desire to change it, we cannot undo the results of its past history. We can at the most splice in new genetic information. A process parallel to this exists in the knowledge field.

Grafting knowledge onto the dominant knowledge tree can be accomplished in two broad ways. One is to splice in, directly, existing material that has demonstrable, direct validity. The second would be to bring in, as metaphors, elements from other traditions that could nudge the imagination and give rise to new concepts. The first attempt would be like the splicing-into the European tradition of, say, the knowledge of the compass, gunpowder, or mathematics, and knowledge of flora and fauna that occurred during the Renaissance and immediately after. The second attempt would be like the many transfers of metaphors that have been constantly used in building the scientific enterprise.

The flow of knowledge across civilizations is not new. In fact the history of the Western scientific trajectory is replete with such examples. So, in the following chapter, I shall offer details of how the background for such exchanges between the East and West existed from very early times, and how cultural traffic did flow from one to the other. This will be an overview and will not go into details of the cultural items exchanged, except rather perfunctorily. Later detailed treatments in the subject chapters provide some of the particulars of this traffic.

The following chapter will therefore set the historical background of the mutual—but selective—percolation between the West (meaning initially Greece, and then Western Europe) and the East (meaning, for our narrow purposes here, South Asia).

THREE

THE BACKGROUND TO CROSSFLOWS: WHERE THE EAST AND THE WEST DID MEET

East is East and West is West,
And never the twain shall meet.

—RUDYARD KIPLING

CLASSICAL CROSSFLOWS

Since the eighteenth century, the earlier South Asian intellectual tradition has been delegitimized. Instead, the Western scientific tradition of the last few centuries has been adopted wholesale and has become the new legitimized knowledge studied in universities and practiced in centers of science and technology. Quite often there is no organic interaction between this new, imported intellectual tradition and the earlier organic knowledge of South Asia. However, the strong bifurcation of the European and the South Asian tradition is only of recent date. A detailed study of the historical growth of the two scientific traditions indicates considerable areas of overlap and mutual influence from very early times. Thus, for example, if the scientific and cultural traditions of the Renaissance looked back to Greek sources for new inspiration, they were in fact looking to Greek sources partly influenced by the South Asians. In this chapter I trace the interactions between these two traditions and indicate some of the areas of overlap and congruence.

The roots of this overlap go back to the seedbed cultures that fed both the European and South Asian scientific traditions. In the case of Greece itself, Greek classical culture (apart from the new evidence of

Egyptian influence as documented by Martin Bernal) was fed initially by the traditions of the Nilotic and Sumerian civilizations, and later by the Aryan-speaking tribes that invaded Greece. The South Asian culture that developed on the Gangetic Plain owes its existence to the Indus tradition on the one hand and, on the other, to cultural influences transmitted through the Indo-European language of Sanskrit and its oral traditions.

One of the civilizations that influenced the making of Europe, through Greece, was Sumer, a civilization with close connections to the Indus culture that was contemporary with it. Some Indus-culture seals have been found in Sumer, and a bust found in Mohenjodaro is an aesthetic sculpture recalling the art of Mesopotamia, the features being Semitic.[1] Such finds indicate that close trading connections and other contacts were maintained between the two regions. At a later period, too, commerce between the region and the Persian Gulf was carried on by the Phoenicians beginning in 975 BC; "ivory, apes, and peacocks" from South Asia were brought to King Solomon.[2]

The language of both the Greeks and the Aryan invaders of the northwest Indian subcontinent, as well as the intervening Persian Empire, had a common source. This common language also very probably carried common cultural traits from its original source and at least some of the experiences of the original speakers. There are thus interesting similarities between the gods of the ancient Greeks and the Vedic gods, as well as similarities between the societies depicted in the epic poems of the Homeric and the Vedic traditions. Both peoples worshiped similar gods, such as Father Heaven (Jupiter, Dyaus Pitar), the dawn (Aurora, Ushas), the sun (Helios, Surya), and Mother Earth. The epic ages described in both the works of Homer and the much larger *Mahabharata* are similar.

A third such process occurred at the time of the Persian Empire, which at its peak stretched from the Mediterranean to the Indus. Edward James Rapson has remarked in *Ancient India* that at no time were means of communication by land more open or conditions more favorable for the interchange of ideas between South Asia and the West.[3] Around 519 BC, the Achaemenidean king Darius I sent Skylax of Karyanda, a seafaring Greek, to explore the Indus, with the intention of later conquering it; Darius I later did just that.[4]

It is after the Greek wars with the Persian empire that the East enters the Western psyche and the myth of a division into Orient and Occident—and the concept of Europe is born. Persia thus became an intermediary between Greece and India. Authors of two standard descriptions of India in pre-Alexandrian times, Skylax and Ctesias, were Greeks employed by the Persians. The latter wrote of bizarre creatures that inhabited the Indian region, and these descriptions played a part in the European construction of India. Recent commentators have noted that Ctesias' accounts of weird beings— with dogs' heads, blanket-like ears, or one foot—are clearly translations from Sanskrit mythology.[5]

Greek historian Herodotus wrote that Darius I (521–486 BC) would frequently called Greeks and Indians together for counsel and discussion.[6] Herodotus also records that in the Persian War (484–425 BC) Indian soldiers were recruited by the Persians.[7] Xerxes thus led an expedition against Greece using, among others, Indian troops.[8] Herodotus also described a meeting between Indians and Greeks at the court of King Susa about the best means of treating corpses.[9] Later, Aristoxenes (350–300 BC) mentions a dialogue on human life between Socrates and an Indian philosopher.[10]

Possibly because of these links, some of the ideas prevalent on the subcontinent from 700 to 500 BC—which are found in the later Vedic hymns, the Upanishads, and in the philosophies of the Buddhists and the Jains—appear in later Greek thought. The parallels are sometimes very striking. The search for one reality in the Upanishads is echoed by Xenophanes, Parmenides, and Zeno, the founders of Greek mathematics who sought the One Reality.[11] Pythagoras, one of the founders of Orphism, is alleged to have traveled widely and been influenced by the Egyptians, Assyrians, and Indians. Pythagoras's thought includes such characteristically South Asian views as transmigration of the soul and the ability to recollect past lives. Pythagoras himself "remembered" having fought in the Trojan War in a previous life.[12] Almost all of the religious, philosophical, and mathematical theories taught by the Pythagoreans, Rawlinson has noted, were known in India in the sixth century BC. Thus Pythagoreans, like the Jains and Buddhists, refrained from destroying life and eating meat.[13] It should be noted, however, that all of Pythagorean mathematics was probably not known in India —for example, Pythagoras's theorem in its theoretical formulation (it was known to South Asian thinkers in practice, however).

The concept of karma, representing his "cycle of necessity," was also central to the philosophy of Plato, a founder of the Western philosophical tradition. According to Plato, rebirth is due to the hand of necessity, men being reborn as animals or again as men, a belief common to all major South Asian religious systems.[14] B. J. Urwick has traced in detail the parallels between Platonic thought and Indian philosophy (1920).[15] Recently Vissilis Vitsaxis (1977), a modern Greek scholar, made a detailed study of Plato and the Upanishads and showed that in structure and method, general approach, and the growth of parallel lines of thought on specific points, the two traditions show common features.[16] Other parallels between the two traditions are indicated by Xenophanes's teaching that God is the eternal unity.

Empedocles (490–430 BC), a disciple of Pythagoras, propounded the four-element theory of matter and the four-humor theory of disease. Empedocles's theory that matter consisted of earth, water, air, and fire has parallels in the earlier *pancha bhuta* concept of *prthvi, ap, tejas, vayu,* and *akasa*—earth, water, heat (fire), air, and emptiness (ether). It should be noted that in the South Asian system, there were sometimes only four pancha bhutas: the Nyaya and Vaisesikas recognized five, while the Jains, Buddhists, and materialist Charvakas recognized only four: earth, water, heat, and air.[17] The later Aristotelian view of the physical world also incorporated the essential pancha bhuta concept of five elements. The Hindu caste categories of the Varnas—*brahmins, kshatriyas, vaishyas,* and *sudras*—also have parallels in Plato's Republic, where guardians, philosophers, soldiers, and the people form the four social strata. The doctrine of four humors was later followed in medicine by Hippocrates (460–377 BC). The theory of humors evokes direct comparison with the parallel *tridhatu* of the much earlier *Rig Veda* and the *tridosha* of Ayurveda.[18]

There are also many parallels between the *Samhitas* of Charaka and Susruta (the classical Sanskrit works on, respectively, medicine and surgery) and the Greek systems. These have been enumerated by Jolly, who cites over fifteen major coincidences. That these parallels would occur by chance is unlikely, given that the odds against so many coincidences would be high. Thus "the Hippocratic treatise *On Breath* deals in much the same way with its pneumatic system as we find in the Indian concept of *Vayu* or *Prana*. Plato in his *Timaeus,* strangely enough, discusses pathology in almost the same manner as the doctrine of the *tridosha*."[19]

While some of the tridosha ideas have earlier South Asian antecedents, as noted above, direct borrowings by the Greeks can only be hypothesized. A possible route is through Pythagoras, the philosopher who most influenced Hippocratic medicine. He is known to have traveled to India, and learned, among other things, the theory of transmigration.[20] On his return to Europe, he taught the theory of medicine and systems of dietetics, along with other ideas. It is very possible, therefore, that there were contacts and transmissions. But the fact that the two regions had differences in theories and practices also suggests at least partial independent growths.[21] The Hippocratic collection also mentions an Indian regimen for cleaning the teeth,[22] as well as listing drugs of Indian origin, some with corrupted Sanskrit names (see chapter 6, below).

Likewise, atomic theories occur in the two systems, appearing earlier in South Asia: "an atomic theory being taught by Pakudha Katyayana, an older contemporary of the Buddha . . . was therefore earlier than that of Democritus."[23] Heraclitus's belief that everything is in a state of flux is preceded in a more sophisticated manner by the *anicca* and *anatta* discussions of the Buddhists. Empedocles's view that love and hate acted mechanically on the five elements is paralleled by the addition of joy and sorrow to the elements by the Buddhists and Ajivakas.[24] Aristotle, in his system of explanation, brought in the doctrine of the mean (in 340 BC), whereas similar doctrines had been taught by the Buddhists and others, several centuries earlier.

After Alexander's encounter with South Asia, there was an explicit dialogue with India; several who traveled with Alexander are said to have met with Indian sages: Onesicritius, the Alexandrian historian, Cynic, and founder of the literary tradition by this name; Democritean; Anaxagoras; and Pyrrho, the radical skeptic. Alexander was himself reported to be in conversation with Indian sages.[25] By the time of Alexander's invasion of India, the Greeks had also tapped Indian knowledge on diet, hygiene, treatment of snakebites, and veterinary science.[26] And Strabo considered astronomy as a favorite pastime of the Brahmans.[27] Magasthenes, who as the ambassador of Seluccus Nicator visited the Mauryan court of Chandragupta in the third century BC, described the parallels between the two traditions as follows:

> In many points, their teaching agrees with that of the Greeks—for instance, that the Deity, who is its Governor and Maker, interpenetrates the whole. . . . About generation and the soul their teaching shows parallels to the Greek doctrines, and on many other matters. Like Plato, too, they interweave fables about the immortality of the soul and the judgements inflicted in the other world and so on.[28]

In the Pali Buddhist literature, an important position is given to the *Questions of Milinda* (Menander), which are the records of religious and philosophical dialogues between the Buddhist monk Nagasena and the Indo-Greek ruler Menander. These summarize much of Theravada Buddhist thought and indicate the continuing intellectual dialogue between Greece and South Asia in the post-Asokan period.

It has also been suggested that Gnostic thought was influenced by Buddhist *Prajnaparamita* literature. Gnostic Carpocratians strongly advocated the idea of transmigration. And at least one Gnostic philosopher, Bardesanes of Edessa (ca. AD 200) had extensive Indian contacts and had even traveled to India. The Persian Gnostic of the third century AD, Mani, spent some time in India, and subsequently incorporated several Buddhist ideas into Manichaeism. Buddha is also referred to by name in these writings.[29]

During the Roman Empire, which continued and maintained the classical European tradition after the eclipse of Greece, contacts with South Asia continued. The Romans traded heavily with South India and Sri Lanka for luxuries, and as Pliny records, four ambassadors were sent from Sri Lanka to Rome.[30] Some South Asians settled at Alexandria and other trade centers; excavations at Pompeii have revealed ivory carvings from the region,[31] and there is evidence of Buddhist statues in Roman remains in London.[32]

In late (classical) antiquity, India was depicted in some debates as the origin of philosophy and religion. It is now alleged that key figures such as Plato, Democritus, Pherecydes, and Pythagoras (as mentioned above) had traveled to India and borrowed ideas. Pythagoras, it is said, got his psychological and soteriological ideas from lines of thought prevalent in the region.[33] Alexander Polyhistor (first century AD), Apuleius (second century AD), and Philostratos (ca. 200 AD) stated that Pythagoras learned many things from the Brahmins.[34] Thus Lucianus wrote in the second century AD that Indians cultivated philosophy before it came to the Greeks.

The founder of the Neoplatonic school, Plotinus, was so serious about learning Indian philosophy that he took part in the military expedition against the King of Persia in the hope that it would bring him into the region.[35] According to one tradition, Plotinus went to India in AD 242 expressly to study its philosophy.[36] There is in fact a strong similarity between Neoplatonism and vedanta and yoga systems, as is suggested by the following quotation from Plotinus, on the absorption of the individual into the World Soul: "Souls which are pure and have lost their attraction to the Corporeal will cease to be dependent on the body. So detached, they will pass into the world of Being and Reality."[37] Neoplatonism also had many features in common with Buddhism, especially abstention from sacrifice and from eating meat. A doctrine similar to Neoplatonism later became part of Christian psychology, as reflected in the writings of the Egyptian St. Anthony, St. John of the Cross, and Meister Eckhardt, among others.[38]

By the second century AD, India had almost replaced Egypt as the origin of thought and learning.[39] The Greek doxographer Diogenes Laertius (third century AD) referred to the role of what were considered as Indian "gymnosophists"—the Persians, Chaldeans, and Egyptians—as influences. These discussions find an echo in other writers, like Clement of Alexandria in the second century AD,[40] who referred repeatedly to the presence of Buddhists in Alexandria. Clement was also the first Greek to refer to the Buddha by name, and he knew that Buddhists believed in transmigration and worshiped *stupas*.[41] On the influence of South Asia on Greek thought, Clement was firm: he flatly declared that the "Greeks stole their philosophy from the barbarians."[42]

During the Roman Empire, Indian philosophers also came to the West. An Indian ascetic called Zarmanochegas (probably *Sramanacarya* in Sanskrit) was in Athens. Under Emperor Antoninus Pius, an Indian delegation visited Europe, and the philosopher Bardesanes received much information on Indian thought from members of this mission. On the other hand, Apollonius of Tyana traveled to India at this time.[43]

According to a tradition in the early centuries of the Christian era, allegedly going back to the time of the Aristotelian Aristoxenes (ca. 300 BC), the churchman Eusebius affirmed that an Indian visited Socrates and asked him about the meaning of his philosophy.[44] *Elenchos*, authored by the Christian cleric Hippolytus in the third century AD, reveals an awareness of Indian ideas beyond what was known in the times of Alexander or Megasthenes.[45] Indian medicines and herbals were referred to by several Roman writers—like Celsus, Scribonius Largus, Pliny, and Discoridas.[46] And evidence of continuing contacts with South Asia in the European Dark Ages, even in the extreme North of Europe, exists in the form of a Buddhist image, dated circa sixth century AD, discovered under the oldest church in Sweden.[47]

In order to highlight the imbalance of current historiographies of science, in the previous examples I have discussed the probable transmission of ideas from East to West at this early era in the very formation of Western culture. However, influences worked in the other direction as well. The Alexandrian incursion brought Greek elements into the sculpture and architecture of the Gandhara region (although the extent of this influence, however, has now been found to be limited geographically in South Asia, contrary to the beliefs propagated during the colonial era). There is also the very well documented transmission of Greek ideas on geometry and astronomy to South Asia.

With the passing of the Classical Age and the onset of the Middle Ages in Europe, contacts between Europe and South Asia continued, this time largely through Arab intermediaries. The Arabs were now to perform the functions earlier performed by the Persians, Alexandrians, and Greeks who had first brought together the ideas of East and West.

THE ARAB CONNECTION

In AD 662, Severus Seboknt, a bishop in a Syrian monastery, complained:

> It will omit all discussion of the science of the Hindus, a people not the same as Syrians, their subtle discoveries in the science of astronomy, discoveries that are more ingenious than those of the Greeks and the Babylonians, their computing that surpasses description. . . . I wish to say that this computation is done by means of nine signs. If those who believe because they speak Greek, that they have reached the limits of science, should be shown these things, they would be convinced that there are others who know something.[48]

In the rule of Khalif al-Mansur (AD 753–774), embassies came to Baghdad from Sind. Among them were scholars who brought South Asian works on mathematics, such as *Brahmasphuta-siddhanta* and the *Khandakhadyaka* of Brahmagupta, which al-Fazari and others translated into Arabic. These two works were used very widely and had a major influence on Arabic mathematics, although there were a number of other works in Arabic on Indian arithmetic.[49] *Aryabhatiya*, by Aryabhata, was brought to the Arab countries by Abul Hasan Ahwazi.[50] Al-Fazari's translation of the Indian work was issued as *Great Sindhind*.[51]

Musa al-Khwarizmi (ca. AD 825), whose importance in the history of mathematics is attested by the fact that corrupted derivatives from his name have given us such words as "algebra" and "algorithm," made his set of astronomical tables based on al-Fazari's translation as well as on certain elements of Babylonian and Ptolemaic astronomy. He also pointed out the Indian origin of the number system. The original of al-Khwarizmi's text is lost, but it lives in a Latin translation, *Algorithmi de numero indorum*, made in 1000 by Maslama al-Majriti, a Spanish astronomer in Cordoba.[52] The basics of modern trigonometry had been developed by Aryabhata I, and they too were transmitted, through translations, to the Arabs, who likewise passed the information into Europe. These theories are described in detail in 1464 by Regiomontanus, in *De triangulis omni modis*.

The Arabs called mathematics *hindsat*, "the Indian art," illustrating their own perceptions of the order of things.[53] The mathematical knowledge borrowed from Indian sources included the Indian method of arithmetical notation and enumeration, the concept of zero, the decimal system, algebra, and trigonometry. In due course, Asian methodology replaced the abacus in calculations used in "commerce, government and technology."[54] Through the Arabs, some Roman works—such as those on astronomy (*The Romaka Siddhanta*) which had been embraced into the South Asian tradition—were returned to Europe retranslated from Sanskrit to Arabic and from Arabic back to the original Latin.

Arabs also translated the major Indian medical works, while Indian nationals actually practiced in Arab hospitals. Later, this transmission was continued further west, when al-Razi or Rhazes (AD 865–925) wrote a major work incorporating much of South Asian medical knowledge that was then translated into Latin by Moses Farachi and became, in the Middle Ages, the standard medical work of Europe[55] and so laid the groundwork for later developments.

Among those who transmitted South Asian mathematical knowledge (as incorporated in Arabic texts) further into Europe was the Englishman Adelard of Bath. Around 1142, he translated al-Khwarizmi's work through its Spanish version by the astronomer Maslama al-Majriti. Other Latin translations found in England included an anonymous fifteenth-century manuscript, in which computations are made for the year 1428 for the geographical latitude of the English town of Newminster.[56]

Other European "onward-transmitters" were John of Seville, Robert of Chester, Alexander de Villedieu, John Sacrobosco, and Leonardo Pisano. Seville was the author of *Liber algorismi*, which was based on Arab texts, including those of al-Khwarizmi. Circa 1141, Chester translated al-Khwarizmi's *Hisab al-jabr wal-muqabala*, which drew on Indian algebra. Villedieu wrote *Carmen de algorismo* (ca. 1200), which followed closely Seville's *Liber algorismi. Carmen* was translated into English, French, Icelandic, and other European languages. Sacrobosco, a contemporary of Villedieu, wrote *Algorismus Vulgaris*, also an arithmetical text. Pisano's classic of arithmetic, *Liber abaci* (1202), the first complete rendering into Latin of Indian and Arabic arithmetic, is considered the beginning of the Renaissance in arithmetic.[57]

Developments in arithmetic in the Renaissance facilitated the rising trade and commerce of the time. The new arithmetic, based on the decimal place value system, was called "algorism," and the sudden appearance of many works on the topic, from the sixteenth century onwards, indicates its importance. These studies include Cardano's *Practica arithmatice et mensurandi singularis* (1501) and Tartaglia's *La Prima Parte del general trattato di numeri e misure* (1556), both Italian; Forcadel's *L'Arithmatique* (1556–57) and Boissiere's *L'art d'arythmetique* (1554), both French; Jacob Koebel's *Rechenbiechlin* (1514), Chistopher Clavius's *Epitome arithmaticae practice* (1583), and Stifel's *Arithmatica integra* (1514), from Germany. England had Robert Recorde's *The grounde of artes, teachying the worke and practise of arithmetike*, which ran to seventeen printings before 1601 and Digg's *Stratiotios* (1579).

THE VOYAGES OF DISCOVERY AND AFTER

At the beginning of the Renaissance, old belief and religious structures were tumbling, and there was a feeling in Europe that technological progress was not only desirable, but possible. In their extensive "Voyages of Discovery," the Europeans possessed a greater array of devices and skills than any contemporary Eastern or past European civilization. Yet this new knowledge included significant borrowings from the East, especially in mathematics and techniques for the newly important arts of navigation, astronomy, and warfare. The South Asian systems of arithmetic, algebra, and trigonometry (transmitted earlier by the Arabs) were rapidly developed and utilized for these purposes.[58]

Thus the Portuguese, through Mediterranean intermediaries interested in astronomy, acquired navigational knowledge in the form of tables, astrolabes, and charts—but these were the fruits of Arabic, Indian, and other Asian ideas. The artifacts and techniques used by European sailors in their travels to the East also included many other borrowings, such as the mariner's compass and the axial sternpost rudder from China, the lateen sail from the Arabs, the sprit sail from India, and the use of multiple masts from many Asian sources.[59] The detailed geographical knowledge that the Europeans required was also obtained from Asian sources, including information needed to cross the Indian Ocean. The Indian Ocean has been crisscrossed over the previous centuries by South Asians, Arabs, and Southeast Asians. Some of the latter —from present-day Sumatra—had by the early centuries AD crossed the entire ocean and settled in large numbers in Madagascar. Some Portuguese traditions held that two voyagers from India had rounded the Cape of Good Hope by the early fifteenth century and had reached a place called Gardin on the Western coast of Africa.[60] By the fifteenth century, Chinese junks had crossed the Indian Ocean, and it is also possible that they too had attempted to round the Cape.[61] According to some sources, Vasco da Gama's navigator across the Indian Ocean to Calicut was either an Arab or an Indian;[62] when he returned to Portugal, he brought back several instruments useful for navigation, called *kamals*. In the Portuguese push beyond Java, Albuquerque relied on several Javanese palm-leaf maps and sea charts for guidance.[63] Except in the crossing of the Atlantic, the genius of the voyagers of discovery was in collecting and applying the scattered geographical knowledge held by different peoples. And after this era, the Portuguese introduced new ideas and products to Europe. These innovations were important, though not as profoundly influential as such earlier Asian imports as algebra, trigonometry, and gunpowder.

In newly discovered lands, Europeans found medicines and herbs unknown in Europe. Here particular mention should be made of the Portuguese Garcia d'Orta, whose descriptions of medicinal plants had a profound influence on the contemporary scientific world.[64] The introduction of these new plants opened the European scientific mind to their possible uses and classification, the latter process reaching its culmination with Linnaeus in the eighteenth century. Languages the Europeans now encountered also encouraged new systems for the classification of languages.[65] The imported knowledge thus opened up fresh vistas in botany, geography, languages, and social customs. The new European strength and ability to combine ideas from different sources in order to create new knowledge only accelerated in subsequent centuries and contributed to the sharp differences between the ongoing development of knowledge in Europe and in Asia.

As described above, part of the enormous corpus of Indian literature on the practice of medicine had percolated to the West in earlier centuries, partly through Greek sources and later through Arab sources, becoming an inherent part of the Western tradition. However, in the eighteenth and subsequent centuries, two important new South Asian contributions arrived in Europe—plastic surgery and, possibly, vaccination against smallpox. Alvares and others have described how plastic surgery techniques worked their way into British practice in the nineteenth century through translations of Sanskrit literature and personal observations of British travelers in India:[66] "Indian [plastic] surgery remained ahead of the European until the eighteenth century."[67] By the late eighteenth century, English commentators such as Dr. Scott had also noted the delicate operations carried out by surgeons in the subcontinent, including the removal of ulcers and operations on the lens of eye.[68]

In the field of technology after the seventeenth century, cultural influences from South Asia continued to reach the West, as documented by C.A. Alvares.[69] He has noted that the first major industry to develop during the Industrial Revolution, the textile industry, owed a great deal to the transfer of technology from India. After a detailed examination of this technological transfer, based on contemporary documents, he sums up: "In England and the Continent, the textile industries were being revolutionized through the study and close imitation of the work of Asian craftsmen. And later, these improvements, harnessed to the machine, would turn the tide of events."[70]

The history of developments in eighteenth-century iron and steel technology, also an essential part of the Industrial Revolution, shows the indirect influence of Indian steel technology. South Asia had long been considered the source of the high-quality steel *wootz*, much admired and used in the eleventh century for the manufacture of Damascus swords, and ninth-century Islamic literature attests to the use of Sri Lankan steels in high-quality sword-manufacture. A joint study by Sri Lankan and British archaeologists of nearly 100 ancient steel-making furnaces in Sri Lanka, published in *Nature* in 1996, reveals a hitherto unknown system of steel-making. Carbon dating of the different sites places them between the seventh and eleventh centuries. The technique used wind power and created high-grade carbon steel, and the sites provide the oldest evidence of the industrialized production of this type of "furnace steel."[71] Indian steel, or wootz, which had been manufactured from the very earliest times, was introduced to eighteenth-century England, where it was highly valued. In the 1790s samples of wootz created intense scientific and technical curiosity in England; after examination, English experts pronounced it the best steel in the world.[72]

Until the eighteenth century, Europe believed that all languages were derived from Hebrew garbled at the Tower of Babel.[73] The key figure in the new discipline of linguistics was William Jones, a judge of the Calcutta Supreme Court who saw some of the interconnections between the Sanskrit and Persian languages on the one hand, and the ancient languages of Europe on the other. By the nineteenth century, the new discipline of linguistics (philology) was on firm ground, assisted by the works of the great fourth-century BC Sanskrit grammarian Panini—who had, in effect, developed and established linguistics as a science.[74] Panini's analysis was formal and descriptive and far in advance of any European work until the nineteenth century. He had analyzed phonology with great accuracy, and South Asia had, roughly at the same time, also developed a system of writing to match this analysis.[75]

There were also significant Eastern elements and traces in the philosophical ideas that influenced the thought of the major scientists of the seventeenth century. Philosophical influences on Newton included "Platonic elements [that] entered into his philosophy and consequently to that of modern science."[76] As I have indicated, there are South Asian antecedents not only for the ideas of Plato but also for the later Neoplatonic school. Being influenced by Platonic ideas, Newton was thus indirectly influenced by ideas which had roots elsewhere. Platonism was not a mechanistic philosophy and was out of tune with the mechanistic science of Newton, yet it is doubtful whether Newton's ideas could have emerged without his Platonism.

Newton's rival and the codiscoverer of calculus was Leibniz. Both developed calculus systems to capture change with the idea of infinitesimals. It is here that one sees another probable influence, as documented by Needham. Thus, the Buddhist concept of time moments, *khanna vada* (other English renderings are "point-instants" or "fleeting moments"), travels from South Asia to China and gets incorporated in the Neo-Confucian synthesis of Chu Hsi (AD 1130–1200). They then become incorporated in the concept of Leibniz's "monadology," which perceives the world as consisting of unextended atomic point-instants.[77]

Leibniz also had an ambitious plan to develop a universal calculus and technical language that could enable one to pursue all scientific inquiries. This was to be a universal deductive science whereby all knowledge would be deconstructed into fundamental, non-overlapping, distinct ideas. Complex ideas would be depicted as combinations of various symbols, as in algebra. Leibniz is, therefore, the Western founder of what would become, in the twentieth century, symbolic logic.[78] But one should note just such a mathematized approach to concepts appeared much earlier, in Jain attempts to outline the permutations of all possible philosophical positions.[79]

Newton's ideas were also to influence the economic and political ideas of his time through Locke and his successor David Hume.[80] These thinkers created a climate of skepticism and fostered a belief in laissez faire that challenged the given order of society and ultimately led to the Enlightenment and, indirectly, to the ideas associated with the French Revolution. Some have pointed to the surprising and detailed similarities between the thought of David Hume and of the Buddha; these commentators include Moorthy, Whitehead, and de la Vallee Poussin (especially in relation to the idea of the self).[81] De la Vallee Poussin summarized the essential Buddhist position as follows:

> According to the Buddhists, no unitary, permanent feeling or think-ing entity comes into the field of inquiry. We know only the body, which is visibly a composite, growing and decaying thing, and a num-ber of phenomena, feelings, perceptions, wishes or wills, cognitions—in philosophic language, a number of states of consciousness. That these states of consciousness depend upon a Self, are the product of a Self or arise in a Self, is only a surmise, since there is no consciousness of a Self outside these states of consciousness. . . . "There are percep-tions, but we do not know a perceiver."[82]

Nolan Jacobson notes that Hume talks in almost identical tones when the latter says:

> There are some philosophers, who imagine we are every moment intimately conscious of what we call our self, that we feel its existence and its continuance in existence, and are certain, beyond the evidence of a demonstration, both of its perfect identity and simplicity. . . . For my part, when I enter most intimately into what I call myself, I always stumble on some particular perception or other, of heat or cold, light or shade, love or hatred, pain or pleasure. I never can catch myself at any time without a perception, and never can observe anything but the perception. . . . I may venture to affirm of the rest of mankind, that they are nothing but a bundle or collection of different perceptions, which succeed each other with an inconceivable rapidity, and are in a perpetual flux and movement.[83]

Jacobson also observes that in both philosophical viewpoints, sepa-rated by over 2,000 years, "there is no thinker but the thoughts, no per-ceiver but the perceptions, no craver but the cravings. . . . The similarity . . . is striking."[84]

Jacobson explores further similarities in the context of the European intellectual climate at Hume's time, pointing out that the years from 1600 to 1769 were the period during which "the orient contributed most to Western thought and they are the years that the very foundations of modern philosophy in the West were being laid."[85] Jacobson strongly suggests that the parallels between Hume and the Buddha were very probably the result of the dissemination of Buddhist ideas at the time. He rejects the notion of an independent discovery by Hume and holds the view that Hume was influenced by the ideas brought from the East to the West, and that these provided the intellectual substratum on which Hume's ideas were based.[86] Jacobson's view, after an examination of the intellectual currents of the time, was that Hume's ideas were shaped by intellectual sources influenced by the Chinese thought pouring into

Europe at the time. And Buddhist ideas were part of that transfer from China.

During the Enlightenment, China was in some quarters the initial point of European interest, especially with regard to China's "practical philosophy," often recommended as a corrective to Europe's archaic traditions. Voltaire, on the other hand, arguing from a radical perspective, declared that India was instead the cradle of world civilization. From this perspective, the differences between popular Hinduism and philosophical Hinduism could also be transferred to a criticism of the prevailing Christianity. Voltaire believed that (Western) religion was derived from what the Indians believed in the distant past. The notion of an Indian origin for Christianity survived as a subject of debate into the nineteenth and twentieth centuries. During the French revolution, too, the concept of India was played against the claims of the primacy and exclusivity of Christianity. Voltaire was also of the view that India also had prior claims in secular learning and worldly culture.

It should be noted, however, that these contradictory images and approaches to the East were based on incomplete and fragmentary information, which fed partisans in their own parochial debates. Later, in the nineteenth century, with more detailed knowledge Europeans attempted grand historical schemes incorporating South Asia.[87] These hegemonic nineteenth-century texts were produced by such figures as James Mill, Hegel, and Marx and gave certain standard views on the subcontinent. Their ethnocentricity paralleled the ethnocentric depiction of Arab countries and culture by Western scholars of the time, so ably documented by Edward Said as "orientalism." Martin Bernal's descriptions of how the concept of "Greece" was socially constructed as the fountainhead of civilization also illustrates the European ethnocentricity of the time. A common theme in the narratives of Mill, Hegel, Marx, and others was that South Asian thought was inherently dreamlike and irrational. Indian perspectives were considered dogmatic and unscientific, whereas the European tradition was considered critical and scientific. More damaging, those South Asian defenders of Indian culture in the late eighteenth century and nineteenth century internalized this view—that "East is East and West is West."[88]

Yet from the nineteenth century, major South Asian texts were being translated directly into European languages by scholars such as Max Müller and Rhys Davids and published in scholarly series like those of the Pali Text Society in London and the Harvard Oriental Series. These translations included key texts in Sanskrit and Pali. For the first time, relatively unfiltered information, what was said and not said in the region, was available in the West.

Partly as a result of these and earlier translations, several nineteenth-century philosophical figures were attracted to South Asian ideas. Schopenhauer identified the central tenets of his philosophy as expressly shared with the Upanishads and Buddhism—partly, one should add, due to a gross misunderstanding of the Buddhist concept *dukkha*, often translated wrongly as "suffering."[89] Nietzsche was attracted to Buddhism in his later period because of his relationship with Schopenhauer and Deussen.[90] Many other German literary and artistic figures of the nineteenth and early twentieth centuries were also influenced by Buddhism, including Richard Wagner, Eduard Grisebach, Josef Viktor Widmann, Ferdinand von Hornstein, Max Vogrich, Karl Ghellerup, Fritz Mauthner, Hanz Much, Herman Hesse, and Adolf Vogel.[91]

Americans had begun to discover Asian thought in the eighteenth century, but this exploration expanded significantly in the nineteenth century: Ralph Waldo Emerson even fulsomely declared, "The East is grand and makes Europe appear the land of trifles."[92] Other admirers included Thoreau, Alcott, and Parker. Arnold's best selling book *The Light of Asia*, the Theosophist Movement, and the World Parliament of Religions in 1893 were highlights in an eventful century of bringing non-Western ideas Westwards.[93] The World Parliament of Religions was meant to be a celebration of Western civilization and its religions, but it introduced, perhaps for the first time in the Western sphere, South Asian thought to a large Western audience. Colonel Olcott, who in the mid 1870s had served as the special commissioner on the investigations that followed Lincoln's assassination, and Madame Helena Petrova Blavatsky together established the Theosophical Society in New York in 1875 with the objective of studying "ancient and modern religions, philosophies and sciences." The movement was a curious mixture of a fascination with the occult and a search for serious knowledge. It attracted many influential persons, including Thomas Edison. After Olcott and Blavatsky had lived for some time in South Asia, further sympathetic transmission of ideas Westwards occurred. They also helped inspire in the local South Asian population a sense of the worth of their intellectual heritage, although because of the Theosophist interest in the occult, their efforts were not always what we would call scientific.

By the end of the nineteenth century the pace of exchange had intensified. Dale Riepe has noted the influence of Buddhism on epistemology, psychology and on ideas of the self in the thinking of William James, Charles A. Moore, Santayana, Emerson, and Irving Babbitt.[94]

As we enter the twentieth century these contacts continue and expand. Interest in Eastern-inspired thinking sometimes leads to the creation of a background for important scientific discoveries. There are thus traces of Eastern philosophical influences and parallels in the realm of modern physics. Let me take the case of relativity and the background of Einstein's theories as seen in the philosophy of Ernst Mach and nineteenth-century discussions of higher spatial dimensions.

The idea that there are spatial dimensions beyond the three we normally experience (the x, y, and z dimensions) was essential to the acceptance of Einstein's idea of a space-time continuum as part of a four-dimensional entity. The publication of the Theosophist Blavatsky's *The Secret Doctrine* inspired interest in extra spatial dimensions. This was not by any means a scientific tract in the usual sense of the word, yet theosophy's antipositivist views were very similar to hyperspace philosophy. Theosophist C.W. Leadbeater equated four-dimensional sights to the idea of "astral vision," and although Blavatsky herself did not believe in a fourth dimension, many theosophists became interested in the idea. More importantly for science, two of the key philosophers of hyperspace, Flaude Boyden and P.D. Ouspensky, got interested in the fourth dimension because of their background in theosophy.[95] In a book with a foreword by Einstein on the recent history of theories on space, Max Jammer highlights the importance of P.D. Ouspensky's *Tertium organum* in sensitizing the new perspectives on space. Ouspensky's first book, *The Fourth Dimension*, had placed him at the front rank of writers on mathematics.[96] Einstein himself mentioned that whether the world was non-Euclidean was discussed extensively prior to the development of relativity.[97]

There was another possible backdoor influence of Eastern ideas on Einstein's thought: philosophical ideas associated with Hume and Ernst Mach, two of the few philosophers Einstein read between 1902 and 1904, immediately before his special relativity paper. Hume's *Treatise on Human Nature*, which Einstein is know to have read, had strong echoes of Buddhism, as noted above.[98]

Mach (1838–1916) was a major philosopher in his day, and Einstein was strongly influenced by his program and methodology. Einstein in fact gave Mach credit for significant influences on his development of special relativity. He also wanted his general relativity to conform to Mach's ideas.[99] Einstein also pointed out in a more philosophical tract written in 1946 that his general theory of relativity provides "a strong support" for some of Mach's ideas.[100]

Einstein used Mach's principle that concepts and statements which could not be empirically verifiable do not have a place in science, finding the simultaneity of two events happening at different places in space to be a nonverifiable notion. This realization led him to special relativity in 1905 and, ten years later, to general relativity. Mach's principle was therefore a heuristic that nudged Einstein into a new way of thinking. Also of use to Einstein was Mach's idea that "the totality of distant masses must be the cause of the centrifugal forces."[101] Of special note: in his obituary on Ernst Mach, Einstein emphasized an interest in the theory of knowledge as necessary for the serious student.[102]

Mach's importance lay in his being an advocate of positivist views and having a presentationalist position, in opposition to theories of representation. By the end of the seventeenth century, a view of the external world had emerged in which to "see" or "experience" implied that sensory impressions did not constitute the external world. "Primary" qualities—such as size, shape, number, and density—now represented the external cause of sensory impressions. Color, sound, smell, and so on were "secondary" qualities, and together with "tertiary" emotional qualities were considered entirely mental, with no analogues to the physical world. This representational view was, however, destroyed in the eighteenth century by Kant, Hume, and Berkeley, and a return to a pre-Galilean presentational position occurred. Sensory impressions or objects were now once again considered parts of the external world, not just representations of it. This presentationalist position was continued in the nineteenth and twentieth centuries through positivism, logical positivism, idealism, phenomenology, pragmatism, and, one should add, Buddhism; all contemporary science follows a presentationalist perspective.

Mach himself was attracted to Indian literature and science, including mathematics, and he also respected all life. Some of his friends were Buddhists, like Paul Carus and Theodor Beer. Many of Carus's ideas were indebted to Buddhism, and his "monism" was closely related to the philosophical positions of Mach, who also contributed to Carus's journals *The Open Court* and *The Monist*.[103]

Mach's first direct appreciation of a Buddhist philosophical orientation, especially in connection with the relativity of the observer, was revealed in his *Analysis of Sensations*:[104] "But to ask that the observer should imagine himself as standing upon the sun instead of upon the earth, is a mere trifle in comparison with the demand that he should consider the Ego to be nothing at all, and should resolve it into a transitory connection of changing elements."[105] Although Mach did not write extensively on Buddhism or on his attitude to the philosophy, in 1913 in an autobiographical fragment, he wrote tellingly:

> After I recognized that Kant's thing-in-itself was nonsense, I also had to acknowledge that the unchanging ego was also a deception. I can scarcely confess how happy I felt, on thus becoming free from every tormenting, foolish notion of personal immortality, and seeing myself introduced into the understanding of Buddhism, a good fortune which the European is rarely able to share.[106]

It was in 1903 that the first references to the resemblance between Mach's philosophy and Buddhism appeared. And in his death year of 1916, a number of articles referred to Mach's restoration of the Buddhist doctrine on the "self" and "ego." The most enthusiastic observer of Mach's connection with Buddhism, Anton Lampa, wrote in his book *Ernst Mach*, published in 1918: "Mach's thought shows a remarkable agreement in its main characteristics with those of Buddha in the exclusion of metaphysics and the concept of substance." Although Einstein (the physicist) was influenced by Mach (the philosopher), Einstein himself, in his own philosophical musings later in his career, rejected adherence to what he called "Mach-Buddhist reductionism."[107]

The other revolution that defined twentieth-century physics was quantum physics, the physics of the very small. Here, the commonsensical world again breaks down and new philosophical insights are required to transcend the classical world of Newton. The key figure in quantum physics, equivalent to Einstein in relativity, is Schrödinger, whose celebrated equation defined the behavior of particles and waves at the microscopic level. If Ernst Mach had Buddhistic philosophical views, then Schrödinger was even more deeply identified with South Asian thought: he was a professed Hindu of the Vedanta philosophical tradition.

Ludik Bass, in a recent review of a biography of Erwin Schrödinger, notes that "throughout his adult life, Schrödinger adhered to a form of monistic idealism which he called the Vedantic view, which denies plurality of minds. His adherence went beyond opinions. Almost all that seems strange about him flowed from his striving to actually live his philosophy. . . ."[108] Schrödinger was clearly influenced by two Western philosophers, Mach and Schopenhauer; the latter had written about his chief work, *Die Welt als Wille und Vorstellung*: "I do not believe that my doctrine could ever have been formulated before the Upanishads." Schopenhauer's direct debt to Buddhism is of course well known.[109] Walter Moore, Schrödinger's biographer, further notes that "the unity and continuity of Vedanta are reflected in the unity and continuity of wave mechanics."[110] Yet Moore, although aware of Schopenhauer's deep commitment to Vedanta, did not see a direct causal link between his physics and his Vedantic philosophy.

More recently, Indian physicist Ranjit Nair has subjected Schopenhauer's philosophical writings and physics to a more detailed examination. Nair finds that Schrödinger's Vedanta philosophy was of central importance to his scientific thought, allowing Schrödinger to occupy a position that normally would have been inconceivable, namely a combination of realism and absolute monistic idealism.[111]

The examples of Ernst Mach and Erwin Schrödinger bring us almost to the threshold of the present, where in the last three decades, South Asian intellectual interactions with the West have increased. There have been exchanges of large numbers of scholars and students, and in the sixties and early seventies, Western youth culture inspired a surge of interest in things South Asian. This interest was in part a passing fad, leaving some spurious remains in the still-flourishing New Age movement. Yet this interest also rang a cord of serious intellectual resonance, with implications in many pursuits. The professional interest in things Asian has also dramatically increased in recent decades. Membership of US professional associations dealing with Asia has increased from a mere five hundred in the early sixties to more than ten thousand by the mid-eighties.[112]

But it is essential to remember that South Asian thought has been much more than a recent fad in European and European-derived cultures. Its presence in Western culture and its inescapable influence on Western ideas through the centuries is attested to over and over again: from the ancient Greek era, the theory of five elements,[113] atomic theory,[114] theories of flux,[115] the doctrine of the mean, and core medical concepts; from the late Middle Ages and after the place system of notation in arithmetic, the concept of zero, algebra,[116] trigonometry, and large segments of the standard European medical text; in the Renaissance, the beginnings of chemistry with the new element, salt, in the work of Paracelsus,[117] and the impetus theory. And then there is surgery in the eighteenth century,[118] linguistics[119] and philosophy in the nineteenth century, and psychology in the twentieth century.[120] Post–Scientific Revolution Western philosophy also borrowed extensively—Descartes' *cogito* has parallels with Nagarjuna, Hume's work with Buddhist ideas,[121] not to mention Schopenhauer, Kant, and Hegel[122] as a few nineteenth-century philosophers whose work echoes the South Asian tradition in various ways, pointing toward the examples of Ernst Mach and Schrödinger in our century.

And as subsequent chapters will show, there is much more yet to be borrowed from South Asian knowledge bases.

FOUR

TRANSFORMATIONS

The story of the growth of modern science has accompanying it a set of events assumed to be unique, a set of unique changes in the attitude to the external world, and a train of unique social changes accompanying the changes in science. In the first category are the various discoveries in science and technology of the last three hundred years or so. In the second category are particular skeptical attitudes to knowledge-gathering. In the last category are the reverberations which these different discoveries had on the body social and politic, such as those associated with Copernicus and Darwin.

The given story of the growth of science is about these "unique" events. This chapter, while recognizing the supreme importance of the scientific discoveries, questions their uniqueness in three realms. Firstly, it gives evidence to indicate that some of the aspects of the key discoveries themselves were not unique. Secondly, it gives data on the existence elsewhere of formal, skeptical knowledge-gathering that parallels contemporary scientific skepticism. Thirdly, it examines those events that shook the Western psyche as the scientific enterprise traversed the West during the last few centuries, and shows that if these events had occurred in South Asia, they probably would not have had the same impact.

KEY TURNING POINTS IN SCIENCE

In the last chapter, we discussed the two-way flow of ideas between the European tradition and aspects of the South Asian tradition. Many of those examples tend to revise the given historiography of science, which centers on turning points that convention says constituted the trajectory that sciences took. Sometimes, as in the more sociologically oriented histories of science, these turning points are embedded in an explicit or implicit theory of history trying to explain societal factors that gave rise to science, as well as explaining the converse—societal influence on key events in science. These key moments occur in many disciplines, from mathematics, to physics, to medicine, to chemistry, to evolution, to psychology, and to changes in industrial and agricultural technology. Let us briefly review some of these turning points.

Among the greatest scientific achievements of the Renaissance were Copernicus's new astronomical system, which challenged geocentric astronomy, and the first anatomy of the human body, mapped by Vesalius: both works were published in 1543. Later achievements included Galileo's discoveries in astronomy and gravitation, Gilbert's discovery of the earth's magnetic properties, and Harvey's discovery of the circulation of the blood.[1]

Tycho Brahe and his pupil Johannes Kepler carried the Copernican system to its logical conclusion, discovering how the solar system actually functioned, with the planets moving in elliptical and not circular orbits, as Copernicus had believed. A further step in these advances was provided by the invention and use of the telescope by Galileo Galilei, who discovered the craters of the moon, the phases of Venus, the rings of Saturn, and the moons of Jupiter with this new instrument, as well as making significant advances in his description of motion.[2]

Science reached maturity with the work of Newton and his contemporaries. Newton's major contribution, *Principia Mathematica*, was a culmination of the work of several earlier astronomers and mathematicians, including Galileo, Kepler, Descartes, Hooke, and Hartley. His work in astronomy helped him discover the inverse square law of gravitation and the law of centrifugal forces and to construct a detailed mathematical model of the planetary system. He also formalized the laws of motion with the three laws that bear his name and made important discoveries in optics; with the German philosopher and mathematician Leibniz, Newton invented calculus. Calculus was a vital tool for physics, as were the telescope and microscope in other fields; several discoveries in mechanics and astronomy largely resulted from its application.

The new emerging science was to be advanced by two of its interpreters, Francis Bacon and René Descartes, who saw the possibilities of the new knowledge.[3] Bacon emphasized the practical nature of the new science and its ability to give a commonsense appreciation of the everyday world, formalizing a system of observing nature and developing the inductive method. Descartes, in addition to his philosophical contributions, made a major advance in mathematics by inventing analytical geometry. Descartes' dichotomy—the strict separation of mind and matter—was to have a profound impact, freeing the scientific worker from the realm of religious interference by protecting the scientific sphere from encroachment by the religious sphere.[4]

Newton's ideas were also to have a strong impact on the economic and political ideas then emerging. This impact was reflected in the philosophy of Locke and his successor, David Hume. Hume promoted an irreverence for the Church's teachings, proclaiming their assertions about faith unacceptable. He remarked, "If we take in our hand any volume of divinity or school of metaphysics, for instance; let us ask, does it contain any abstract reasoning concerning matter of fact and existence. No. Commit it then to the flames: for it can contain nothing but sophistry and illusion."[5] Locke and Hume, and their successors, contributed to the climate of thought that shaped the Enlightenment and thus, ultimately, the French Revolution.

Let us take some of these and other key points in the modern Western trajectory and examine them from the South Asian comparative perspective. This examination reveals that many turning points are not what they claim to be: discoveries thought unique to the West had in fact occurred earlier, elsewhere, in part or fully developed. Enumerations of such parallels is important because these discoveries are normally assumed to have arisen from the Western tradition. In the chapters following, I will describe in detail many such instances, but here I mention briefly the more obvious ones in mathematics, physics, chemistry, medicine, evolution, and psychology, as well as industrial and agricultural technology. We will see their partial or full parallels in another knowledge trajectory, that of South Asia.

Mathematics

The concept of very large numbers appears quite early in the South Asian tradition, with figures of up to 10^{53} in the middle pre-Christian centuries.[6] In modern astronomy, large numbers open up a view of an almost infinitely large universe. This expansiveness is in contrast with the medieval European concept of a closed universe, as exemplified by Dante's *Divine Comedy*.[7] Similarly, the mathematics of very large numbers is necessary to particle physics in the twentieth century.

At the other end of the spectrum, it took Europe a long time to accept the zero, "a mere nothing," as a separate number with its own arithmetical logic.[8] What came to be known as irrational numbers in the West had, however, long been accepted in the South Asian tradition. Even in the sixteenth century, these numbers were not considered legitimate in the West, both Pascal and Newton holding that the square root of three, for example, could not be realized as genuine arithmetical numbers but only as geometric magnitudes.[9] Likewise, negative numbers—well accepted in the Asian tradition—were regarded by sixteenth- and seventeenth-century European mathematicians as "absurd." Even in the nineteenth century, some influential European mathematicians, like Kronecker, attempted to rid mathematics of such concepts as irrational numbers, imaginary numbers, and transfinite numbers.[10]

In chapter 3, we explored the transmission of arithmetic and algebra from South Asia through Arab intermediaries to Europe, noting how the place system of numerals and the concept of zero were transmitted, along with various computational methods. Just before and during the Renaissance, many of these South Asian techniques of arithmetic found their way into books written not only in Latin but also in the local European languages, and found a ready audience. Some of these works, as already noted, ran into several editions. In fact, the ease of calculation that the new techniques brought was one of the facilitators of the Renaissance. Various techniques and approaches in algebra and trigonometry were also transmitted and included methods for solving different equations.

We have already seen how the co-discoverer of calculus, Leibniz, was influenced through his Chinese contacts by the idea of point-moments, infinitesimals of time, drawn from Buddhist theories of time.[11] We will also later see how aspects of calculus appeared in such earlier Indian mathematical approaches as slicing a sphere into minute pyramids to get at the sphere's volume.

Irrational numbers were used from the time of the Vedas, as in the *Sulbasutras*. The use of very large numbers and other numbers considered irrational, imaginary, or in any other way strange in the West (like transfinite numbers) appears early, as does the study of permutations and combinations and the so-called Fibonacci series, along with other solutions in trigonometry and algebra. In algebra, the usage of symbols for unknown quantities was developed. Indeterminate analysis, the Pellian equation, the inverse cyclic method—associated with names like Fermat, Euler, Lagrange, and Galois; concepts of infinity and transfinite numbers; aspects of the binomial theorem; Pascal's triangle: all appear in South Asia before the times of those to whom they are now conventionally attributed. Post–Scientific Revolution developments like the Taylor and Gregory's series and Lhuilier's results are more recent European rediscoveries following the South Asian tradition.[12]

The Physical Sciences

Theories of motion constitute another turning point in the history of Western science: the development of a theory of impetus, followed by the views of Galileo and others, and culminating in Newton's three laws of motion and theory of universal gravitation.

Earlier South Asian parallels in these fields (see chapter 7) include the Vaisesika texts' descriptions of *vega* (impetus), the Jain's description of a motion in a straight line (Newton's first law), Brahmagupta's and Varahamihira's views on how the earth and sky are held together by gravitation "like a piece of iron in the grip of magnet," and Kerala astronomers development of a heliocentric mathematics similar to that of Tycho Brahe.

In the seventeenth century, Gassendi, Boyle, Newton, and Huygens introduced the atomic perspective into modern discourse. Atomic theories, first associated with Democritus in the Western tradition, were further elaborated by Lucretius in the first century BC; then they virtually vanished from intellectual view until resurrected 1,600 years later. However, atomic theories had surfaced earlier in South Asia than in Greece; they were known, for example, to an older contemporary of the Buddha, and therefore predated Democritus.[13] The theories, common to the Vaisesikas, the Buddhists, and Jains, persisted through subsequent centuries.[14] As were all theories at that early age, these were speculative in that the existence of atoms could not be proved through, for example, chemical reactions. Instead, there were various permutations and combinations on the common theme. Thus, the Jainist concept had atoms combining together because of inherent qualities such as attraction and repulsion.[15] The Vaisesika theory of atomism described atoms as eternal and spherical in form. The disintegration of a body, according to the Vaisesika view, results in its breaking down to constituent atoms. If a solid block like ice or butter melts, this process is explained as a loosening-up of the atoms due to heat, giving rise to fluidity—a concept, one should note, that predated by over two millennia the modern heat-based view of fluidity.[16]

Relativity and Quantum Physics

Relativity and quantum physics were of course not postulated anywhere else before this century. But as we have noted earlier, there are tantalizing philosophical commonalities in the South Asian tradition. I have mentioned Mach, the greatest philosophical influence on Einstein, and Mach's appreciation of radical Buddhist views on the role of the observer in reality. I have also noted that some of Theosophist/Buddhist-inspired discussions laid part of the background for the appreciation of multiple dimensions and hence of a space-time continuum.

In the case of quantum physics, a class of thorny philosophical questions arise, including whether a particle is actually "there" or not. In chapter 9, I describe the parallels between these questions and Buddhist logic, as noted by Western scientists such as Robert Oppenheimer, the head of the Manhattan Project. I also mention the many parallels that have been noted between the interconnectivity of matter, as portrayed by modern physics, and similar views in some South Asian systems.[17]

The physicist who took quasi–South Asian ideas of reality the furthest in physics was the late David Bohm, who postulated that an underlying implicate order existed in nature, of which the overt behavior of micro-physics was only the surface manifestation. These views are of course not universally accepted. But Bohm's own thinking had many obvious and direct parallels with South Asian philosophical thinking, as witnessed in, for example, his joint publications with Krishnamurti, an Indian philosopher at Oxford University.[18]

So, in this comparative perspective, the impression that twentieth-century physics inspired a gestalt switch is tempered by an analysis of the cultural background of South Asia.

Chemistry

The modern chemical tradition began with Paracelsus,[19] considered the founder of modern chemistry. As its foremost practitioner, he epito-mized Chemistry in the Renaissance. Through his efforts and those of other like-minded people, the chemical laboratory at the end of the Renaissance became similar to present-day laboratories with their fur-naces, stills, retorts, and balances. But some of these apparatus appear in both the Arabian and South Asian traditions earlier.

Philippus Aureolus Theophrastus Bombastus von Hohenheim (1492–1541) called himself Paracelsus to indicate that he was superior to Celsus, a predecessor of ancient times. In keeping with the prevailing spirit of disrespect for the past, he burned the books of Galen (131–201) and Avicenna (980–1037), and drawing on the old traditions of alchemy transmitted by the Arabs, gave the subject a new direction.[20] Alchemy appeared in India more for prolonging life than for con-verting base metal to gold. In the West, iatro-chemists', and especially Paracelsus', major contribution was the view that the human body was a chemical system consisting of mercury, sulphur, and salt. Sulphur and mercury were already known to the alchemists; the third element, salt, was introduced by Paracelsus. This theory of *triaprima* went against the four-humor theory of the Greeks advocated by the authority of Galen and Avicenna. If the body was constituted of mineral substances, then in times of disease, went the view of Paracelsus, it could be cured by chemical means.[21]

One should note here that Greek alchemical texts did not show an interest in pharmaceutical chemistry, in marked contrast to India. In the *Atharva Veda* (eighth century BC) there are references to the use of gold for preserving life. The transmutation into gold of base metals by con-coctions using vegetables and minerals is discussed in the Buddhist texts of the second to the fifth centuries AD. In the Arabic literature one finds, for the first time outside the Chinese and Indian cultural realms, an emphasis on medical chemistry. In fact, many Indian alchemical ideas had been transmitted to the Arabs.[22] A salt-based alchemy had already been taught in twelfth-century Saudi Arabia by a South Indian alchemist named Ramadevar. This probably percolated further West through the usual Arab intermediaries and helped to nurture the European alchemical tradition.[23]

Anatomy and Surgery

The father of modern Western anatomy, Vesalius, is known to have visited burial grounds to study bones. However, bones were studied, classified, and numbered in South Asia from the time of Susruta, the practice definitely dating to at least a few centuries before the Christian era. One of the required early meditation practices of Buddhist monks was to go to grounds where the dead had been abandoned and observe the decaying body as its bones were revealed. Buddhist records give a count of the main bones and enumerate them up to a figure slightly higher than three hundred. Buddhist monks, it is known, also indulged in medical practice, and even wrote several special texts. While these early exercises in anatomy studies have not left us with sophisticated diagrams like those of Vesalius, they point to detailed efforts in a detached study of anatomy.[24]

Parallel to the limited knowledge of anatomy in Europe prior to Vesalius, was the low prestige attached to surgery—which was practiced only by barber surgeons—until the eighteenth century. In India, however, surgery was an honored calling, from at least the pre-Christian time of Susruta.[25] The social acceptance of surgeons extended to the highest social strata: a fourth-century Sri Lankan king, Buddhadhasa, was also noted for his surgical feats.

Circulation of Blood

The modern mapping of the circulation of blood has been conventionally credited to William Harvey (1578–1657), yet this attribution has stirred controversy among medical historians. Early Greek ideas on the functioning of the heart and blood vessels were rudimentary. Galen, without any idea of circulation, held the view that blood moved in a flow and ebb process. Avicenna, on the other hand, viewed that blood was continually created in the liver and carried to other organs. This view was also accepted by Vesalius, in spite of his pioneering the field of dissection. Ibn Al-Nafis (1010–88) criticized Galen and Avicenna and discovered the lesser circulation. Others, like Servetus (1511–53) and Colombo (1516–53), continued this new tradition, until Cesalpino

(1524–1603) explained general circulation and Harvey came up with the complete experimental exposition of circulation.[26]

The circulation of blood is maintained by three factors: contraction and relaxation of the heart muscles, the heart's working as a pump, and the movement of the blood throughout the body. Sastry, reviewing the classical South Asian literature, points out that all three factors had been explicitly recognized in works dating from the early pre-Christian centuries to circa 1000 AD. The *Susruta Samhita* had described the muscular structure of the heart; the *Satapatha Brahmana* and *Yogavasishtha* described how the heart works like a pump; and the *Charaka Samhitha* described the circular motion of blood for the body's nourishment.[27] Harvey's descriptions are not there in detail, but the major elements are outlined.

Geography

The voyages of discovery helped expand the European geographical perspective. New lands became known, as did new plants and new customs. But how unique was this knowledge? Europeans on the Atlantic seaboard may not have known much of the world outside. Some still believed in the mythical descriptions of South Asia derived from Greek sources, while others relied on the more scientific classical sources, such as the maps of Ptolemy or the later descriptions of Marco Polo.

But if we shift our focus away from Europe, we will come across several cultural areas which had a geographic perspective as large as those of Europe after the age of discovery, and very much larger than that of the Middle Ages. One set of Asian travelers, Buddhist pilgrims, is known to have covered the whole of Asia; some Buddhists from China went as far as the Americas, according to American archaeological research. Buddhist artifacts have been found as far as Rome, London, and under the oldest church in Sweden.[28]

Shifting the focus to a geographically small country, my own Sri Lanka, illustrates the early widespread geographical knowledge possessed by inhabitants of South Asia. (I use this example not out of ethnocentricity but because I am familiar with the material, and so can reel off details easily.) Sri Lanka is represented in relative detail on the earliest Greek map, that of Ptolemy; in fact, it appears as large as the Indian mainland. This cognitive distortion is probably the result of extensive trade contacts with the country at that time. Pliny records visits of four Sri Lankan ambassadors to Rome, while Roman coins have been found throughout Sri Lanka. Sri Lankan Buddhist missionaries went far afield to different parts of Asia, particularly China, including fourth-century Sinhala nuns who helped establish a higher order there. Important travelers also passed through the country, including the Chinese Buddhist Fa Hsien in the fifth century, Marco Polo in the thirteenth century, and the Arab Ibn Battuta in the fourteenth century. There are recorded connections with Persia, many other parts of South Asia, and Southeast Asia, including intensive cultural contacts with the countries known today as Burma, Thailand, Cambodia, Laos, and Indonesia. Because of its position on global trade routes, Sri Lanka acted as a way station, its familiarity with other countries exceeding that of western Europe in the Middle Ages.

But Sri Lanka's geographical and intellectual reach were common to other South Asian peoples, whose knowledge was eventually tapped and collated by the European voyagers who showed Europe its other, South Asia.

Biology and Evolution

Chapter 1 documented the results of anthropological research on the intellectual curiosity of the simplest of groups living in very small communities in forests. Like modern personnel using the Linnaean system, these earlier peoples classified elements of their world through the similarities and differences of what they studied. The difference between a Linnaeus and a local forest dweller is that the former had a larger variety of items to work with. Linnaeus's large collections came from stores of tropical plant types built up in Europe during the voyages of discovery— first by the Portuguese and then by the Dutch, both having their sources largely in South Asia.

The idea that all life is interrelated was part of received mythology and ingrained in all South Asian cultures; humans were not separated from the animal kingdom by a creation myth. In some of the religious traditions there is an evolutionary perspective that goes back to very early times and, in its treatment of lineages, echoes current concerns.[29] Lyell's discovery of sea fish fossils on Mt. Etna sparked nineteenth-century debates on the age of the earth, debates that would not have occurred in South Asia, where the earth's vast age was accepted as a matter of course.

Psychology

Several pioneers in the field of psychology, especially in the last century, moved European knowledge out of its set ways. In chapter 8, I outline a whole set of results from South Asia that predate by several millennia Western research, including work now coming out of Western universities and research centers.

Industry and Agriculture

So far, we have examined turning points in science. But science and technology have often worked together, especially in later centuries, where developments in science fed those of technology, and vice versa. And the greatest initial impact of Europe on the rest of the world was through the Industrial Revolution, which rested on a prior Agricultural Revolution.

The Industrial Revolution emerged in Britain. But without detracting from the major achievements of the British, it should be noted that some of the technological achievements of the Industrial Revolution resulted from the combination of elements drawn from other traditions. There were also direct borrowings, too. Claude Alvarez has noted that the textile industry, the first to develop during the Industrial Revolution, owed a great deal to the transfer of technology from India. After a detailed examination of this technological transfer based on contemporary British documents, Alvarez sums up: "In England and the Continent, the textile industries were being revolutionized through the study and close imitation of the work of Asian craftsmen. And later, these improvements, harnessed to the machine, would turn the tide of events."[30]

The history of iron and steel technology in the eighteenth century, an essential part of the Industrial Revolution, shows at least some influence of Indian steel technology. As mentioned in the previous chapter, Indian steel, or wootz, which had been manufactured from the very earliest times, was introduced to England in the 1790s and was considered the best steel in the world.[31] Dharmapal, commenting on an English study of wootz, notes that

> Its qualities were thus ascribed to the quality of the ore from which it came and these qualities were considered to have little to do with the techniques and processes employed by the Indian manufacturers. In fact it was felt that the various cakes of *wootz* were of uneven texture and the cause of such imperfection and defects was thought to lie in the crudeness of the techniques employed.[32]

This view was revised only a quarter of a century later. An earlier revision was not possible because the fact that "iron could be converted into cast steel by fusing it in a closed vessel in contact with carbon" was yet to be discovered. It was only in 1825 that a British manufacturer took out a patent for converting iron into steel by exposing it to the action of carburetted hydrogen gas in a closed vessel, at a very high temperature, by which means the process of conversion is completed in a few hours, while by the old method it was the work of from 14 to 20 days."[33] The British Royal Society now described wootz as being of "harder temper than anything we are acquainted with."[34]

But the Industrial Revolution went hand in hand with an agricultural revolution which, by increasing the productivity of the land, freed laborers to become part of the new industrial proletariat. The prior South Asian parallels here are very revealing, and I am indebted here to the excellent work of the PPST group in Madras, who has brought together the material from British sources.

The *Encyclopedia Britannica* lists three major innovations that transformed British agriculture in the "era of Improvement" in the eighteenth century: Jethro Tull's invention, in 1731, of the drill plough, "whereby the turnips could be sown in rows and kept free from weeds by hoeing thus much increasing their yields"; the introduction of crop rotation in 1730–38 by Lord Townshend; and the selective breeding of cattle, introduced by Robert Bakewell in 1725–95. It has been said that through crop rotation, "the average yield of wheat in England increased from 8 bushels per acre in the early eighteenth century to 20 bushels per acre within a few decades prior to 1840, and it established England as the school of agriculture for the entire Western world."[35]

But all of these "innovations," according to British records, had in fact been common to South Asian practice from early times, although perhaps not all in the same place and time. The use of the drill plough in India was noted by a British colonial administrator, Captain Thos Halcott, in 1797.[36] R. Wallace, a British agricultural scientist who had traveled extensively in India, concurred in an important book on agricultural conditions in the country, *India in 1887.* Wallace described various kinds of drill ploughs used in India "from time immemorial."[37] In another authoritative source, *A Text-Book on Indian Agriculture* (3 volumes, 1901), J. Mollison of the Bombay Agricultural Department— the first Inspector General of Agriculture in India—confirmed the use of conscious breeding in cattle, "for a very long time."[38]

A form of crop rotation had been practiced in Vedic times (first to second millennium BC), when in the same field rice was grown in summer and pulses in winter. Roxburgh, considered the "Father of Indian Botany," noted that "the Western world is to be indebted for this system of sowing."[39] Based on his own observations on rotation of crops and mixed cropping, Wallace noted that "an extraordinary variety of rotations" was practised in India and that these systems were "much more varied and numerous than in England."[40]

Dr. John Augustus Voelcker, Consulting Chemist to the Royal Agricultural Society of England, was commissioned in the late nineteenth century by the British Government to make enquiries on Indian agriculture, published as *The Report on the Improvement of Indian Agriculture* in 1893.[41] Voelcker noted a form of rotation of crops that involved interplanting a leguminous crop with a cereal crop, an arrangement that could yield many complicated patterns and permutations.[42] Mollison gave further evidence that crops were systematically rotated with leguminous crops.[43] These techniques were not known in the Western scientific tradition until late in the nineteenth century. Only in 1880, after considerable, drawn-out controversy which lasted for thirty years, did Western science finally accept as fact that pulse crops enriched the soil.[44]

THE SEARCH FOR VALID KNOWLEDGE, REASON, AND BLIND FAITH

The scientific approach and methodology is assumed to consist of a set of skeptical procedures that would yield verifiable results on our internal and external environments. But all Indian knowledge systems were interested in valid and correct knowledge, through means described in texts from the many different schools. I have already mentioned the formal methodologies for obtaining valid knowledge in medicine and surgery. Such approaches were common in many intellectual systems. Many emphasized skepticism, observation, and reason. There were of course others that emphasized faith and a strong allegiance to a particular point of view. What follows are some brief examples of South Asian philosophical attitudes to knowledge-gathering that parallel those of the scientific enterprise; there are several others not mentioned here.

Thus the *Anguttara Nikaya*, a Buddhist canonical text, says explicitly:

> Do not put faith in traditions even though they have been accepted for long generations and in many countries. Do not believe a thing because many repeat it. Do not accept a thing on the authority of one or another of the Sages of old, nor on the ground that a statement is found in the books. Never believe anything because probability is in its favor. Do not believe in that which you have yourselves imagined, thinking that a god has inspired it. Believe nothing merely on the authority of your teachers or of the priests. After examination, believe that which you have tested for yourselves and found reasonable.[45]

The orthodox tradition of Purva-mimasa accepted the reality of the world. For its followers, the correct knowledge (*pramana*) is obtained through six methods: perception (*pratyaksa*), inference (*anumana*), verbal testimony (*sabda*), analogy (*upama*), presumption (*arthapatti*), and nonexistence (*abhava*). Some variants accepted only the first three methods. Still later, other schools accepted only five, having excluded nonexistence.[46] The Nyaya-Vaisesika system, a scientific system of logic outside of religious beliefs, had sixteen categories of evidence. The later Vaisesika had six means of arriving at correct knowledge: substance (*dravya*), quality (*guna*), motion (*karma*), universal (*samanya*), differential (*visesa*), and inference (*samavaya*).[47] Many traditions had sophisticated lines of exploring how to arrive at testable truth.

THE SCIENTIFIC REVOLUTION AND SOCIAL LIFE

The Scientific Revolution's aftermath rocked the Western collective psyche in several important ways. There was a campaign against those who believed in a nongeocentric universe—as exemplified by the trials and tribulations of Copernicus and Galileo. The church would not allow its imagination to admit anything more than a reading of the Bible for explanations of the workings of the universe. Similarly, in the biblical version of events, humans were created by a God, and at a very recent time. These literal readings were still dominant nearly two centuries after the onset of the Scientific Revolution.

"God" in this rendering had created man in "His" image, and also man's home, the earth. This act of creation took only a few days, and it happened only in the very recent past: 4004 BC to be exact, according to the detailed calculations of Bishop Usher in the seventeenth century. Man, being created in God's image, was set apart from all other creatures in the animal kingdom because they had been created for him by God. Similarly, the earth itself had a special position as the abode of God's image; the rest of the heavens had to circle around it. Time itself begins only a short time ago, with the "Creation." However, Copernicus eventually challenged the idea of earth and man's central position, Darwin was to challenge the special position of man in the animal kingdom, and subjects such as geology challenged the narrow concept of time.

But in South Asia, these revisionary theories would not have had the impact they had in Europe. None of the religious or philosophical South Asian traditions give man or his abode the separate role or location designated in the Judaeo-Christian tradition. The horizons in all South Asian traditions are much wider, more imaginative, and more sophisticated. In the more popular nonphilosophical traditions, man is one of a variety of beings, physical and nonphysical, and may be born as one of these other creatures, and vice versa. All, including the gods, are subject to *rta*, universal law, and karma, causal engine. In this mythical scheme man is related not only to monkeys but also to elephants, deer, and eagles; humans could be born as animals, and animals as humans. Living creatures also included unseen, microscopic life forms, the chain of life extending from the very small to the very big.

Further, in South Asia, large time spans comparable with those found in modern astronomy were part of mythology. This same world view

held the earth as only one world out of many abodes of sentient beings—thus the Jains belief in two suns and two moons, a strong nongeocentric position. And in the Buddhist tradition there are three world
systems: minor, middling, and major. A minor world system has hundreds and thousands of suns and earths and "higher worlds." The middling world system is a hundred thousand times as great, and the major
system one hundred thousand times still greater.[48] These three systems
are not the result of modern observation through the powerful tele-
scopes. They are the results of observations, with the naked eye, of a
large numbers of stars in the sky and the imposition of a speculative
conceptual framework on those observations. The old Buddhist framework does not necessarily coincide with the modern one, however, but it
is much larger than similar imaginative constructs in the pre-modern
Western traditions.[49]

The universe and the process of its coming into being are therefore
viewed from a perspective entirely different from that of the Judaeo-
Christian tradition: "The universe is very old, its evolution and decline
are cyclic, repeated ad infinitum."[50] To indicate the length of time involved, one should mention that the longest cycle of evolution within
the Buddhist conceptual framework is 311,040,000 million years.[51]
The Christian perspective, which places the creation of life a few thousand years ago, is insignificant in comparison. So is the idea of one
single abode for creatures—apart from hell and heaven—in comparison with the innumerable universes of South Asia's mythical schemes.
So a theory such as Copernicus's, which denied the exclusiveness of the
habitable earth, would not have created the controversy in South Asia
that it did in Europe. It could well have been accepted easily.

The same could be said of Darwin's theory of evolution. Another
Anglican bishop, William Wilberforce, ranted in the nineteenth century
against the evolutionary theories of Darwin. Echoes of these fundamentalist traits continue in today's demand, by American Christian fun-
damentalists, that creationism be taught alongside evolution, a demand
which found much sympathy with former President Reagan.

In contrast, most South Asian schemes have evolutionary cosmogonies. Thus, although the Buddha advised against the search for ultimate
causes or ultimate beginnings, Buddhist literature gives examples of
evolution, beginning with the remote past and extending into the distant future, with time frames on the order of magnitude referred to in
the previous paragraph. This Buddhist evolutionary scheme, *parinama*,
was affirmed by certain groups to mean "that nature is a unitary entity
which evolves in varying forms, including minds [here regarded as distinct from souls]"[52] The interrelationship between all forms of life constitutes a central element of all South Asian belief systems. All higher
animals were considered sentient beings in most traditions, differing

only in degree from humans. According to these belief systems, Darwin's view of evolution would not have produced the trauma in South Asia that it did in Christian Europe.

The ideas of thinkers like Hume, who were also influenced by the zeitgeist of the Scientific Revolution, brought new insights into the human condition in Europe and an irreverent look at existing dogmas. Some of Hume's formulations, we now know, paralleled statements of the Buddhists, probably from direct or indirect borrowing.

At the early period of the Scientific Revolution, Descartes postulated his dichotomy of mind and body, a separation that allowed a God to exist. This principle unwittingly shielded the scientific domain of enquiry from direct influence by God, yielding unto his God largely a ceremonial, and for all everyday purposes, a nonfunctional role. This Cartesian dichotomy is today breaking down and yielding in several disciplines a different view of the subject-object, mind-body relationship. Descartes' division of mind and matter was also accompanied by a mechanical view of the body: "If there was a machine that had organs and the external features of a monkey, or some other animal, we would have no way at all of knowing that it was not, in every aspect, of the very same nature as those of animals."[53] Descartes did not fully include humans in this category because, true Judaeo-Christian he was, he thought that only humans, because they possessed the speech facility, had the power to communicate and apprehend complex thoughts.

But the machine metaphor itself had been used to describe the body in the South Asian tradition at least one thousand years previously, albeit from a different perspective and with entirely different assumptions of the relationship between mind and matter. In this case, the state of humans was discussed from the idea of "no self." This Buddhist perspective denies the separation of the body and mind and is philosophically both different from and more sophisticated than the Cartesian dichotomy.[54] Let me give as illustration two selections from the writings of the fifth-century commentator of the Sinhala tradition, Buddhaghosa:

> The mental and material, both are here in fact,
> A human substance though cannot be found,
> Void it is, set up like a machine,
> A mass of conflict, like a bundle of grass and sticks.[55]

Buddhaghosa quotes the verse from the "ancients," the Buddhist writings in the Sinhala records from the third century BC to the fourth century AD (which Buddhaghosa and other commentators translated and codified in the fifth century AD into the more universal language of Pali). Buddhaghosa continues describing the "self"-less, soul-less (the standard Buddhist description of the true human entity) human machine further:

> As a puppet walks and stands through a combination of wood and strings, although it is empty, without life, without impulse, so this contraption of mental and material factors, void, without soul, without free will can walk and stand, as if it had will and work of its own.[56]

Another important challenge to European orthodoxy, one which shook the Western psyche, came from Freud's and his followers' insights into psychology. Freudian theories are now questioned, but some of their key elements—like the existence of a subconscious—remain as basic principles of psychology. But the importance of Freud lay not in the particulars of his theories but in the jolt to the European psyche that was the result of thinking of mental phenomena as capable of everyday analysis.

But the South Asian theories of mind, personality, and the human condition encompassed much larger horizons than those of pre–twentieth-century Europe, with a multitude of theories adopting a wide variety of intellectual positions regarding the human psyche. The detailed psychologies in South Asia, as pointed out in chapter 8, were all observational. Some schools believed in an unconscious; some schools (like varieties of Tantra) had a place for sexual diagnosis of the human condition, including "therapeutic" practices in which a ritualistic celebration of sex occurred; other formulations took the opposite view on sex. Similarly, the mind was analysed, and many layers beyond the immediately conscious were assumed, and there were many theories dealing with the unconscious. The psychology of the region was more varied, richer, and more sophisticated than simpleminded Christian dogma. The theories of a Freud or a Jung could easily have found a place, though perhaps not a prominent one, in this broad spectrum of intellectual positions.

Other aspects of science changed conventional world views, as did the perspectives created by relativity and quantum physics, further releasing humans from the uniqueness of a single observational platform. These were less debated on the Western public platform, because by the twentieth century, Christian orthodoxy had lost its earlier hold. While by no means wishing to detract from the intellectual strength of twentieth-century physics, one should again note that South Asian cosmogonies would not have had difficulties in adjusting to this discipline. I have already noted Mach's observation that the Buddhist world view was even more radical than that of Einstein. (Let me hasten to add that this congeniality of world views does not by any means imply that Buddhists had an inkling of Einstein's theories of relativity.) Quantum physics also brought changes, with non-commonsensical theories of the nature of the very small. Some of these views have their parallels in multiple logics and arguments for the interconnectedness of phenomena, both of which occur in several South Asian discourses. Varieties of multiple logics inhabited different schemes, like those of Jains and Buddhists.[57] And in a popular book published nearly twenty years ago, Fritsjoff Capra brought to attention some of these seeming parallels. I have shown elsewhere that some of these parallels were spurious because Capra had misunderstood some Eastern positions in his essentially romantic, New Age journey.[58] But all the same, he did point out some of the parallels, indicating again that the New Physics would not have had a major world-view-challenging position in South Asia.

Science indirectly overcame the religion-based dogmas of the European Middle Ages. But the word "religion" has a heavy load of connotations carried over from the Judeo-Christian roots of the English usage. The word also describes South Asian belief systems, but they comprise a variety of systems which differ widely from the Judeo-Christian; some, like the Charvaks, were out-and-out materialists. Some, like the Buddhists, are atheists. All have heavy doses of philosophy as a central element of their belief systems. Some, such as the Jains, found mathematics an important ally, in fact mathematizing some of their belief systems and developing important mathematical findings on the way. (This exercise was similar to Newton's trying to find God's order through his work in physics and mathematics.) Several belief systems developed important psychological observations which are now being corroborated in laboratories.[59]

It is clear that the key figures (Copernicus, Darwin, Freud) who freed the West from religious orthodoxy and created a scientific spirit would have challenged the South Asian orthodoxy only mildly, if at all. South Asians were not necessarily more scientific than Europeans; rather, the wider range of beliefs in South Asian culture only raises wider questions concerning the social context of science, the relationship between history and science, and what constitutes science. But before leaving this discussion, I want to touch briefly on another important point of comparison: the stagnation and regeneration of science. According to the conventional view, European science stagnated between the classical period and the Renaissance, and only then began its modern flowering. This simple view of the intervening Dark Ages was a useful myth created by the Humanists of the Renaissance. Yet even if one were to debunk the myth of the Renaissance and acknowledge the fact that there was a mini-Renaissance a few centuries earlier, one would still have to admit that Western science stagnated from the classical period at least until the twelfth century.[60]

Now no such stagnation occurred in the South Asian context in the period corresponding to the European Middle Ages. In fact, most of the major advances in mathematics and astronomy took place in this period (Aryabhata and Buddhaghosa lived in the fifth century, Brahmagupta in the seventh century, Mahavira in the ninth century, and Bhaskara in the twelfth century). Although the advance of science slowed down, relatively, after the twelfth Century, it continued until the seventeenth century, as Rahman has demonstrated.[61] The "Dark Ages" in South Asian science, in relative terms, really only begin then, with the colonial incursion. Clearly the key points of the Western trajectory, if they had occurred in South Asia, would not have led to the problems and controversies that occurred in the West.

The aim of these comparisons is to highlight the ethnocentricity, conscious and unconscious, of traditional historiography. This ethnocentricity is partly the result of the colonial period, when most of the histories of science were attempted.

ON EUROCENTRIC BLINKERS AND CONTRARY EVIDENCE

The eleventh-century Arab astronomer and historian of science Said al-Andalusi, in his survey of nations *Kitab Tabakat al-Umam* ("The Categories of Nations"), placed India first in its development of science at the time.[62] This reading—from a representative of an ethnic group that helped kindle the Renaissance—is a precursor to some of the facts arising from current historical records that we present here. These revisionist facts suggest that elements of turning points in the Western scientific tradition appeared, as discrete intellectual pieces of the puzzle, in other traditions, but they did not come together to create particular amalgams until the Scientific Revolution.

But when facts about South Asian scientific knowledge came to light in the early decades of this century, they were explained away under the prevailing ethos of a strong Eurocentricity. Thus, when Arab sources pointed to the Indian origins of modern sine functions, a typical turn-of-the-century reaction was to argue that the knowledge had in fact come from the Greeks, via transmissions to South Asia, a position held by Paul Tannery.[63]

Similar attitudes were shared by the early British historians of Indian science. In *Indian Mathematics*, G.R. Kaye wrote

> The achievements of the Greeks in mathematics and art form the most wonderful chapters in the history of civilization, and these achievements are the admiration of western scholars. It is therefore natural that western investigators in the history of knowledge should seek for traces of Greek influence in later manifestations of art, mathematics in particular.[64]

And so Kaye devoted considerable time to showing the derivative nature of Indian mathematics on linguistic grounds. His knowledge of

Sanskrit for this comparative exercise in philology was such that he depended solely on local pundits for translations. M. Berthelot (1827–1909), a leading chemist as well as historian of the subject, formulated a similar response. Having come across detailed descriptions on the preparation of alkalis in the *Susruta Samhita*—compiled a few centuries BC—Berthelot suggested that these descriptions were in fact a later interpolation.[65]

Another misconception advanced by Eurocentric historiography is that forums for rigorous intellectual discussion developed only in the West. Thus J.D. Bernal noted that Plato's Academy provided a central forum for discussion until the sixth century and was a precursor of today's universities. But forums for deep intellectual discussion existed in India from early times, examples being the great universities at Benares and Taxila (certainly known in the sixth century BC) and Nalanda. Among the well-known alumni of Taxila were fourth-century BC grammarian Panini, Kautilya (author of the renowned economics and political text the *Arthashatra*), and Charaka, one of the two great names of Indian medicine. According to A.L. Basham, "It seems that veritable colleges existed at these 'university towns.' Thus we read of an establishment at Benares with 500 students and a number of teachers, all of whom were maintained by charitable donations."[66] The fifth-century Chinese scholar-traveler Fa Hsien reported large monasteries in Sri Lanka, each with nearly 10,000 inmates in vigorous scholarly debate and discussion.

These remarks, which aim to correct the historiography of science and remove its Eurocentric bias, should not divert us from the fact that the development of science in Europe since the Renaissance has been a liberating and meaningful process that has helped expand human consciousness and remove the blinkers of bigotry and ignorance. My aim is to problematize the Eurocentric imbalance, in light of recent historical research. European achievements are but part of a larger context. One aspect of that context is that some of the social and cultural challenges that science has posed to Western orthodoxy would not have occurred to the same extent in South Asia.

Eastern precursors or parallels did in fact exist, in whole or in part, at different (earlier) times and in different places. In Europe, however, the different strands came together into the particular combination of ideas—Western science—that we are all heir to. But was this the only possible scientific tradition? In a different civilization, at a different coming together of institutional and other factors, could a qualitatively different "scientific revolution" have been possible? Could we have been left with different chains of thought and ideas that would have led to different lines of exploration—and entirely different cognitive maps of physical reality?

SECTION 2

EXAMPLES OF MINING FOR CONTEMPORARY SCIENCE

FIVE

INDIGENOUS KNOWLEDGE:
COMING INTO ITS OWN

The origin of science is without any doubt in the very
anthromorphic necessity of man's struggle for life.

—Schrödinger[1]

When Columbus traveled to the Americas, Vasco da Gama sailed to
the Indies, and Magellan circumnavigated the globe, they were engaging
in two forms of discovery. One was the geographical discovery of
lands that they had heard of and knew in relative detail, as they did Asia.
The other was the accidental discovery of lands that they stumbled
upon, as happened with the Americas. But even in the case of the latter,
further discoveries depended on acquiring detailed knowledge from
locals, just as in their travels to Asia they depended on detailed knowledge of oceans and lands held collectively by people bordering the
Indian Ocean and the West Pacific. In addition, they were also using in
their voyages still other forms of knowledge—such as the compass, the
lateen sail, and the axial rudder, and arithmetical calculations—which
in previous centuries had been transmitted from Asia.

This traffic in knowledge from other cultures also initially helped the
major burst in science which followed the voyages of discovery. The development of some newer sciences, such as the biological and human
sciences, was strengthened with the knowledge of a variety of life and
social forms revealed in the newly discovered worlds. Yet in spite of
occasional inputs from across Europe, the development of science since
the Scientific Revolution occurred largely within the internal dynamics of European civilization. Propelled by new social factors, including

specialized institutional arrangements for scientific pursuits, scientific knowledge grew geometrically.

In this growth, Europe drew on its current knowledge system as well as selectively on the past European knowledge tradition. The spread of the resulting knowledge (as the only legitimated one across the globe since then) has also meant the curtailment of earlier knowledge traditions in many parts of the world. Some of these traditions, as in West, South, and East Asia, were as deep and varied as the Greek-derived European tradition, which selectively fed the knowledge system leading to the modern scientific tradition. In the course of the spread of modern science, these other traditions were often delegitimized and cognitively "lost" to science. Yet just as the Greek tradition presented many pathways that could be selectively traveled in the pursuit of modern science, there are many conceptual elements in Asian traditions that could provide useful fodder for science.

This book is about uses of this non-European conceptual and empirical fodder. It takes as its starting point a variant of the question which Joseph Needham, principal author of the monumental series on the history of science and technology in China, posed a few decades ago: why did science not develop in ancient China, even though prerequisites were present? Our variant question would be, what if science developed in China? If science, with its array of procedures, tools, verification systems, and interlocking institutional arrangements developed in a non-European civilizational base, what would its nature be? What avenues would it explore? What conceptual elements would it selectively accept? Clearly, these would be different from the European elements and would result in a qualitatively different knowledge tree.

Today, institutional arrangements for science exist in most parts of the world, so that if the barriers to the possibilities in past knowledge traditions are overcome, one could possibly tap vast conceptual stores. A parallel to such a general civilizational project exists today in the fully legitimized search by Western science for knowledge of tropical flora and fauna held by shamans, pastoralists, farmers, and non-Western medical practitioners. Much of this knowledge is far less formalized than the civilizational knowledge systems of Asia. Yet over the last few years, as both tropical rain forest acreage and the numbers of practitioners have shrunk, new, intense efforts are being made to capture this knowledge. This endeavor has led to several imaginative efforts to link this traditional knowledge base to modern ones (especially those dealing with biotechnology) before it vanishes. The result has been the partnerships of several biotechnology firms with local knowledge groups to identify and creatively use this knowledge. Such linking also raises profound commercial questions of ownership, to which we shall also refer.

This book is about using a similar approach to "civilizational" knowledge stores—that is, the formal knowledge systems of regional civilizations. For reasons of space and personal familiarity, I will restrict myself to the region of South Asia. Through examples drawn from several fields I will also indicate that a further search within the Eastern traditions could reveal a rich range of useful facts, metaphors, concepts, and solutions that could still be fruitful.

But such examples presume that the impulse toward formal, verifiable, replicable knowledge is a universal trait in all human societies. The present knowledge trajectory since the Scientific Revolution of seventeenth-century Europe has formalized this thirst, made considerable changes in approaches to knowledge (and even changed definitions of what "useful" and "real" knowledge is), made possible the wide dissemination of knowledge, and brought about key institutional changes. It has through some very remarkable men, and at a suitable historical juncture, fashioned a legacy which is today's scientific enterprise. But what has been so fashioned is in part a formalization and an intensification of a common human impulse.

To further illustrate the universality of the scientific enterprise, later chapters offer detailed evidence, through the histories of different disciplines, that what have been considered key turning points in the history of modern science have also occurred elsewhere, partly or fully. Perhaps the real importance of the recent European story in science is the fact that a unique set of circumstances brought many factors together, accompanied by a process institutionalizing the search for scientific knowledge. Ever since, that accumulated knowledge has grown at such a rapid rate that according to the calculations of Price, it doubles every eleven years.[2]

The different disciplines that constitute the family of sciences use different approaches to define what constitutes an interesting problem, a solution, and verifiable fact, and "truth." Some of the theoretical issues of what constitutes science and how it progresses, seen from a nonethnocentric, non-Eurocentric, long-range point of view, will be discussed in a later section. But at this point, I will turn to some of the technologically simplest of societies for examples.

Recent work of anthropologists on these small social groups, so called "primitive" peoples, reveals that the impulse to be scientific is universally present. I focus on these societies deliberately, because the Scientific Revolution began after the voyages of discovery, and aspects of both projects interacted with each other, namely the search for science and the search of the Europeans' "other." The imperialist perspective that accompanied both events soon began to assert the superiority and exclusivity of everything considered European. Soon, these attitudes crystallized, to varying degrees, into a view that other cultures were inherently incapable of the intellectual work that goes today under the rubric of "science." And this perspective has colored subsequent views on knowledge, views that only over the last two decades or so are being gradually rethought.

Let me begin the revision with the simplest and oldest of human groups. It had been estimated that Stone Age man had access to a vast array of valid knowledge about nature. This lore was of such quantity that studies on similar contemporary hunter and gatherer peoples suggest that "if all their knowledge about their land and its resources were recorded and published, it would make up a library of thousands of volumes."[3] Such a complex corpus of oral knowledge could not have been acquired and retained without a highly developed system of classification and codification of the individual items of knowledge.

The claim to a systematic status for this knowledge—although less sophisticated than the modern scientific or the past civilizational traditions—is best illustrated by the following words from the introduction to a textbook on taxonomy: ". . . the most basic postulate of science is that nature itself is orderly. . . . All theoretical science is ordering and if systematics is equated with ordering, then systematics is synonymous with theoretical science."[4] If these criteria are followed, then one finds that systematic classifications relevant to their environment have been made by many little traditions, as is indicated by examples taken from a number of simple cultures. Thus Amerindians have taken pains to observe and systematize scientific facts concerning lower animal life.[5] These animals do not afford any direct economic benefit to this culture; the interest in classification has been purely intellectual. Referring to another simple culture, B.E. Smith notes that "every plant, wild or cultivated, had a name and a use, and every man, woman and child knew literally hundreds of plants. . . . [My instructor] simply could not realize that it was not the words but the plants which baffled me."[6] In yet another culture, any child could identify

> The kind of tree from which a tiny wood fragment has come and furthermore, the sex of that tree, as defined by kabiran notions of plant sex, by observing the appearance of its wood and bark, its smell, its hardness and similar characteristics. Fish and shell fish by the dozen are known by individually distinctive terms, and their separate features and habits as well as the sexual difference within each type, are well recognized.[7]

Other instances of systematic classification and accumulation of rational knowledge of the environment have been supplied by several writers, for example for Hawaiians,[8] the Hanunoo,[9] and the Filipino Negrito.[10] But what should be emphasized is that this systematic thinking and process of classification are not static but are based on the constant

expansion of existing knowledge and on experimentation. As Ackernecht has forcefully noted, "in my investigations, I have found that there
is more than enough evidence to indicate that experimentation in the
old fashioned trial and error system is not unique to Western European
societies or even to civilized societies," but exists also in the simplest
ones.[11]

That this experimentation is ongoing is indicated by the fact that
many simple societies continually develop, from local plants, medicines
that have been estimated as at least 25–50 percent effective.[12] Further,
these isolated cultures have often developed medicines for diseases introduced by Westerners, a notable example being the Amerindians'
development of a cure for malaria using the cinchona bark. The manner in which this process goes on at the tribal level is best indicated in
the following observation by Fox regarding the Negritos of the Philippines:

> The Negrito is an intrinsic part of his environment and what is still
> more important, continually studies his surroundings. Many times I
> have seen a Negrito, who, when not being certain of the identification
> of a particular plant, will taste the fruit, the leaves, break and examine
> the stem, comment upon its habitat, and only after all of this, pro-
> nounce whether he did not know the plant.[13]

Further, if a plant introduced to the environment is found to be
useful, "it will be quickly utilized. The fact that many Filipino groups,
such as the Pinatubo Negritos, constantly experiment with plants has-
tens the process of the recognition of the potential usefulness, as defined by the culture, of the introduced flora."[14]

Even after recent acknowledgment of these extensive classification
systems, their objectivity has been questioned. There has been a dispute
about whether humans all over the world perceive classification of biological organisms in the same way.[15] One view holds that classification
systems differ because of variations in the information and traditions
passed down from the elders; according to this view, such ordering systems are not, in fact, based on objective observations of animals and
plants. The other view holds that the natural order determines folk biological classifications, because objective observations based on similar criteria take place all over the world. Hence, folk classifications of
different groups are similar to the extent that the different natural orders are similar. The objective patterns of similarity among organisms,
according to this perspective, determine classifications. Much recent
anthropological research (as, for example, by Berlin, Breedlove, Raven,
Boster, O'Neill, and Bulmer)[16] supports this view.

This objective impulse is further corroborated by field studies of attempts at classification by American students who had no formal train-

ing in biology and did not know beforehand the biological specimens they were asked to classify. They all came to similar classification systems based on observed criteria. It therefore appears that "people have come to share an understanding with others by directly observing the world rather than by learning from others."[17] Perceivers independently observe the environment and come to similar conclusions.

But why do peoples of the forest search out and classify these plants? There have again been two broad opposing positions in explanations of indigenous peoples' biological classification systems. One, the utilitarian—which has lost ground recently— assumes that there is no special innate curiosity in indigenous people for deciphering nature and that their classification systems are purely motivated by the need to know the biological items that will be used for practical purposes. The second, on which an increasing body of knowledge is fast accumulating, holds that indigenous people have a natural thirst for knowledge and that they engage in classification largely out of an intellectual urge. Thus field anthropological research has shown that in these classifications, people are attempting to get at the structure and inherent order of the biological system, irrespective of any mundane, practical reasons.[18]

The different organisms on earth number in the millions, and *no* system of classification, including the contemporary scientific one, can be entirely comprehensive. Indigenous people, who sample a relatively small area and for whom the larger part of the world's flora and fauna remain unknown, will necessarily classify a small number of organisms.[19]

Just like the Western-trained scientist, an indigenous biologist observes many animals and plants and develops an implicit system of classification. As field studies have shown, he or she uses exactly the same types of clues as the Western scientist to recognize a particular organism and then categorize and name it as belonging to a particular grouping. When the Western scientist and the indigenous classifier see the same species, both classify based on the differences or similarities between species. And so if they both are classifying the same animals in a given area, their grouping should closely correspond to each other, and the two resulting taxonomies would then correspond. Further, in such a case the scientific system could be used to predict the classification of organisms in a given area in the folk system, and vice versa.[20] Berlin has indeed tested out the latter hypothesis in the field and shown that it holds true in actual cases.[21] The difference between on the one hand a Linnaeus, eighteenth-century founder of the modern classificatory system, and on the other hand a folk classifier, becomes partly one of degree. Linnaeus had access to a wider store of plant samples, created through European expansion into the rest of the world.

The world, it appears, is thus littered with indigenous starting points for potential trajectories of knowledge—trajectories which, if they were developed, would have led to different explorations of physical reality. The existence of all this anthropological evidence does not solve the problem of Western ethnocentricity or of the distinctive rise of Western science, but it does help to further problematize them.

Now as the nature of the technologically simplest groups' knowledge systems has been rethought, there have concurrently been moves to recognize the practical worth of that knowledge and to use it for contemporary needs. This trend has appeared in recent work on advice about problems in general development as well as in the narrower subject of biotechnology.

Some of these uses are of a technological nature, but it should be noted here that technological developments and those of science are not in separate baskets, as was once simplistically assumed. Science and technology are closely intertwined in many ways, mutually helping each other. Thus a technological development like a telescope, microscope, or particle accelerator revolutionizes sciences in their fields—in this case, respectively, astronomy, biology, and the physics of the very small. Further, certain subjects' developments of high theory are intimately tied to very down-to-earth problems. A case in point is thermodynamics and the concept of entropy, one of the most fundamental of scientific developments. As Cardwell has shown, thermodynamics would not have emerged without the practical interest in the early industrial age on how to squeeze more energy out of heat engines.[22]

A few recent examples indicate the recommended uses of indigenous knowledge by contemporary commentators within the Western scientific sphere.

CONTEMPORARY USES OF INDIGENOUS KNOWLEDGE

Forest-dwelling peoples, Posey has noted, know, classify, and use over 99 percent of the rich biological diversity of tropical floral and faunal species. In contrast, entire tropical forests are being devastated by modern techniques in order to exploit less than 2 percent of the available species. The local knowledge of the forest peoples can thus be both an inexhaustible storehouse for new potential commercial products and the fountainhead of new ideas on sustained natural resource management. These potential new products include natural colorings and dyes, medicines, natural insecticides and fertilizers, cosmetic products, and new foods.[23] Indigenous knowledge systems of potential use in natural resource management also cover a wide field—agronomic practices, pastoral management, agroforestry, and genetic resources to name a few promising areas. The disappearance of these practices, on the other hand, has been characterized by several negative features, among others, erosion of genetic resources, drops in the fertility of the soil, vulnerability of the ecological system, and the destabilization of the soil-water-plant relationship.[24]

Ethnoecology, which constitutes indigenous ways of perceiving, managing, and using resources, has been recommended for countering the social and ecological destruction of Amazonia. Posey and his coworkers explored six areas of local ecological knowledge—cosmology, agriculture, aquaculture, gathered products, resource units, and knowledge of the interrelationships between these categories. The researchers came to the conclusion that development based on these local knowledge bases would lead to more ecologically sound production which, at the same time, would be commercially viable.[25]

This trend has been noted elsewhere. Indigenous knowledge has also been advocated for social forestry activities in Thailand in contrast to the previous practices of importing foreign experts who transfer unsuitable knowledge gained in colder, more northern latitudes.[26] Similarly, local traditional knowledge and cultural practices are being used in environmental education and development efforts among nomads in Africa. Hasty measures and instant, foreign-imposed solutions which aim at quick results, dependence on foreign aid, and then subsequent abandonment of a threatened community are now being replaced by respect for the local knowledge system.[27]

In Africa, a reappraisal of agriculture—the most foundational of human activities—is under way. Top-down, external, expert-driven agriculture has failed, and current recommendations now advocate going back to the earlier knowledge base. It is now urged that existing indigenous knowledge should be taken into account before introducing any innovations. Further, and equally important, it is suggested that this existing knowledge system should provide the *basis* for new practices and innovations.[28] Uses for such indigenous farming knowledge have been found all over the globe. For example, most traditional farmers employ a variety of techniques to control pests, techniques that can be divided into two broad approaches. One uses cultural, mechanical, physical, and biological practices and nonchemical systems of pest control that reach the problem directly. The second relies on the control mechanisms inherent in the traditional farmers' complex farming systems and utilizes biotic and structural diversity.[29] The cultural practices used for controlling pests manipulate crops in space and time and include overplanting, selective weeding, manipulation of crop diversity, and adjustments of planting and harvesting times. In addition, farmers manipulate other variables, such as the microclimate, soils, crop genetics, and the chemical environment. The result is a rich variety of techniques that rely more on knowledge—that is, on detailed knowledge of the agricultural environments—than on commodities in the form of expensive inputs. These techniques thus become very useful elements in agricultural development work.[30]

One should note parenthetically that traditional farmers' reliance on intensive knowledge has something in common with practices of the emergent postindustrial information world in industry. Here again, the emphasis is on the manipulation of information for products, in contrast to the blunderbuss approach of the typical mass production and smoke stack industries of the first industrial revolution. The use of chemical inputs in mass agriculture is the equivalent of the mass production of the industrial age, and so it appears that traditional farmers are nearer to the practices of the information age than the so-called modern practices that traditional approaches are replacing.

In Asian and Pacific countries, the suggested use of indigenous knowledge has been extended beyond agriculture and forestry, to disaster preparedness. The first director of the Asian Disaster Preparedness Center, headquartered in Thailand at the Asian Institute of Technology, has explicitly stated that one of the objectives of his organization is fostering the application of indigenous knowledge and skills in risk reduction.[31]

Brief mention has already been made of the use of ethnobiological knowledge in biotechnology.[32] Developing new biotechnology material requires access to a variety of useful genes from plants, yet only 1,100 of 265,000 plant species have been thoroughly studied. Of these, 40,000 probably have medicinal and nutritional applications for humans.[33] In fact, plants that are unknown in developed countries have many uses which have been identified as useful over the centuries by farmers, pastoralists, and traditional healers across the globe. Over three-fourths of the 121 plant-derived compounds in the (developed) world's pharmacopeia have been derived from plants.[34] This knowledge is now being gathered by multinational corporations and the plants and their properties identified; later, the particular gene responsible for a desired property is isolated, to be incorporated into a new, genetically engineered plant. The scientist who finds and transplants the specific gene is then rewarded with a patent—and the social group which through generations had identified the properties of the plant and developed it in the first place would receive no reward whatever. Further, the patented plant or seed could then be fed back as a commercial product into the agriculture of these same social groups (possibly also to the same people who identified the desired trait in the first place), replacing the existing flora.[35]

The knowledge that has been patented comes, therefore, from two sources: the original knowledge of the farmer that had over the centuries identified the plant's useful properties, and the multinational corporation that isolated and incorporated the gene. Yet under present arrangements, only the multinational corporation is rewarded. This problem has been a subject of deep debate in several United Nations organs, including the World Intellectual Property Organization (WIPO).[36]

However, novel models seem to be emerging that jointly exploit local "traditional" knowledge and that of multinationals, in a less exploitative and less ecologically harmful manner—and these cooperative projects have the potential to further increase the world's store of scientific knowledge. One imaginative solution to the tension between biotechnology creation in multinational firms and local knowledge hopes to maximize benefits for both parties. In an experiment termed "chemical prospecting," drug multinational Merck and Costa Rica have teamed up to rationally exploit the rain forest in a potentially mutually rewarding relationship. This partnership was the result of a recognition of enlightened self-interest on both sides. Merck, on its part, had woken up to the fact that a vast reservoir of knowledge of potential drugs in the tropical forest was fast disappearing. Further, because of rapid screening techniques, it was now possible, for a relatively modest investment, to yield thousands of extracts in a brief time. In return, the company would provide funds for the National Biodiversity Institute of Costa Rica, helping the latter's ecology-driven program of chemical prospecting.[37]

Programs on similar lines are being pursued by a number of organizations, including other multinationals like Monsanto. A smaller company, Shaman Pharmaceuticals, and Conservation International (an environmental group) have teamed specifically to collect material from tribal groups that have knowledge of the uses of tropical plants. The US National Cancer Institute supports programs in chemical prospecting using ethnobotany and is developing models of ownership that, it is promised, would share profits with local governments.[38]

But these solutions are not without their ethical questions. On one hand the current holders of traditional knowledge will probably be the last of their kind, as their children embrace "modernization" and ignore traditional callings; on the other hand, the multinationals that extract their knowledge are guided largely by the profit motive. The extracted knowledge would not vanish but in a modified, albeit commercialized, form would be spread around the world.

The above examples are suggestive of the universality of the scientific impulse among all humans, yet, they do not touch some of the more abstract, more sophisticated sciences. These we shall deal with in much more detail in forthcoming chapters. But to indicate that the impulse to find out applicable knowledge—the ultimate essence of science—is present in a very nascent form even in the higher animals, and so is possibly biologically driven, I will refer to current anthropological research. It has been documented that animals find and use plants for therapy, and thus a new subdiscipline, zoopharmacognosy, is now reported on in regular anthropological forums. The findings have led to a further understanding of plants as medicine.

For example, researchers have observed that certain birds, like the hoatzin, are not effected by the toxic plants that they eat, like poison oak and poison ivy. Apparently there are molecules in the bird that neutralize the poison. Chimpanzees eat a bitter plant, aspilia, when they are sick; the plant contains a compound, thiarubrine-A, that is toxic to nematodes, fungi, and some viruses. There are, in fact, many reports of animals using medicines for their ailments, including howler and muriqui monkeys, Kodiak bears, and coatis. Gorillas seem to eat plants to rid themselves of intestinal parasites, and orangutans are also reported to use medical plants. Chimps use plants for stomach illnesses, for killing parasites, and possibly to prevent diseases. About fifteen plant species have already being identified as being used by apes for medicine.[39] These studies, it is believed, could also give clues as to what plants, of the thousands in the tropical forest, could be candidates for medicines.[40]

But of course the use, by some animals, of efficacious plants does not by any means constitute more than the barest beginnings of science. But then today nonhumans, in the form of "intelligent" computer programs, perform many functions that in the eighteenth century (or for that matter even a few decades ago) would have been considered equivalent to science. So science is not some mysterious thing—a theme touched on earlier when I briefly discussed the *necessity* for the artificial intelligence of discovery machines, used in the highest levels of today's sciences.

But even shorn of its aura of mystery, sustained efforts in science cannot be made without means of recording, institutional settings, and other prerequisites. These requirements, of course, emerged in abundance after the seventeenth century in Europe. So even if forest dwellers, in their systematization, approach in many ways the efforts of modern scientific biologists, they do not exhaust other fields of scientific enquiry, let alone that of biology. They demonstrate instead that the will to knowledge is common among all humans.

"Civilizations," on the other hand, are better candidates as reservoirs of scientific knowledge. They are spread over larger geographical settings and have specialized divisions of labor, with knowledge systems attached to each such division. Specialized groups are specifically charged only with knowledge production; these were initially priests and scribes in the riverine civilizations of, say, the Nile, the Sumer, and the Indus. But within such religious pursuits themselves there arose and developed a curiosity and search for more mundane matters. Further, civilizations had means for recording what they garnered so that there could be a reliable transmission belt of knowledge from generation to generation. There were also institutional arrangements, centers of learning, and also interregional and intraregional travel that allowed for exchange of ideas, debate, and discussion. So the natural universal curiosity about nature was not only kindled but developed and formalized in every major civilizational area. Each could therefore have provided, if the conditions were right, an independent starting point for science. There would have also been elements of commonality between civilizational groups, which would have faced similar scientific problems and at times come to similar answers.

These similarities could be due to independent discoveries in the different regions or due to cross-transmission of ideas and cross-fertilization between regions. Both have occurred, as we shall notice in the more detailed subject chapters. But how did cross-traffic occur in such widely separated regions of the world? In fact, as we showed in chapter 3, there is considerable historical research to indicate that, contrary to the nineteenth-century imperialist refrain of the British poet Rudyard Kipling, the East and the West did meet, and in historical terms, quite often at that.

The examples of how indigenous knowledge sources feed sciences suggest the potential for the inflow of a variety of concepts from regional civilizations to the modern science enterprise. A contemporary search within the Eastern tradition could reveal a rich array of techniques, metaphors, and intellectual solutions that could advance in knowledge. It is to such exercises that we now turn.

SIX

MEDICINE: AYURVEDA

The validity of an abstract subject like mathematics has not been contested since the ancients. Two plus two remains four—and the theorems of Euclid are still used today in schools, as are the algebraic discoveries of Aryabhata and Bhaskhara. The ancient work may be added to or enlarged as, for example, in the development of non-Euclidean geometries; or entirely new subjects may be developed, as in topology or the study of fractals.

In the physical sciences, the situation is somewhat different. A lever may still be used to shift a large load, a principle applied in the building of large structures like the pyramids or stupas, structures of comparable height. The fine tuning of the use of a lever may be accomplished today through applied mechanics and the theoretical knowledge of the strength of materials. Yet, in spite of these continuities, sea changes have occurred in our understanding of almost all of the sciences, pure and applied. Medicine is one such area.

In the Western world, most of the medical assumptions held sacred into nineteenth century have now been eroded, replaced by testable theories of the causation of disease. Ancient explanatory systems like that of the three humors theory have been supplanted by explanations based on microorganisms, malfunctioning of organs (like the heart), or the malfunctioning of cells (cancer). This modern medical knowledge is also related to basic discoveries not only in biology, but also in other sciences, such as physics, chemistry, and mathematics. Yet even with this knowledge there are vast gaps between the universal wish for health and medical reality.

It is here that expanding the existing knowledge base to incorporate verifiable elements from other traditions would be beneficial. The tradition we explore in this chapter is India's classical medical tradition, Ayurveda, with its attendant pharmacological and surgical practices. Ayurveda is by far the dominant non-Western medical tradition in India.

AYURVEDA METHODOLOGY

Ayurveda is perhaps the oldest continuously practiced system of medicine in the world. As the name suggests, it has roots in the Vedas, the oldest literature of South Asia, some of which dates from the second millennium BC (2000 BC). Ayurveda had been formalized as a coherent system of medicine by the pre-Christian era, the classical collections (based on, and essentially being a compilation of, earlier work) being completed during the pre-Christian period. The most famous of these collections, the *Samhitas*, were written by Charaka (on general medicine) and Susruta (on surgery). There were other notable writers, another pre-Christian one of note being Atreya. This literature was added to over the centuries and constitutes a formidable body of work.

As in other pre-modern medical systems, there are elements that from today's perspective are clearly unscientific. For example, Ayurveda has elements in its doctrines relating to "medical astrology" that are clearly untenable scientifically. (But then in a supposedly more scientific time and more rigorous area of discourse, Western doctors in the 1950s held that the typical meat-ridden, fat-dripping, vegetable-short diet was the best, which we now know is scientific nonsense.) Also untenable are the pancha bhuta (the five elements of the ancient South Asian systems) and tridosha (three humors) systems of causation of disease, which latter has close affinities to the Greek humoral tradition, the Ayurveda probably being the older of the two. The pancha bhuta and the tridosha are related to each other in the Ayurveda system, in which disease is said to be the result of imbalances in the tridosha. Even though ultimately based on spurious classifications, the Ayurvedic system could, however, include in its idea of balance elements that have, through empirical observation, been determined necessary for good health.

But it should be noted that until the beginning of the nineteenth century, medical practice in the West was itself "traditional." The "scientific" methodology that had come to physics in the seventeenth century, to biology in the eighteenth century, and was only then informing practices in chemistry and thermodynamics, had yet to penetrate medicine. When it did, it brought to diagnosis the search for single causes and cures, an active principle in pharmacology. This single magic-bullet approach is undergoing more sophisticated changes today.[1] In its stead, a multifactorial causation for disease will probably emerge, incorporating ideas of balance and equilibrium.

Shorn of these foundational, philosophical questions, Ayurveda has very objective approaches within its system of diagnosis and treatment with which a modern doctor could find much to agree. Descriptions of such approaches appear in classical texts, as well as in anthropological observations of Ayurvedic practice.

Ayurveda divides medicine into two broad schools, those of physicians and surgeons. It has eight specialties: Internal medicine (*kayachikitsa*), pediatrics (*balachikitsa/kaumarabritya*), psychological medicine (*grahachikitsa*), otorhinolaryngology and ophthalmology (*urdwangechikitsa/shalakyatantra*), surgery both general and special (*shalyatantra*), toxicology (*damshtrachikitsa/agadatantra*), geriatrics (*jarachikitsa/rasayanatantra*), and eugenics and aphrodisiacs (*urishyachikitsa/vajikaranatanta*).[2]

The Ayurvedic physician, according to the classics, conducts two types of examinations, one of the patient (*rogi pariksha*) and one of the disease (*roga pariksha*). The physician examines the patient as a totality, noting the bodily state (including tissues, site of disease, digestion, general metabolism, pulse, urine, faeces, tongue, eyesight, auscultation, response to tactile stimulation, and body structure) and the constitution of the patient (age, dietary habits, and so on), as well as the season.[3] After careful examination, treatment is prescribed: it consists of avoiding the causative factors and taking medicines, having a suitable diet, following a regimen that includes suitable activity for restoring the balance of the body, or if need be, having surgery.[4]

The treatment aims at not only curing any given disease but restoring the body to its proper equilibrium. Ayurveda thus exemplifies a philosophy of total health care, as the name itself signifies—namely "the Veda of long life." It also works to increase resistance to disease, recommending daily, nightly, and seasonal routines, as well as an ethical routine. In daily living, it prescribes for a regulated diet, sleep, and sex.[5] From its inception, Ayurveda assumed that the body and mind are intertwined, that all diseases are in fact psychosomatic, an approach that Western science has now adopted. In Ayurvedic examination and diagnoses, treatments of the body go together with treatments of the mind.[6]

Diseases due to deficiencies—or, the reverse, of excesses—in nutrition were recognized early in the South Asian regimen. Some essential nutrients are in fact described in terms of traditional uses and practice. Anemia was treated with iron as early as the fourth century AD; iron deficiency was identified as such in the West only in the sixteenth century.[7]

According to Ayurveda, disease can also be caused by small organisms, *krimi*, which can be either visible or invisible. These krimis, as described in the classical pre-Christian Ayurvedic literature, include a variety of organisms; some, like insects, can be easily seen, while others are invisible to the naked eye. From the second to the eighteenth centuries, dozens of books covering various aspects of krimis in disease appeared, describing ailments caused by krimis that infect the eyes, nose, teeth, intestines, ribs, and head. They are given as one reason for miscarriage, abortion, and sterility in women and are alluded to as a cause of the infectious spread of *kushta*—that is, skin diseases.

The visible (*drsta*) krimis include flies, mosquitoes, and intestinal worms. It would be tempting to suggest that invisible (*adrsta*) krimis correspond to modern bacteria and viruses, but that would be an over-extension of the evidence, because bacteria were identified only after the discovery of the microscope. It would not be incorrect, however, to say that Ayurvedic practitioners hypothesized the existence of invisible organisms—through formulations based on what they observed—to explain diseases and their spread. This is not an inappropriate assumption: both atoms and viruses were surmised to exist long before anybody could directly see them through instruments. Susruta described how infectious diseases are spread by a variety of means, including inhalation, ingestion, body contact, sexual intercourse, and the use of infected objects. These observations, of course, predate those of Lister. Susruta also connected the formation of maggots to the activities of flies.[8]

The Ayurvedic tradition possesses sophisticated methods of preparation and standardization in its repertoire.[9] Many of the classic Indian texts written before the tenth century detail different pharmaceutical processes. These texts—*Caraka-Samhita, Susruta-Samhita, Astagasamgraha, Astangahrydaya, Cikitsakalika*, and several others—describe alcoholic preparations (*Arista* and *Asava*), preparations made from different fats, pharmaceutical processing, special methods for powdering drugs, and different types of infusions.

Susruta, the classic authority on surgery, describes three phases in the surgical care of the patient: the preparation, the actual operation, and the postoperative phase. He outlines six different types of surgery and details methods of diagnosis, different types of operations, bandages, and four methods of arresting hemorrhage.[10] He also describes various types of enemas, four kinds of sutures, and three types of needles.[11] The surgical practices he outlines include setting of fractures, operations for hydrocele, reduction of hernia and ruptures, laparotomy (opening the abdominal area), Cesarean section, extraction of the fetus after first crushing it, crushing stones (lithotripsy), removing stones (lithotomy), limb amputation, plastic and rhinoplastic operations, extracting foreign bodies, and removing piles and fistula. Some of these techniques have not changed since they were first described by Susruta.[12] Other operations in his catalog include those for treatment of dropsy, abscess, tooth extraction, and removal of foreign matter.[13] Susruta describes details of operations for the conditions of ascites, obstructions in the rectum, and for removal of a dead foetus without killing the mother, considered a very difficult procedure.[14]

Susruta also details a considerable repertoire of operations in orthopedic surgery. He classified bone injuries into different types, with six kinds of skeletal injuries, twelve types of fractures, and two types of dislocations. Clinical features of dislocations and fractures are important to practice. A fracture, according to Susruta, is characterized by swelling, tenderness to the touch or in movement, loss of functioning, varieties of pain, inability to find a comfortable position, and the presence of crepitus: all of these symptoms fit well with modern practice. Further, he details complications arising from bone injury, corrections for malunited fractures, and instructions on how to deal with compound fractures. He even specifies medicine to be taken internally to help in the healing process.[15]

Susruta's texts also cover plastic surgeries, such as repairs to the nose and severed lips. For example, he catalogs fifteen methods of repairing ear lobes, indicating the depth of detailed knowledge and practice at the time.[16] The section on plastic surgery for the nose is illustrative, and following is Krishnamurty's translation of the relevant passage:

> I shall tell you exactly the method of joining of a nose that has been cut off. Take from a tree, a leaf of the same size as that of the nose concerned. Place on the patient's cheek in such a way that it stays supported, while you trace the cut portion of the nose precisely on it. Remove now a portion of the skin as per this tracing from the cheek nearby by slicing; do so in such a way, that a small part of it remains still attached to the cheek. Place this deflected portion of the skin on the injured nose and then suture it quickly on any pattern that is suitable to the case concerned, with all possible care. Then raise up, the two alae around the noses by inserting two sticks of castor plant and see that the final form of the nose would thus (i.e., by your raising) look shapely and natural. Then tie and dress by the healing powders of *patanga* and the *anjana* of *yasti* and *madhuka*."[17]

There was also considerable knowledge of dentistry, including extraction, treatment for toothache, and the making of dentures. In 1194, the king Jai Chandra, when beaten in battle, was recognized by his false teeth.[18] Other cosmetic practices in fields allied to medicine included use of hair dye, as the thirteenth-century Sinhala text *Saddharma Ratnavaliya* testifies.

Ayurveda also had strict ethical procedures, formalized at least two centuries before the Hippocratic oath was formulated. The elaborate oath required of disciples of Ayurveda contained more than seventeen do's and don'ts for the medical practitioner. In addition, the texts were peppered with additional ethical instructions.[19]

KNOWLEDGE ACQUISITION TECHNIQUES

Civilizational systems of knowledge such as Ayurveda, although traceable to a long historic past and sometimes to paradigmatic positions set very early, are not static. A conventional dichotomy holds that such systems are oriented only to the past and, further, are not competitive, but consensual. In contrast, modern science is viewed as having a more open epistemology, which is the reason for the higher quality of modern science. Yet studies on classical texts and actual case studies of Ayurvedic medical practitioners reveal a different, more varied picture. They are shown to exercise a high degree of dynamism in their knowledge system.

The Ayurvedic practitioner both accepts the importance of the past and recognizes the provisional nature of the knowledge thus obtained. Consequently, the theoretical system can take into account other theories that challenge it. These challenges (it was observed in careful field studies) are not ignored, the practitioner being well aware of their effects on the system. The tradition now appears as a more dynamic, open system, allowing for growth and refinement.[20] The effects of this openness are seen in the regional variations of Ayurvedic practice across South Asia, as well as in the selective adoption of Western systems, as documented in considerable evidence from both theory and practice. Sometimes this epistemological openness arises from encounters with different patterns of disease, when theory has to be revised.

On the other hand, there are different regional variations in the available minerals and flora from which most Ayurvedic medicines are made. A particular region would have some medicinal plants which are more effective than the ones described in the classical texts. Or the latter may not exist in the particular region, necessitating a search for alternatives. The result is a dynamic body of knowledge which gives attention to changes in the nature and number of diseases and the efficacy of medicines. These changes may not be so rapid or dramatic as the changes in modern Western medicine, especially those occurring in this century, but changes occur all the same. Here, we should note, there are parallels to "indigenous peoples'" knowledge acquisition processes (that is, knowledge acquisition by small social groups, often at the hunter-gatherer level). These groups act virtually as "research groups," moving into different ecological niches and testing for and discovering new medicinal plants.

Further, knowledge in Ayurveda is formally described as dependent on *pratyaksha* (direct evidence), *anumana* (inferential evidence), and *yukti* (logic), all of which are considered more reliable than *shabda pramana* (textual narrations).[21]

Consequently, various forms of experimental techniques were used in the classic ayurveda tradition. Animals were used to test for poisons, for example, some animals displaying disturbed behavior at the sight and smell of certain poisons. Tests were also performed on human subjects—in one instance on a patient who had the same symptoms as those of a king who was ill. The man was first rendered unconscious; then a physician opened the abdomen and observed a long, round worm in the intestine. Later, after suturing, various medicines were used to try to kill the worm. Ultimately, an effective one was found: it killed the worm, which was then observed to have passed through the stool. The same medicine was then tried on the king, and he too passed the worm and was cured.[22]

The classical writers Charaka and Susruta also advised medical practitioners to ask those who spent much of their time in the forest— shepherds, goatherds, cowherds, fowlers, and ascetics—about useful plants and drugs.[23] This, we should also recall, is the exact methodology being followed today by pharmaceutical companies when they search for effective cures among indigenous peoples in the Amazon forest.

The importance of number and measurement in correct practice is also found very early in South Asian intellectual traditions. The *Charaka Samhita* devoted eight chapters to the basic idea of measurement.[24] Different body parts were subject to measurement, the unit being the subject's own finger-breadth. (Early measurements were based on anthropometric standards, as witnessed in the English use of the "foot" as a measure.) The system developed various standard, desired measurements for various parts of the body, based on this unit. These anthropometric measurements, it was held, were of prognostic value. Desired measurements, in the views of the different authors, were associated with longevity, physical and mental strength, and ability to resist disease.[25] Exact criteria are, of course, given in the preparation and dispensation of different drugs. One does not have to believe in the theory behind the measurements or their efficacy and outcomes to appreciate these attempts at quantification.

This openness and experimentation is also seen in surgery. Susruta emphasized observation and dissection, and if what the practitioner found did not square with the received tradition, he advised changing it.[26] Susruta, therefore, was probably the first in the world to advocate dissection for obtaining firsthand information. He denounced persons who obtained knowledge only from books, comparing them to warriors who have never been on a battlefield and so run away at the first taste of combat.[27]

For Susruta, theory and practice go together like "two wheels of a cart." The good surgeon, he believed, should be the experienced one, and he advocated students' use of models for the practice of surgery. These models ranged from fruit, to leather bags, to bladders or other parts of animals, each providing in turn the consistency and feel of the human organ to be operated upon.[28] Susruta laid down elaborate procedures for dissection of the human body, which was compulsory. To quote an illustrative passage:

Therefore, he who desires to acquire the correct knowledge of anatomy, free from doubt, should process a cadaver and thoroughly examine (by dissection) all the various parts of the body. What is observed directly should then be compared with the standard text; [interaction] will lead to further increase of knowledge.[29]

The Greeks, it should be noted, did not dissect the human body: "The anatomical knowledge of Hippocrates was derived chiefly from dismemberment of animals, experience in slaughtering and sacrifices, and from the observation of surgical cases."[30] Just prior to the Scientific Revolution, Vesalius and other European anatomists of the sixteenth century articulated views similar to those of Susruta. Vesalius, coming from a long line of medical men, conducted dissections after his appointment in 1537, at age 23, as Professor of Surgery and Anatomy in Padua. His observations were published in his *De Humani Corporis Fabrica* (1543) and showed the many errors in the "truths" passed down from Galen, including mistaken assumptions about the lower jaw, the sternum, and the heart.

Gananath Obeyesekere, a professor of anthropology at Princeton, has reported on the actual daily practice of Ayurvedic physicians in Sri Lanka. He observes that they constantly experiment, partly through variations on classical prescriptions. Thus, although an Ayurvedic doctor may depend on written material in either Sanskrit or Sinhala, he or she may vary treatment if the first attempt fails, either by mixing two existing prescriptions or by deleting elements from (or adding to) standard prescriptions. In this way clinical experimentation often occurs in a physician's practice. These effects are not achieved through guesswork but through a theoretical understanding of the disease from the Ayurvedic perspective. When a physician encounters a foreign disease, then ad hoc experimentation is resorted to. But here, there is no theory (based on the formal Ayurvedic knowledge) to guide practice.

Falsification is also built into the practice. If a treatment does not work, it is dropped from the physician's repertoire. This happens whatever the source of the treatment, be it age-old ancient text or the physician's own invention, and as a result, Ayurveda produces a large repertoire of treatments. A patient may also change physicians, in which case the patient takes along the old prescriptions so that the new physician has a record of past treatment. A new approach now ensues with the new doctor.[31]

There are classical examples of such changes in practice. In Sri Lanka, after the abandonment of its classical irrigation-fed cultivation systems in the thirteenth century, parts of the country succumbed to malaria. Malaria was largely eliminated in the 1940s by the use of insecticides, but before that, local villagers had to cope with endemic malaria. On the basis of surviving informants and available records, K.T.

Silva has reconstructed the coping mechanisms of the pre-DDT era: Ayurvedics responded to the epidemics with local herbal traditions, which used a combination of antiparasite and antivector approaches.[32] And as for the ancient knowledge of how the disease itself was spread, Sir Henry Blake—British Governor of Sri Lanka, who delivered an address to the Royal Asiatic Society in 1905—pointed out, on the basis of texts translated through Buddhist scholars, that malaria had been recognized as a mosquito-borne disease. This realization had occurred over a millennium before the work of Manson and Ross, who are now credited with this discovery. For nearly two thousand years, Sri Lanka had a flourishing irrigation-based civilization, and would have known the causation of the disease which was so closely tied to water cycles and the breeding of mosquitos.[33] Theory and practice has continued to change with the coming of Western medicine into South Asia, and some Ayurvedic practitioners do incorporate Western medicine to supplement their repertoire of practices.

Ayurvedic practitioners have traditionally learned either from older individual practitioners or in an institutional setting. The latter included the vast Buddhist monastic universities populated with thousands of students, some dating to the fourth century BC. Increasingly the current institutional settings are in the form of Ayurvedic colleges with some institutional practices taken from Western-style universities. Especially in the latter, some Western-influenced medical knowledge is also taught, notably contemporary anatomy and physiology. There the Western system is flowing into the Ayurvedic one. No detailed studies exist of how this penetration is taking place, but there is no doubt that the phenomenon also points to the epistemological openness of traditional systems.

TRANSFERS

Through the ages, there were cross-transfers between South Asian medical knowledge and that of Europe, beginning during the time of the Greeks. Similarities abound, as well as records of direct transfers, in both theory and practice.

As discussed in chapter 3, the Greek theory of four elements and four humors evokes direct comparisons with (and is anticipated by) the Ayurvedic pancha bhuta (the five elements) theory and the Indian tridosha theory, which has its antecedents in the Brahmanic, Upanishadaic, Jain, Buddhist, and even Vedic sources. The humoral concepts are first mentioned in the *Tridhatu* of the *Rig Veda*; the three humors appear in the *Atharva Veda*, and a clear exposition occurs in the last period of the Vedic literature in the *Atharvaveda-parisista*. The ayurveda literature *Bhela Samhita* has the humors as a comprehensive theory, while the later *Charaka* and *Susruta Samhitas* contain the complete theory.[34]

In the Western tradition, the philosopher-physician Alcmaeon of Crotona (fifth century BC) extended Empedocles' theory of four elements (see chapter 3, above) into the domain of health. He maintained that the human body consisted of these four substances—earth, air, fire, and water—and that when they were in balance, one was healthy, and when there was imbalance, one was sick. Corresponding to the four elements were four qualities (hot, cold, moist, and dry) and four humors (blood, phlegm, and yellow and black bile). Imbalances among the humors gave rise to disease.[35] The doctrine of four humors was later followed by Hippocrates (460–377 BC).

Later, the Greco-Roman physician Galen (AD 129–199) developed a system of medical diagnosis that used earth, water, fire, and air—the stuff of the universe. He linked the four elements with four humors (melancholic, phlegmatic, sanguine, and choleric) and with the organs supposed to be associated with them (spleen, brain, heart, and liver). Thus a system was developed that could through its theory attempt explanations of many of the ailments.[36]

The Indian system starts with the pancha bhuta of vayu (air), prithvi (earth), teja (fire), yapa (water), and akasa (ether). There are three Indian doshas (vayu/air, pitta/bile, and kapha/phlegm), although Susruta includes a fourth dosha, blood, a concept which later fell into disuse. Blood is an important humor in the Greek system, although in the Indian system it is considered a *dhatu*, a constituent of the body that is capable of producing disease. In the Greek system, air is not one of the Greek humors, while in the Indian system, it is the most important one, vayu.[37]

There is also a key difference in the way the two systems believe humors cause diseases. According to the Greeks, the imbalance of the humors themselves cause the disease, but in the Indian system the doshas do not directly produce disease—they have to be disordered through their corresponding causative factors, their respective *nidanas*. Once the disorder takes place, doshas influence the dhatus.

The *Bhela Samhita* and the *Agnivesa Samhita* appeared before the Hippocratic texts, which were written between the fourth century BC and third century AD. The theory of a vital organic breath as the cause of all motor activity in the body appears in these earlier Indian texts, described in clear and detailed terms. A similar idea appears, in a less precise form, in the Hippocratic texts. In the later *Charaka* and *Susruta Samhita* compendia, these ideas are elaborated to a more sophisticated extent, indicating a natural growth of ideas from the earlier local sources.[38]

Concepts of the creation of the embryo, found in many Indian traditions, are repeated in the Hippocratic text *On the Nature of the Child.* In Indian sources, the embryo is the result of a coming together of the parental semen, the maternal seed, the vital breath, and the eternal soul. The concept of a maternal "semen," unusual in the Greek tradition, is also found in *On the Nature of the Child*, pointing to an Indian source. *The Embryo of Eight Months* includes the concept of the mother's vital breath circulating in the embryo's body; yet, other details that appear in the Indian texts—the embryonic heart and embryonic blood-circulation and nutrition—are absent from Hippocratic texts. This points to a Westward transmission that was incomplete.[39]

There are certain books in the Hippocratic collection which are closer in their treatment of their subject matter to Ayurveda than to other Greek ones. One example is the *Peri phuson* (*About the Winds*).[40] The Hippocratic collection also reiterates in detail an Indian medical process for cleaning the teeth, and mentions Indian drugs such as pepper, peperi, akoras, and sesamon,[41] some with corrupted Sanskrit names: kardamomen (cardamom), shringavera (*Zingiber officinale*), maricha (*Piper nigrum*), kinnamonas (cinnamon), amomon, *Sesamum indicum*, Nardostys Jatamanshi, and guggul (*Boswelia thuriferia*), to name a few.[42]

Because of these antecedents, direct Greek borrowing can be hypothesized. One possible route is through Pythagoras, who is known to have traveled to India and learned, among other things, the theory of transmigration;[43] on his return to Europe, he taught the theory of medicine and systems of dietetics, along with other borrowed systems. It is also very possible that there were contacts in these fields aside from Pythagoras.[44]

The many parallels between the *Samhitas* of Charaka and Susruta and the Greek systems have been enumerated by Jolly, as follows:

Pathology as due to humors

Stages of fever having a raw, ripening, and a ripe stage

Remedies divided into hot and cold and dry and oily

Healing by applying remedies of opposite character;

Seasons having an influence on dietetics;

In development of the body all parts form simultaneously;

Twins are born by semen being divided into equal parts;

The foetus is viable in the seventh month but not in the eighth;

A dead foetus is dismembered and extracted by using a hook in the eye sockets;

Body movement used to advance the placenta;

Lithotomy method as used in surgery;

Treatment of hemorrhoids by branding, cauterizing, and cutting;

Many common surgical instruments;

Similar techniques in operating on cataracts of the eye. . . .[45]

That such parallels could occur by chance is very unlikely. A few of the similarities between the Ayurvedic and Hippocratic texts may be due to common borrowings from the Babylonian tradition. For example, there are many medical concepts in an ancient Mesopotamian document of circa 2000 BC that reappear in both the Indian and Greek texts.[46] But to explain the bulk of the commonalities one would have to assume there was some transmission of ideas.

Other early influences are apparent in Indian exports to Greece, such as pepper, cardamom, indigo, cane sugar, and musk-root plant, all of which were used as medicines.[47] After Alexander's invasion of India in the third century BC, the Greeks' stock of Indian knowledge on diet, hygiene, snakebite treatments, and veterinary science increased further.[48] Alexander's army, which marched as far Eastward as the northwest of the subcontinent, had no knowledge of conditions peculiar to India, such as snakebites. For these, Indian doctors were called in, and some of them accompanied Alexander's army back to Greece, thereby directly introducing South Asian practices.[49]

With the passage of the classical age, the medicine practiced in Europe after the seventh and eighth centuries was partly derived from Latin translations of Arab texts. The latter in turn had drawn heavily from Sanskrit texts.[50]

In the eighth and ninth centuries, Baghdad was the center of the Arabic world. Its ruler was Harun al-Rashid (786–814), and a family of Indian physicians, the Barmecid, occupied a prominent place in his court. This family was descended from the chief monk (*Pramukh*; in Arabic, *Barmak*) of the Buddhist temple in Balkh. Some of the Barmaks became Muslims and translated many Indian texts to Arabic and Persian. These translated texts included almost all the standard Indian texts, such as those of Charaka and Susruta, as well as other treatises dealing with such subjects as diseases of women, drugs, poisons, snakebites, mental diseases.

There are many descriptions of how Indian physicians in Baghdad were able to cure diseases that confounded the Greek-derived system. Records attest to how the caliph himself, suffering from a serious disease which could not be cured by Greek practices, sent for an Indian physician who effected a cure. There are many similar stories of the prowess of Indian medicine, including another court incident in which a person given up for dead was "resurrected" by an Indian physician who found him only to have fainted from a condition that he could cure.[51] Many Indian doctors worked in hospitals including one who was made a director of a hospital. Several Arabian scholars of the ninth and tenth centuries affirm that Indian medicine and physicians were held in high esteem.[52]

The Abbassid caliphs encouraged the translation of Indian texts, including the *Susruta Samhita* and *Charaka Samhita*. The transmission further westward continued when al-Razi, or Rhazes (865–925), produced a comprehensive book incorporating Indian medical knowledge. This book, *Kitab-al-hawi*, was translated into Latin as *Liber continens* by Moses Farachi in the thirteenth century and in the Middle Ages became the standard medical work in Europe.[53]

The next significant transfer package occurred after the voyages of discovery—without intermediaries, through Europeans themselves. A principal raison d'être for the voyages was a search for sources of spices. Small wonder, then, that there was also an interest in remedies that used exotic plants, in addition to the usual traditional treatments. Key figures in such interests were those closely allied to the colonial enterprise. Some of that transmitted information also fed the development of the biological sciences because now, a larger range of specimens from all over the world was available for study, in contrast to the limited variety existing in cold Europe. Thus, for example, the knowledge brought by Garcia d'Orta fed the initial interest in botany, as did the larger shipments sent by the Dutch East Indies Company, which were used by Linnaeus and the still later systematic efforts of the Britishers through their worldwide establishment of botanical gardens.[54]

Because of the Portuguese presence in South Asia from the sixteenth century on, there was much two-way interaction. Yet in the field of medical science in the sixteenth century, Europe was more influenced by India than was the reverse.[55] Many leading Portuguese had high praise for local doctors,[56] and studies of Portuguese writers of the time throw much light on the nature of this interaction.[57]

Virtually no Portuguese in the sixteenth century were aware of any Indian language beyond the rudiments required to, say, trade in the bazaar. Up to the end of the sixteenth century, they were even unaware that a language called Sanskrit existed, let alone it was the medium of a vast scientific literature. Only in the seventeenth century did a formal study of local languages begin among Europeans. Hence whatever was written by the Portuguese about India did not come from the formal literature but from observation and through dialogues using interpreters.[58]

Tome Pires, one such observer, wrote a famous letter in 1516 to the King of Portugal from Cochin describing the drugs he found in India. Among other observations, he reported surprise at the Indian custom of not eating when having a fever. He mentioned that wounds were treated with warm coconut oil, in contrast to the prevailing treatment by hot iron in Europe. He also reported on a new disease hitherto unknown in Europe, filaria.[59]

Garcia d'Orta's classic *Cloquios dos simples e drogas da India* (*Colloquies on the Drugs of India*) was published in Goa in 1563, five years before his death, and was very influential in the European acquisition of local knowledge. The fact that it was the third book printed by the Portuguese in India attested to its importance to them.[60] The work has fifty-seven chapters, each devoted to one drug, usually of plant origin. The plant's appearance is described, as well as where it grows, its local name, and its therapeutic uses. D'Órta thus described parts of the *materia medica* in use and its associated botany, as well as medicine in general. He was acquainted with local physicians, but there is no mention of any local text on medicine or of any of the local seats of medical learning. So his view of local knowledge was largely of its customary use; he did not know that an old and highly articulate formal body of knowledge lay behind the practice.[61]

He says that in case of disease in his stay in Goa, he first tried the European cures he knew, and when that failed, he turned to local remedies. He describes local cures for dysentery which he finds better than those in use by the Portuguese (during this period widespread dysentery was a major medical problem in Portugal), as well as treatments for fever and inflammation.[62]

D'Orta, on his part, was a man with an open mind who never considered the physicians of Greek antiquity as the final authority. He wrote, "the testimony of an eye-witness is worth more than that of all the physicians, and all the fathers of medicine who wrote on false information." Just as Vesalius had corrected Galen's views on anatomy through his dissections, d'Orta thus corrected other ancient authors, including Dioscorides. D'Orta's work had a special influence during the Renaissance because it was taken up by Charles de l'Ecluse, a Dutch physician who translated the work into Latin from Portuguese. The book became very popular, running into five editions in a few years' time. It was also later translated into other languages.[63]

The other Portuguese authority who significantly facilitated the transmission of Indian knowledge and practice was Cristovao da Costa, through his book *Treatise of drugs and medicines of East India*. Like d'Orta before him, he had no idea of formal Ayurveda or that there was a written literature of medicine such as existed in Sanskrit. He says that what brought him to India was his interest in finding "in several regions and provinces learned and curious men from whom I could daily learn something new; and to see the diversity of plants God has created for the human health." In this pursuit he consulted many physicians of different nationalities in India.[64]

The author of *Breve Relacao das Escrituras dos Gentios da India Oriental* described the comparative efficacy of the local medical traditions as follows: "Several times I heard him [the Chief Portuguese Medical Officer—Physico-more] say that the *Brahamans* called *Pandits* treat patients much better than the Europeans who practice medicine in India.[65] John Huighen Linschoten, a Dutch traveler who lived in Goa from 1583 to 1588, confirms this observation in his book *Voyage to the East Indies.* He notes that the Portuguese upper strata in Goa—viceroys, archbishops, and the aristocracy—preferred to be treated by Indian doctors.[66] The governor of Goa, Barreto, referred to the pundit who was his physician while in 1548, and the Jesuit College was being treated by a Brahmin physician.[67] And in Sri Lanka, another South Asian region where they had a foothold, the Portuguese used local Ayurvedics to cure dysentery and snakebites among their soldiers.[68]

C.R. Boxer, a modern historian of the Portuguese empire, puts the comparison as follows:

> From the perusal of Garcia d'Orta's *Coloquios* . . . it seems obvious that the native Indian physicians were on the whole decidedly more advanced than their European colleagues. It is also significant that the local Viceroys preferred Hindus as their medical attendants rather than trust themselves to the tender mercies of the European Physico-More, and this, despite the promulgation of decrees condemning the practice.[69]

This last refers to the ban on treatment by local physicians after the coming of the Inquisition: in one instance, the governor's wife was fined in 1589 for using local medicine.

Generally, South Asian knowledge contributed more to the Portuguese practice than the other way around. The influence occurred not only in the introduction of unknown treatments, but also in providing an opportunity to persons like d'Orta to correct Greek myths through firsthand observation, and so contribute in other ways to the development of post-Renaissance medicine. Thus, according to a recent Portuguese commentator Antonio Sergio, "If Garcia d'Orta had not left the European environment he would not have dared get rid of the superstition of the authorities, and turn from the attitude of the *Homo credulus* to the attitude of the critical spirit."[70]

The impact of Eastern medicine on European practice is exemplified by the drug import pattern of Britain, which would not be untypical of Europe as a whole. In 1588, only 15 percent of drugs imported into England came from outside Europe. This increased to 48 percent in 1621 and to 70 percent in 1669. The majority of these imported drugs came from South Asia and the East Indies.[71]

The center of European expansion now moved from the Iberian

peninsula to the more northern latitudes. The Dutch and the British now acted as key transmitters of medical and associated botanical knowledge. Thus the Dutch in the coastal regions of Sri Lanka used in their hospitals both local medicines as well as local physicians in addition to their own.[72] Hendrik Van Reede published *Hortus Malabaricus* in twelve volumes between 1678 and 1693. It was probably the first comprehensive compilation by a European of medicinal plants, belonging to 740 species, that were used in the local medical treatment. The text contains approximately 700 line drawings and plant names in three local languages—Sanskrit, Urdu, and Malayalam—as well as in Latin. He speaks of his local helpers in the following words:

> I was often seized by the desire to explore and examine the leaves, flowers, bark, and roots of those plants. And then I found that they frequently had a very sweet smell and a penetrating taste. And when I asked natives who accompanied me on my journeys whether they knew anything about these plants, they not only disclosed the name, but also knew very well their curative virtues and uses. I have often witnessed this on the way, when one of our companions suffered from some complaint, either an internal or an external one, although they [those Indians who accompanied Van Rheede] were not endowed with either medical or botanical knowledge.[73]

The westward transfer of knowledge continued after the Dutch ceded supremacy to the British in the late eighteenth century and nineteenth century. But during this period medicine in Europe was maturing and the flow in the reverse direction, from Europe outward, increased very rapidly. Still, there were transfers from South Asia. Dharmapal has collected several illuminating accounts by Britishers on Indian medical practices in the eighteenth century. Thus, J.Z. Holwell (FRS) gave a detailed report on the widespread practice of inoculation against smallpox to the president and members of the College of Physicians in London, as follows:

> Inoculation is performed by a particular tribe of Brahmins who are dedicated annually for this service from the different colleges over the various provinces. They travel in small groups of three or four from place to place so as to reach various places some weeks before the usual onset of disease. . . . When they begin to inoculate they pass from house to house and inoculate. . . . Usually they inoculate outside of the arm, and between the wrist and elbow for the male, and between the elbow and the shoulder for the females. Before the inoculation, the inoculator takes a piece of cloth and rubs the part to be inoculated for eight or ten minutes. He then makes a small prick with a small instrument till a drop of blood appears. After that he opens a linen double cloth and takes from there a small piece of cotton charged with variolous matter which he measures with a few drops of Ganges water and applies it to the wound with a bandage.

Dr. Holwell continues: "When the before-recited treatment is strictly followed, it comes as miracle to hear that one in million fails for receiving the infection or of one that miscarries under it."[74] Later, the British banned the practice. The high incidence of smallpox epidemics after this in the nineteenth and early twentieth centuries could probably be assigned to the banning of this practice before the new Jenner vaccination system could become fully widespread.[75]

In the late eighteenth century, several British observers remarked on the South Asian practices in surgery. In a letter written circa 1790–1801 by Dr. Helenus Scott to Sir Joseph Banks, President of the Royal Society, London, we find a description of an operation for cataracts of the eye: "They practice with great success the operation depressing the Chrystalline lens, when become opake and from time immemorial they have cut for the stone at the same place which they now do in Europe." Dharmapal also notes the following remarks of one Colonel Kyd, on Indian surgery: "In Chirugery (in which they are considered by us the least advanced) they often succeed, in removing ulcers and cutaneous irruptions of the worst kind, which have baffled the skill of our surgeons, by the process of inducing inflammation and by means directly opposite to ours, and which they have probably long been in possession of."[76]

A Viennese doctor, Niccolao Manucci, was the first to describe in relative detail the reconstructive surgery in India for a severed nose. If enemies were not killed, often their noses were cut off; new surgery was thus required. The surgeons

> cut the skin of the forehead above the eyebrows, and made it fall down over the wounds on the nose. Then giving it a twist so that the live flesh might meet the other live surface, by healing applications, they fashioned for them other imperfect noses. There is left above, between the eyebrows, a small hole, caused by the twist given to the skin to bring the two live surfaces together. In a short time, the wounds heal up, some obstacle being placed beneath to allow of respiration.[77]

The English-language *Madras Gazette* of 1793 reported a rhinoplasty by a Maharatta surgeon near Poona as witnessed by two British medical officers. This case was later described in London's *Gentleman's Magazine*. These reports helped spread the technique in Europe and America. A book by Carpues entitled *An Account of Two Successful Operations for Restoring a Lost Nose from Integument of the Forehead* (1816) created considerable interest in the subject. In Germany, Carl Ferdinand von Graefe performed the first total reconstruction of a nose in 1816 and used the term "plastic surgery" for the first time. In America, the Indian method of rhinoplasty was adopted by Jonathan Mason Warren in 1834. In the *British Medical Journal* of 1897, a Captain Smith published an article "Notes on Surgical Cases—Rhinoplasty," with suggestions for improving the technique. Later, in 1900, Keegan reviewed the improvements on the Indian technique that had subsequently been made in the West.[78] By the end of the eighteenth century, the British had also studied the Indian surgical procedures for skin grafting. Grafting skin to correct for deformities of the face then became "the starting point for the modern specialty of plastic surgery."[79]

As Krishnamurty points out, "Many of the ideas re-discovered and appearing from time to time as new and original, can be found to date back to the ancient Sanskrit works on the art of healing, and particularly surgery."[80] Among different surgical techniques, one also finds bamboo splints recommended in Sanskrit texts; rattan cane splits were adopted in Western practice in nineteenth-century surgery to keep the affected parts in alignment. Seutin, Diffenback, and others cite improved versions of apparatus used in surgery for fractures by Greek and Arabian surgeons—apparatus described in ancient Sanskrit texts centuries before those of the Arabs and the Greeks.[81]

Early on, the subcontinent also used hypnotism, in the nineteenth century referred to as "mesmerism," after its alleged discoverer, Mesmer. The various techniques used subsequently in the West for hypnotism —such as staring fixedly at an object, looking steadily into a subject's eyes, revolving mirrors, and mesmerizing passes—were all known on the subcontinent much earlier.[82] Max Neuberger, a historian of medicine, has noted that

> India must also be credited with at least an indirect influence upon the spread of hypnotism, as the empirical practice of suggestion was there more developed than elsewhere. To mention only one fact, it was certainly no accident that it should have been in Calcutta that the English surgeon Esdaile hit upon the idea of performing numerous operations induced by hypnotism.[83]

Generally speaking, there were two broad phases in the British encounter with local medical practices in India. During the first period, up to about 1860, there was a willingness to learn from local practices, especially those regarding local hygiene. Subsequently, the hardening of racial divisions in the course of the century saw India as unable to contribute anything valuable. This period also saw changes in European medical doctrine, particularly with the arrival of the germ theory. This was less conducive to some of the earlier European theories derived from Greek sources with affinities to the Indian ones.

When cholera broke out in Europe in 1830, opinions of Indian medical practitioners were avidly sought. European views on the causation and prevention of the disease, which were partly derived from indigenous sources, had an impact on British medicine at the time, reflecting Enlightenment views on openness of inquiry. The respected Dutch physician Jerome Gaub, in his *Adversariorum Varri Argumenti*, for example, advocated the use of local medicines in the Indies, carrying out investigations on the medical properties of local roots. Others compiled local pharmacopoeia and described properties of Indian drugs and herbs. Some recommended using vegetarian food and warned against the heavy consumption of animal food. One book by James Johnson, the most influential writer on tropical medicine, considered health in terms of the different phases of life, where certain activities were considered appropriate to each age—a position clearly analogous to Hindu ideas on the topic. Later however, after the 1860s, the notion that one system had to be superior took hold.[84]

CURRENT RESEARCH ON LEGITIMIZING

Now, the background we have indicated should suggest that indeed the Ayurveda system should have elements worthy of exploration and resurrection. But a preliminary word is in order here. Knowledge of cures that we have in mind "become cures" in a scientific sense only if they are published in a Western-oriented journal, following established experimental criteria. In the absence of this series of steps, one may have the most efficacious and miraculous cure, but it is not considered as such in the modern universe of discourse.

There is another factor peculiar to the South Asian case. In the nineteenth century, South Asia suffered a strong cultural and intellectual division. One track continued to follow the older intellectual traditions, while another branch dumped them altogether to adopt wholesale the Western system of knowledge, down to detailed copying of syllabuses, texts, and even the concept of the university (initially London University, in particular). The spirit of this attempt at wholesale Westernization is conveyed by an oft-quoted remark by a British education reformer of the time, T.B. Macaulay. His notorious and influential minute summed up the European views of the intellectual tradition of the region: "A single shelf of a good European library is worth the whole native literature of India and Arabia." Macaulay discouraged the training of Arab and Sanskrit gurus, emphasizing that they gave "artificial encouragement to absurd history, absurd metaphysics, absurd physics and absurd theology."[85] The colonial self-interest that determined this policy is best expressed in his own words:

> We must at present do our best to form a class who may be interpreters between us and the millions whom we govern; a class of persons, Indian in blood and colour, but English in taste, in opinions, in morals, and in intellect. To that class we may leave it to refine the vernacular dialects from the country, to enrich those dialects with terms of science borrowed from the Western nomenclature, and to render them by degrees fit vehicles for conveying knowledge to the great mass of the population.[86]

Although some voices were raised against this dichotomy and there were attempts at merging the two traditions, the division has persisted. Prestige and class factors are involved, apart of course from the sheer intellectual power and practical efficacy of the post–Scientific Revolution knowledge package from the West. In medicine, too, this dichotomy continues, although many South Asians, including those of the upper, Westernized classes do at times and for particular types of ailments patronize Ayurvedics.

But in contrast, the dichotomy did not develop to this degree in China, where there was a continuous interaction between the Western and local systems because of a user-oriented health care system which utilized whatever resources were available. In South Asia, on the other hand, there has long been a strong cognitive gap between Ayurvedic and Western practices: the two operated in two worlds. But with some world trends recognizing aspects of non-Western thought and practice, some of this antagonism is now dissipating.[87] As a result there are now greater efforts from the westernized sector to find what is useful in the non-Western sector and incorporate it in health practice.

The examples below stem from recent explorations into Ayurveda that have been reported either as already scientifically proven remedies or as potential ones. Almost all are in the realm of medicine as opposed to surgery, because that is where the search has occurred most frequently. Yet in Eastern surgical practice, too, there are potential elements for resurrection, as indicated by just one example—surgical treatment of hemorrhoids. In the West, this condition is treated by cutting off the growth and suturing. But a surgical technique, found from the earliest times in Ayurveda, uses a thread and is less invasive, with a faster healing time, than the common Western technique. It has been now evaluated on Western scientific criteria and is entering practice.[88] Let us now return to the examples from medicine proper.

In 1968, S.K. Jain gave a list of medicinal plants from the Ayurvedic tradition. The study had screened the relevant published literature for thirty years prior to the sixties. It was a conservative collection and only included those plants that have been verified by pharmacological and other experimental work within a Western frame of reference. The list of approximately one hundred efficacious medicinal plants varied from Indian acalypha and aconite, through ishabgul and psoralea, to tylophora and ashvagandha. This collection devotes a chapter to each plant. Apart from such data as the description of the plant and where it is found, there is information on the properties of the drug and its efficacy. The latter details are taken only "from very authentic publications, and only those uses of medicinal herbs are described which have been recognized in the British *Pharmaceutical Codex* and/or United States *Dispensatory*, or whose properties have been shown experimentally on animals or in clinical trials."[89] One of the features that struck Jain was how few of the medicinal plants available in the Ayurvedic tradition had been subjected to pharmacological experiments or clinical trials.

Since 1968 subsequent research has attempted to shift some Ayurvedic medicines to the Western cognitive field. Below are details of a few of these attempts, indicative of further possibilities. First are treatments for common ailments, followed by treatments for more chronic illnesses, especially those that come with advanced age. The selection covers a list of difficult-to-treat diseases, like cancer and diseases of the nervous system and the cardio-vascular system, as well as more everyday health problems like diarrhoea, wounds, and ulcers. In addition, I give names of ailments whose Ayurvedic cures have been the subject of several studies in Western science.

These latter lists are from several journals that specialize in testing non-Western drugs on the basis of laboratory and clinical tests, using controlled criteria such as double-blind tests. These studies are referred to in regular publications of abstracts. One, *Medical and Aromatic Plants Abstracts*, includes abstracts from a list of nearly ninety journals, to take a sample recent issue (1993). The list includes *Advances in Plant Sciences, Biotechnology and Bioengineering*, the *Chinese Medical Journal, Indian Drugs*, the *Indian Journal of Experimental Biology*, the *Indian Journal of Medical Research*, the *Journal of the American Chemical Society*, the *Journal of Ethnopharmacology, Naturwissenschaften, Plant Cell Reports, Phytochemistry, Planta Medica*, and *Scientia Pharmeceutica*, and suggests a broad geographical and disciplinary coverage. These abstracts cover different areas, such as pharmacology and toxicology, clinical studies, and patents, the last referring to patents obtained for traditional medicines. There are many entries pertaining to South Asia, probably the largest number from one geographical area, closely followed by entries from East Asia.

To give a flavor of this tested store of medicines, I give below a selection of successful treatments for various diseases using Ayurveda-derived drugs and techniques. After a few detailed descriptions from sources outside the abstracts, I give additional information that has appeared over the last few years in these abstracts under the section on clinical studies—that is, those studies that show effectiveness for the various ailments. Because of space considerations (the list could go to tens or even hundreds of pages) I am not listing the authors, the titles, or the original journals here, only the page references to *Medical and Aromatic Plants Abstracts*.

Wounds and Ulcers

A study done by Chitatampalli and Mulcahy noted that an old Ayurvedic approach using honey and sugar as treatments for wounds and ulcers was also being followed in contemporary biomedicine. In their literature survey, they came across 33 works that attested to topical clinical uses of honey and sugar for their microbiological properties. These included both clinical as well as in vitro studies. Susruta, however, had detailed descriptions for the topical use of honey and sugar in the lancing of suppurated lesions, ulcers, removal of darts and arrows, abscesses and tumors, and cutaneous eruptions of ringworm.[90]

The plant *Centella asiatica* (gotu kola) has been shown to have an effect on skin afflictions (including eczema, psoriasis, lupus, lepra, and scrofula) and leprosy.[91] Another plant, *Psoralea corylifolia* (in Sanskrit, *Vakuchi*) has been used in Ayurveda for leucoderma or viligo. An oral treatment derived from this, Psoralen, is now used widely throughout the world. Its efficacy as a treatment for depigmented patches has been found to be "very satisfactory."[92]

Abstracts lists dozens of successful Ayurvedic treatments for wounds, aterial and fungal wound sepsis, burns, hemorrhoids, fistula, fractures, inflammation, inflammatory wounds, psoriasis, severe burns, acne, dermatitis, partenium dermatitis, infective and allergic dermatoses resistant to conventional therapy, and nonspecific dermatitis.[93]

Diarrhoea and Stomach Diseases

Ayurvedic texts describe a number of preparations to combat diarrhoea as well as complications arising from it. Of the very many approaches to the disease in Ayurveda, only a few have been tested under formal scientific criteria, including three specific preparations made from different recipes and given to patients under controlled conditions. All performed well as anti-diarrhetics.[94] A French medical doctoral thesis on the treatment of acute intestinal amoebiasis using Ayurvedic medicines has demonstrated their efficacy. Researchers using double-blind tests in a hospital milieu cleared up patients' amoeba-based dysentery.[95] The *Abstracts* citations list a number of successful treatments for a range of additional stomach diseases: diarrhoeal dehydration, nonspecific childhood diarrhoea, hyperacidity, helminthic worm infections, irritable bowel syndrome, dyspepsia, duodenal ulcers, gastrointestinal disturbances, peptic ulcers, malabsorption syndrome, malnutrition, and anaemia.[96]

Liver Ailments

Jaundice and diseases of the liver are referred to regularly in the Ayurveda literature, with several prescribed medical preparations. In a controlled trial with the modern drug Prednisolone acting as control, some of these drugs have been scientifically evaluated. The four drugs under trial were *Katuki, Daruharidra, Katukyadi-Yoga,* and *Kumari Asava.* Standard monitoring indicators such as serum bilirubin, serum thymol turbidity, serum alkaline phosphatase, serum protein, serum albumin stercobilin, etc. were used. The study showed that patients suffering from hepato-cellular jaundice had complete relief after treatment with Katukyadi-Yoga. Roughly 35 percent of those suffering from chronic hepatitis were cured with the same treatment, while the remaining 65 percent improved their condition. The effect of this drug was comparable with that of the modern drug and, in the case of hepato-cellular jaundice, performed better. Of the other Ayurvedic remedies tried, Katuki scored the next best in its efficacy. Other drugs tested showed mixed results. On the basis of the controlled tests, the authors recommended the use of Katukyadi-Yoga for both types of liver diseases.[97] Another traditional drug, *Piccrohzia kurroa,* has also shown very promising liver-protection properties in experimental studies and now awaits further clinical studies.[98] Alcoholics, who often experience liver damage, are one class of candidates for such drugs.

Other related diseases that have been effectively treated, according to the *Abstracts* list, are hepatitis, infective hepatitis, viral hepatitis, and diabetes; Ayurvedic treatment has also been used in nonsurgical lithotomy of the kidneys and urinary bladder.[99]

The Reproductive System and Contraception

There are occasional references, going back to the earliest literature in Ayurveda, on preventing conception, although the subject is notably absent in the classical Ayurvedic literature of Susruta and Charaka. But references to oral as well as locally applied contraceptives are available in the literature from the eighth century on. Fourteen important medical texts with references to the theme have been found since that date, detailing medicaments applied locally to prevent conception, as well as oral contraceptives for both males and females. Thus, a World Health Organization study observes that the roots of the plant *Plumbago rosea,* mixed with alcohol from the plant *Madhuka indica,* have been used in parts of India as a postcoital contraceptive, orally self-administered.[100] Dash and Basu have identified over thirty drugs similarly used for oral contraception, some still in use today. These, however, have not yet been subjected to clinical testing.[101]

Abstracts lists scores of studies on subjects such as male contraceptives, birth control, premature ejaculation, sexual dysfunction, low sperm count, impotence, sterility and menstrual irregularity, urinary infections, gynecological disease, urinary tract infection and obstructive uropathy, urolithiasis, cervical spondylosis, ureteric calculi and urinary tract infections, medical termination of pregnancy, lack of lactation, fertility, postpartum care, inadequate lactation, pregnancy-induced hypertension, and menopausal symptoms.[102]

The Nervous System

The term *chittodvega*, mentioned in Ayurvedic literature, could be translated roughly to mean "anxiety neurosis." In the Ayurvedic sense, the condition arises from both mental and physical causes, and a number of Ayurvedic drugs are used for its treatment. While these drugs have not yet reached the "world" pharmacopoeia, modern scientific tests have been carried out on a combination of them: *Centella Asiatica* (*Mandookaparni*), *Glycyrrhiza glabra* (*Yashti*), and *Narodostachys Jatamansi* (*Jatamansi*), combined and suspended in *Ksheerabala Thailam* (an oil), another drug for anxiety neurosis. Patients of both sexes suffering from generalized anxiety disorder were involved in the test. Standard psychological and physiological parameters were established before and after the treatment. These included Taylor's Manifest Anxiety Scale, the Max Hamilton anxiety rating scale, perceptual retention, fatigue rate, heart rate, pulse rate, alpha activity, and various tests on urinary excretions. The patients received, either a placebo, diazepam (Valium), or the Ayurvedic drug, prescribed to patients at random. The treatment lasted for 45 days. The same patient received the other drug after an interval of seven days. The findings were that the Ayurvedic combination was more effective than the other two control drugs in enhancing perceptual and psycho-motor discrimination and in controlling somatic and psychic anxiety.[103]

Another study evaluated ancient Ayurvedic drugs with a capacity to modify and rehabilitate intelligence. One such drug, *Bangiya Brahmi*, is believed to be of great psychotropic value. When given to children, the drug helped them perform much better mentally than a placebo used as control. Under the drug's use, various indicators changed for the better, including vasomotor function (maze learning), pattern of immediate memory (as exhibited by the digit span test), perceptual organization and visual motor perception (as measured by the Bender Gestalt Test).[104] *Vacha* (*Acorus calamus*), another common South Asian drug, contains an essential oil that has been shown to have marked sedative properties.[105]

One of the drugs mentioned earlier, *C. asiatica*, is commonly marketed in US health stores as a brain booster under its Sinhala name gotu kola. It has demonstrated many remarkable properties. Animal experiments have shown that extracts of the plant yield good sedative and cardiotonic activities. Other experiments on animals reveal cardiodepressant and hypotensive actions and psychotropic effects. Gotu kola's effects on the nervous system resemble those of reserpine and chlorpromazine.[106]

With South Asian traditional concerns on mental phenomena, it would be natural to expect a large number of compounds for psychological diseases in its pharmacopoeia. It is useful to conclude this section by noting that the first anti-anxiety and hypertension drug was reserpine, derived from *Rauwolfia serpentina*, used in South Asia for treatment of insanity. It has a calming effect on the mind, and Gandhi is reputed to have used it. The first Western description of the drug came from Garcia d'Orta in 1563, but Western physicians ignored its stated effectiveness until 1952.[107] It is very probable that many such psychotropic compounds are waiting in the background to be discovered.

Medical and Aromatic Plants Abstracts lists many studies dealing with other mental functions and related areas: cures for low intelligence, mental backwardness, and other intellectual dysfunctions in children; stabilization of visual acuity; psychotropic effects of anticonvulsion drugs; pain and analgesics; mental diseases; migraine, stress, nervousness, depression, epilepsy, and behavior disorders.[108]

Cancers

One hundred years ago, it was recognized in the West that arsenic could be used in the control of blood counts for patients with chronic myeloid leukemia. But in Ayurveda, an arsenic-containing compound has been used to control blood counts of patients having hematological malignancies as well.[109] *Tylophora asthmatica* W and A are quite well known in the Indian medical system and have already been recorded in *The Hindu Materia Medica* and in the *Pharmacopoeia of India*. The British in India used it with great success as far back as 1885 to treat troops suffering from dysentery. Its use by the British as an emetic is also recorded. In the traditional literature, it has also been recommended for a variety of other ailments. The plant has been tested for its active principles as a treatment for cancer and has shown activity against certain types of cancer.[110]

Ayurveda uses sesame oil regularly as part of its normal repertoire, both topically on the skin as well as (seasonally) in the colon. Its use, Ayurveda claims, improves physiological balance, longevity, and vitality. Sesame oil has relatively high levels of the essential polyunsaturated fatty acid linoleic acid, which has been shown to selectively inhibit and kill a number of tumor cells, both human and animal, an effect demonstrated both in vitro and in vivo. There has been no significant effect shown in such studies on normal cells. A study done in vitro by John Salerno for a doctoral dissertation indicated that both free linoleic acid and undigested sesame oil strongly inhibited the malignant melanoma and colon adenocarcinoma cell lines in comparison to the corresponding normal cell lines.[111] Several studies show clinical efficacy in the treatment of benign enlargement of the prostate, general cancer, oral cancer, and leukemia, and in adjuvant therapy for cancer patients.[112]

Heart, Lungs, and the Cardiovascular System

Apart from cancer, other common diseases of old age are those relating to the respiratory and cardiovascular systems. Here too several Ayurvedic treatments have been validated according to Western standards. *Tylophora asthmatica* W and A is often used in Ayurveda. In clinical evaluation as a treatment for bronchial asthma, it was found to be very effective; after one week, all symptoms had disappeared.[113] Ginger has been used in Ayurveda to treat arthritis, pain, fever, and blood clumping. These are all diseases connected to the metabolism of arachidonic acid, which affects various eicosanoids. In a Danish study, ginger reduced thromboxane production by almost 60 percent, thus confirming the Ayurvedic remedy in its anti-aggregatory properties.[114]

Hoechst, the West German multinational, used an Indian botanist, Virbala Shah, to identify medicinal plants. Using Ayurvedic literature, she identified a number of useful plants, and among the medicines yielded by her research was Foskolin, derived from the root of the turnip-shaped *Coleus foscoli*. This medicine acts as a hypertensive and has the ability to intensify heart contractions. Another medicine came from *Stephania glabra*, out of which an active ingredient was isolated that could dilate blood vessels and help prevent heart attacks.[115] Similarly, the herb *tulsi* is an old Ayurvedic drug going back to the times of Charaka, one of its announced uses being for heart disease. On animal models, this preparation has been shown to cause dramatic drops in cholesterol levels.[116]

Other relatively common ailments of the heart and lung not necessarily associated with old age are also treatable with Ayurvedic medicines. This list includes the common cold and cough, chronic tonsillitis, sinusitis, respiratory diseases, upper respiratory infections, asthma, bronchial asthma, tropical eosinophilia, broncho constriction, tuberculosis, pulmonary eosinophilia, effects of chronic smoking, and infections of the upper respiratory tract.[117] Conditions more often associated with later age that have also been effectively treated through Ayurveda include cardiac neurosis, cardiovascular disease, bad lipid profile, bad lipid profile of diabetics, blood coagulation and fibrinolysis, coronary heart disease, heart disease, hyperlipidemia and hypercholesteremia, high serum lipids, high blood pressure, serum blood cholesterol, myocardial infarction, ischemic heart disease, hypertension, hypertensive heart disease, obesity, obesity and high blood pressure, and pulmonary treatment of hyperlipidemia and obesity.[118]

Rasayana therapy, a classical branch of Ayurveda, aims at the general promotion of strength and vitality and broadly advancing healthful longevity. A controlled study evaluated the effect of Rasayana compounds consisting of six such drugs from over forty mentioned in the classical Ayurvedic literature. These six were selected because of their particular actions on the different systems of the body. Thus Bala's (*Sida cordifolia*) acts on the cardiovascular system, *shatavari* (*Asparagus racemosus*) on the neuromuscular system, and *punarnava* (*Boerhavia diffusa*) on the excretory system. Extensive assessments (in clinical tests and in biochemical, psychological, and anthropometric measures) of the effects of the aging process were done on the subjects in the study. The results before and after a six-month trial period were compared. The clinical studies indicated that blood pressure had come down significantly to normal levels and that hand grip and pulse rate had improved significantly. The drugs had toned up the cardiovascular and respiratory systems and had improved physical stamina. On the psychological side, statistically significant increases were seen in finger dexterity, work output, and visual reproduction. There was also increased eye–hand coordination, improved short-term memory, and mental performance. On the biochemical side, there was a significant drop in lipids, and hence fat. Uric acid also decreased, suggesting an effect on gout (a result of excessive uric acid).[119]

Medical and Aromatic Plants Abstracts, in its detailed treatment of age-related diseases, lists (among others) treatments for chronic degenerative arthritis, general arthritis, osteoarthritis, osteomyelitis, and rheumatism,[120] as well as for diseases of the eyes, teeth, and skin. These include ophthalmic conditions, refractive error and cataracts, senile macular degeneration of the eye, conjunctivitis, allergic conjunctivitis, and dermatological and dental infections.[121]

Indicative of the trend toward shifting drugs from the Ayurvedic frame to the Western was the fact that by the early 1980s, over 200 promising Indian medicinal plants were being tested every year.[122] Further, the Indian Council of Medical Research, which conducts Western-oriented research, was by the early nineties supporting research in traditional medicine on certain widespread diseases, especially those conditions for which cheap medicines are lacking in the Western system: anal fistula, urolithiasis, viral hepatitis, bronchial asthma, and diabetes mellitus.[123]

If these are fragmentary examples of successful recent dredging of the Ayurvedic reservoir for medicine legitimized in the Western sphere, do we have some measure of what lies in store? What is the potential for finding such drugs? Can we give some estimates?

ESTIMATING THE RESERVOIR

Plants, because they cannot run away from animals that devour them, have evolved a host of chemical defenses, becoming "experts" on animal biochemistry. These defenses are a vast reservoir for medicine.[124]

There are an estimated 250,000 to 750,000 species of higher (flowering) plants in the world, and many have not been described botanically. Norman Farnsworth has estimated that of this number, roughly 10 percent—25,000 to 75,000—have been used in traditional medicine across the globe. And of that number, only 1 percent—just 250–750—species have been acknowledged through scientific studies to have therapeutic value. Almost all therapeutic uses have been identified based on knowledge of their medical uses in non-Western traditions. These figures indicate both the overall Western ignorance of potential medicines as well as the immense possibilities for remedies.[125]

Yet worldwide, only about 1–2 percent of higher plants have been adequately screened for biological activities. Even alkaloids, chemical compounds that have yielded several therapeutic compounds, are virtually unexplored; less than 2 percent of the estimated 250,000 species of flowering plants have been examined for the presence of alkaloids.[126]

But by far the strongest biodiversity—the presence of many different species—is found in the present developing world. There are many telling indicators of this gross nature-endowed bounty of the tropics. The British Isles have far less biodiversity than a tiny island off Panama; the whole of Denmark sustains only half as many species as exist on one hectare in Malaysia; a 15-acre area in Borneo has more species of wood than all of North America. Taken as a whole, more than two-thirds of the world's plant species thus come from the southern hemisphere.[127]

In the modern Western pharmacopoeia, at least 7,000 medicines are derived from plants,[128] and it has been noted that 74 percent of compounds that are extracted from plants for use as traditional medicine have the same or closely related medical uses in modern medicine.[129] This gives a hint of the potential in traditional medicines.

At the start of the nineties, the total worldwide sales of pharmaceuticals stood approximately at US $130 billion; of this, conservatively $32 billion were based on traditional medicines.[130] As a corollary, it was estimated in the mid-eighties that one medicinal plant lost in the rain forest could mean a loss of more than $200 million.[131] As the first issue of the *Journal of Ethnopharmacology* put it, "almost any classical pharmacological agent is derived from a classical botanical source originally employed as a native remedy." Some of the medical compounds that

have already come out of indigenous sources include, among others sali-
cylates and scopolamine (analgesics); cocaine and curare (anesthetics);
morphine and its analogues (painkillers); quinine (antimalarial); col-
chine and podophyllotoxin (antimiotic agents).[132] The problem would
then be how to identify similar candidates in the vast reservoir of plant
compounds.

Because of the costs of searching for useful compounds from scratch,
the field of "ethnobiology" is becoming a cost-effective means of identifying useful plants. The ethnopharmacological method selects plants
that are known to have therapeutic uses identified by local groups. Of
the three current search methods—randomized, chemotaxonomy, and
ethnopharmacology—the latter is therefore the most efficient. Chemotaxonomy would lead to new sources of closely related chemical
compounds, but the last method will lead to the discovery of new medicines. Such ethnopharmaceutical data, it should be noted, not only
identifies useful drugs but also discovers their preparation.[133]

Breakthroughs in these hidden medicines are more likely in the formal medical traditions of India and China. Several European, American, and Japanese multinationals are therefore entering into research
agreements with Indian (and Chinese) drug companies to commer-
cialize these potential products. This interest has now been hastened
through biotechnology companies who have suddenly discovered natural products as a shortcut to viable medicines.[134]

References to curative plants in South Asia go back to the Rigvedic
period (3,500–1,800 BC) in what are probably the largest collections of
information on medical plants in antiquity. Some of these plant references have been fixed with certainty and include the *Semal, Pithvan,
Palash,* and *Pipal.* The *Rig Veda*'s few references to plant remedies increase in the later *Atharva Veda.* It is, however, the appearance of the
two collections, the *Charaka Samhita* and the *Susruta Samhita,* that bring
to light large banks of information. The two compendia are actually
summaries of earlier works. *Susruta Samhita,* for example, deals with
about 700 drugs, some of them outside the subcontinental region. Later,
through the centuries the number of medicinal herbs mentioned in
the formal Indian collections increased to 1,500.[135]

But this number is not particularly large considering the fact that
within the great agroclimatic variations on the subcontinent there are
over 15,000 species of higher plants.[136] According to another estimate,
there are about 15,000–20,000 plant species in the Indian region that
have medicinal properties. Yet only about 100 are being now exploited
by (Western) pharmacists, which gives an indication of the riches in
store.[137]

As recorded by the historian of medicine Neuberger, a translated
passage from Susruta stimulated the growth of modern plastic surgery

in nineteenth-century Europe. But as Krishnamurty points out, this is only what amounts to a stray reference among the many procedures described by Susruta. It had the fortune of catching the imagination of a Western expert. There could very well be, Krishnamurthy observes, many other similar descriptions that could be rediscovered for modern medicine. He also notes that *Rauwolfia serpentina*, source of the first tranquilizer, reserpine, was only referred to in a stray manner in Susruta as a plant used to treat mental disease. D'Orta, as mentioned above, was the first European to describe it. But this information lay dormant in Europe for four hundred years after that. There are in fact other uses of the plant also described in Susruta (including as a remedy for rat poisoning). Under remedies for mental diseases, Susruta gives a very large list of further plants. It is possible that screening of these plants could give rise to a much larger set of useful remedies.[138] Yet only "a small fraction" of the vast reservoir of plant remedies in the Ayurvedic system has been so far subject to investigation by Indian (Western-oriented) agencies.[139]

India is first in the world in use of herbal drugs, with nearly 540 plant species appearing in different medical recipes.[140] In Ayurveda there are over seventy standard books, which contain around 8,000 recipes. In addition there are a large number of as yet unpublished recipes in everyday use.[141] In India alone there are over 460,000 traditional medicine practitioners. Of these about 270,000 are registered with the state, the vast majority being Ayurvedics.[142] Each is a reservoir of information.

However, the vastness of the South Asian collection of medical knowledge has to be tempered with some facts about the tradition. Sometimes, a claim for efficacy is supported as a certainty without going back to original evidence. Thus one finds a repetition of reasons either for or against a particular treatment in subsequent publications that are simply either translations or adaptations of the original source.

There are many publications on the subcontinent's medicinal plants, and the books vary in length and quality. Some run to several volumes of hundreds of pages each. These have been written to a variety of audiences, such as the Ayurvedic physician, botanist, pharmacist, or pharmacologist. The literature is also so vast and scattered that an exhaustive treatment is not possible. So any examination of the subject must mean that useful references would inevitably be left out.[143]

Sometimes, apart from some core uses, the same plant has different properties attributed to it in different parts of the subcontinent. (A modern parallel would be medicines known to be effective for one ailment being found good for others. Aspirin, for example, earlier used for headaches and fevers, is now used to thin blood as a preventive for heart attacks.) The variety of properties associated with some medicinal plants has increased the number of plants listed in the regional languages. Sometimes there are exaggerated claims with references to miracle drugs, a common occurrence in all ancient medical literature. Apart from this formal written literature there is, of course, another large store of knowledge in the ethnobotanical literature, that of pre-literate tribes living in different parts of South Asia.[144]

In India alone there are also 108 undergraduate teaching institutions in the traditional medical sphere offering four-and-a-half-year courses of study. There are also two institutions and 21 departments awarding postgraduate degrees and doctorates. The government has also established a Central Research Council under which about 50 research institutes and 200 units are doing clinical and drug research.[145] These are all reservoirs of important knowledge, and there has consequently been modern scientific evaluation of Ayurveda in several centers in India. One such center is the Institute of Medical Sciences at Banaras Hindu University.

If the above are a rough indication of the extent of the reservoir of knowledge available, it is opportune to close this chapter by giving some idea of the research already under way.

The Potential: A List of Plants from the WHO

As recorded by the WHO, the Medical Faculty at Varanasi University, India, has compiled a carefully selected list of plants that could be used for different medical conditions. These are all plants in use in the traditional system that could become test candidates for the Western study.

The following table covers a range of treatments—anti-diarrhoeals,

TABLE 1. MODEL LIST OF PLANTS

Gastrointestinal Tract Remedies	Carminatives
	Cinnamon zeylancium
Anti-diarrhoeals	*Elettaria cardamomum*
Acacia arabica	*Matricaria chamomile*
Acacia catechu	*Mentha spp.*
Berberis aristata	*Ocimum senctum*
Commitfera mukul	*Origanuni spp.*
Punica granutum	*Thymus vulgare*
	Umbelliferous fruits, anisem caraway
Laxatives	*Coriander, cumin, dill, and fennel*
Aloe ferox	*Zingiber officinalis*
Cassia acatifolia	
Chicorium intybus	Spasmolytics
Glycyrrhiza glabra	*Atropa belladona*
Plantago ovata, P. psyllium	*Datura spp.*
Rhamnus frangula	*Hyoscyamus spp.*
Ricinus communis	*Solenestemmea argel*

Stomachics
Rheum officinalis

Anti-emetics
Atropa belladona
Hyoscyamus spp.
Mentha spp.
Zingiber officinalis

Remedies for Upper Respiratory Diseases

Adhata vasic
Allium cepa
Althea officinalis
Ammi visnaga
Cassia fistula
Cinnamonum zeylanicum
Ficus carica
Glycyrrhiza glabra
Hibiscus sub darrifa
Linum visitatissmum
Mentha spp.
Nigella sativum
Ocimum sanctum
Prunus domestica
Psidium guarjara
Tilia tomentosa, T. ulmifolia
Urginea maritima
Zingiber officinalis

Remedies for Skin Diseases

Aloe vera, A. herbadnse, A. ferox
Ammi majus
Azadiracta indica
Ficus carica
Fumara officinalis
Lausoria alba
Lupinus termis
Matricaria chamomile
Nymphaea alba
Santalum albam

Anthelmintics
Albizia anthelmintica
Artemesia cina

Anti-pyretics
Allium capa, A. sativum
Fagonia arbica

Analgesics and Anti-Inflammatory Agents

Lactuca sative
Matricaria chamomile
Peganun harmala

Anti-allergics
Cydonia oblonga
Zuzyphus vulgaris

Remedies for Urinary Infection

Ammi visnaga
Balantis aegyptiaca
Cucumis sativum
Cympopogon proximus
Nymphaea alba
Raphanus sativum

Remedies for Arthritic Conditions

Capsicum minimum. Cannum
Commifora mukul
Withania somnifera

Remedies for Eye Diseases

Berberis aristata
Rosa domascena

Treatment for Burns, Scalds, Wounds, Abscesses, and Swellings

Aloe vera, A. barbadense, A. ferox
Lawsonea alba
Lenium usitalissium
Punica granutum

Treatment for Snakebites and Scorpion and Insect Stings

Aloe spp.
Azadiracta indica
Heliotropeum stringosum

laxatives, carminatives, spasmolytics, stomachics, anti-emetics, medicines for upper respiratory diseases, skin disease cures, anti-helmintics, anti-pyretics, analgesics, anti-inflammatory agents, anti-allergics, urinary infection remedies, remedies for arthritis, eye medicines, treatments for burns, wounds, abscesses and swellings, and snakebite and scorpion sting cures. This wide range of candidates has been culled from prior screening of possible remedies.[146]

A 1992 WHO overview of Ayurveda notes that there are hundreds of candidates for such drugs, and it goes on to list a selection of thirty plants that show the most promise. Almost all of these plants are in use today in Ayurveda and hence no extensive animal studies on toxicity would be required.

The list in Table 2 gives the areas for which the plants are now prescribed as well as reference to existing work already done on them in a Western framework. Because they are all in use today, clinical testing can continue without much expenditure or delay. The candidate drugs are those used to treat the liver, cardiovascular disease, hypertension, inflammation, cough, malaria, asthma, dysentery, gastric acidity, filaria, leishmaniasis, and diabetes, as well as those to be used as contraceptives and to improve the central nervous system and the general quality of life. The report states that a careful evaluation of these candidate drugs should yield "at least two or three important additions to [the] therapeutic armamentarium of modern medicine."[147]

TABLE 2: PLANTS FOR POSSIBLE STUDY*

Picrorhiza kurrooa	To protect the liver (Handa and Kapoor, 1989)
Terminalia arjuna	To treat cardiovascular disease (Handa and Kapoor, 1989)
Morlnga olelter	To treat hypertension (*Medicinal Plants of India*, Vol. 2, 1987)
Curcuma longa	An anti-inflammatory agent (Handa and Kapoor, 1989)
Anfrographis paniculata	To protect the liver (Handa and Kapoor, 1989)
Ocimum sanctum	To treat cough (Handa and Kapoor, 1989)

Centella asiatica	To improve the quality of central nervous system responses (Handa and Kapoor, 1989)
Daucus carotus	To prevent implantation and as a contraceptive (*Medicinal Plants of India*, Vol. 1, 1976)
Plumbago rosea	To prevent implantation and as a contraceptive (*Medicinal Plants of India*, Vol. 2, 1987)
Artemesia annua	To treat malaria (Handa and Kapoor, 1989)
Dichroa febrifuga	To treat malaria (*Medicinal Plants in China*, 1989)
Xanthium strumarium	To treat malaria (*Medicinal Plants of India*, Vol. 2, 1976)
Phyllanthus amarus	To protect the liver (*Medicinal Plants of India*, Vol. 2, 1976)
Albizzia lebeck	To treat asthma (*Medicinal Plants of India*, Vol. 1, 1976)
Adhatoda vesica	To treat respiratory problems (Handa and Kapoor, 1989)
Ancistrocladus heyneanus	To treat dysentery (*Medicinal Plants of India*, Vol. 1, 1976)
Croton oblongifolius	For gastric acidity (*Medicinal Plants of India*, Vol. 1, 1976)
Azadirachta Indica	An anti-inflammatory agent (Patnaik and Dhawan, 1986)
Nyctanthes Indica	To treat leshmaniasis (Patnaik and Dhawan, 1986)
Streblus asper	To treat filaria (Patnaik and Dhawan, 1986)

Hamiltonla auaveolans	To treat diabetes (Patnaik and Dhawan, 1986)
Gymnema sylvestre	To treat diabetes (*Medicinal Plants of India*, Vol. 1, 1976)
Momordica charantia	To treat diabetes (Handa and Kapoor, 1989)
Withania somniferum	To improve quality of life (Handa and Kapoor, 1989)
Boerhaavia diffusa	To improve quality of life (Patnaik and Dhawan, 1986)
Tinospora cordifolia	To improve quality of life (Patnaik and Dhawan, 1986)
Asparagus racemosus	To improve quality of life (Handa and Kapoor, 1989)
Tripterygium wilfordii	Male contraceptive
Bacopa monniera	To improve the quality of life
Hibiscus rossasinensis	Female contraceptive

*After Chaudhury, WHO 1992.

In 1963 the Institute of Medical Sciences at Banaras Hindu University evaluated pediatric practices, among others, in Ayurveda. In 1987 it brought out a volume giving the more recent of these studies as well as a summary of the studies done over the previous twenty-five years.[148] This publication has listed particular Ayurvedic approaches, which in controlled tests conducted at the Institute, have been found effective: treatments dealing with intestinal parasites; gandmala; respiratory disorders such as bronchitis; marasmus; giardiasis; H. nana infestation; Indian childhood cirrhosis; eosinophilia; aplastic anemia; ascariasis, a common helminthic infestation; ancylostoma duodemale infestation, a common worm infestation; intestinal amoebiasis; and intestinal cestodes and nematodes. The studies summarized in this book entailed nearly two hundred different Ayurvedic medicinal plants and recipes.[149]

This study represents the systematic and scientific evaluation of Ayurvedic drugs in one institution in one particular area, namely pediatrics. Although conducted according to Western criteria, these studies—under the normal rules of the game—have to be replicated many times over. Then the drugs in question have to be further analyzed for active ingredients (if we adopt the magic-bullet view of medical therapy) and refined, and mass-market drugs then evolved. But the important thing to note from exercises such as these is that the large reservoir of non-Western classical medical treatments and drugs can be subjected to relatively fast and comprehensive studies.[150]

A Regional Survey of Possibilities

In South Asia, the traditional medical corpus, apart from a common core, varies from region to region. These regional variations increase the available spectrum of medicines. A survey of four regions in India —Uttar Pradesh, Madhya Pradesh, Bihar, and Orissa—reveal this diversity. In the 1990 study *Medicinal Plants and Folklore*, V.K. Singh and Abrar M. Khan provide data on the plants in these local regions and the medical claims made for them. The data on medical uses come from local herbalists as well as those practicing in the Ayurvedic and Unani traditions.[151]

In the Aligarh district of Uttar Pradesh, for example, 183 taxa of medicinal plants, belonging to 65 families, had not been reported earlier. Prescriptions for various ailments and modes of using the preparations are also noted.[152] The medical uses of the plants listed cover ailments such as gonorrhea, ulcers, bladder stones, diarrhoea, dysentery, cuts and sores, malaria, syphilis, insanity, epilepsy, fevers, coughs, rheumatism, skin disease, tuberculosis, piles, swelling, eruptions of skin, spleen enlargement, cholera, asthma, cardiac problems, leprosy, inflammation, urinary complaints, tumors of the abdomen, and hysteria.[153]

Another region of Uttar Pradesh yielded about 110 taxa of medicinal plants. From this, a total of 81 claims for medical uses were obtained from local practitioners.[154] The Gwalior forest division of Madhya Pradesh yielded 149 hitherto unrecorded claims for medicinal plants. The range of uses are as wide in the case of the Aligarh district.[155] In the Ranchi and Keonjhar forest regions of Orissa, 230 claims for medical uses previously unrecorded were found, comprising 156 taxa.[156] In recording their findings, the authors note that these flora with medicinal properties remain virtually unexplored.[157]

The German historian Karl Sprengel has pointed out that several modern European remedies were rediscoveries of earlier techniques known to the Greek medical tradition. Verneuli Belloste (1654–1730), considered a minor prophet of modern surgery, in fact often looked into "forgotten old books."[158] Some of these recovered medical techniques, referred to in the Greek text *Dioscorides*, are the application of wool-fat to wounds, the use of male fern for tapeworms, castor oil as an external application, elmbark for eruptions, use of horehound for phthisis, and lac for ulcers. A later rediscovery was the use of carbolic acid in surgery, conventionally credited to Lister but actually anticipated by Bottini (1837–1903).[159]

A search in forgotten old books from another tradition has been documented here, a search bound to be as fruitful as a search among the European past's medical knowledge—if not more fruitful. This search is, however, also beset by commercial and regulatory regimes that vary from country to country. Thus in the US it takes nearly a decade and several hundreds of millions of dollars before a drug is approved by the FDA. But when these drugs come from an established traditional source and not from a laboratory, there is the question of the degree of proprietary claim for the pharmaceutical company that spends money on the approval procedure, thus limiting its incentives. The growth of the health movement in the US is whittling away at some of these restraints. In Europe, on the other hand, there is a less torturous route for getting drugs approved, and a recent market study on Ayurvedic-type drugs concluded that "it is a buoyant, prosperous market that continues to grow at a rapid pace."[160]

LIMITS AND CAUTIONS

But a search for efficacious health and medical systems from the South Asian tradition should not boil down to a romantic/scientific version of *Roots*. The search has to be carried on with the fullest rigor. The bottom line is that the elements of therapy mined from this tradition should work. They should alleviate suffering, make existing faculties better, and prolong life. There can be no other criteria. The evaluative process should be hardheaded, and when the chips are down, these are the only elements to be judged by. When cures come with explanatory systems that are very doubtful, we should jettison the explanatory systems and factor in only the cures.

Especially in certain segments of the US population since the sixties, matters Eastern have been in vogue. This interest has helped popularize some Eastern practices as well as introduce them to serious researchers who could then either validate them on objective criteria or falsify the claims. But this opening to ideas from the East has also sometimes led to the easy path of fads accepted uncritically. These New Age efforts may sometimes yield good music or evocative paintings, but they may also lead to bad science. Twenty years ago this happened when Capra's *The Tao of Physics* achieved near-cult status. In spite of some useful insights, the book was a partial misunderstanding of both East and West. The same thing could happen in Ayurveda, giving good scientific efforts a bad name among more conventional scientists. Hence sometimes indiscriminate acceptance of Ayurveda is as bad as its blanket rejection. After all, Ayurveda is an early system of knowledge in which, apart from the possibilities of new findings, one could see many failings.

Deepak Chopra is an Indian-born, Western-trained doctor who "discovered" Ayurveda as a useful formal body of information on health after he had been fully trained in Western medicine. He has now written many popular books in the US, including some—like *Ageless Body, Timeless Mind*—which have been on the nonfiction best seller list in the US for a number of months. Some of his recommendations are "commonsensical," summarizing much current state-of-the-art thinking in, say, aspects of aging. But a reader searching the contents of Chopra's books for detailed research findings on the efficacy of Ayurveda along the lines of Benson's research on meditation (which also led to popular books) will search in vain.[161]

In fact, Chopra's *Perfect Health: The Complete Mind/Body Guide,* which describes the Ayurveda system, is disappointing because it accepts without question the Ayurvedic explanation of disease. This framework, based as it is on vata, pitta, and kapha, is guided by the Indian humoral basis of disease, the tridosha system. This, as we noted earlier, is similar to the Western, pre-scientific humoral theory of disease. With the rise of modern scientific medicine, however, these explanations have been given up as inadequate. This system's description of body types, which has parallels to discarded Western descriptions (of a "phlegmatic" personality, for instance) is also accepted without any qualms.

Further, Chopra's recourse to farfetched concepts like "quantum healing," which are part of the more esoteric teachings of Maharishi Mahesh Yogi and appeal to the instincts of New Agers, hardly create the impression of scientific rigor. Dubious and far-out theories are not required for justifying the efficacy of the system if its success can be clearly demonstrated.

There have also been failures in the search for traditional medicines. Thus the search for an anti-fertility drug based on an Ayurvedic plant traditionally used for the purpose, *Vicoa indica*, was given up after considerable research, although it showed significant promise in primates. Later, post facto writings list a host of possible reasons for the failure, including over-commitment of resources to just one such plant and the fact that the shrub was used differently under local conditions and under laboratory conditions.[162]

Sometimes these failures stem from use of inappropriate techniques borrowed from the field of synthetic medical compounds research. In the search for synthetic drugs, drug candidates are arrived at from a large pool of synthetic compounds, which are then tested for activity. Through gradual and successive elimination, active components are then isolated. This approach, a WHO study notes, cannot be used for all medicinal plants. If the activity of the plant decreases with further fractionation then the technique is totally inappropriate. Similarly, if the plant has two or more ingredients that operate together for their efficacy, then too, the method breaks down. Instead of blindly screening hundreds of plants, as has often happened in the past, it has been suggested that it is much more rational to screen plants whose activities have been well documented in the traditional sphere.[163]

We end this chapter, which has documented areas of potential and promise, with these cautionary notes, because like mining for precious metals, much useless debris has to be thrown out before finding treasure. But the effort would be well worthwhile.

SEVEN

MATHEMATICS

All cultures, even pre-literate ones, have notions of mathematics. They develop systems of keeping track of numbers, spatial arrangements, and other systems of symbolic manipulation. In her fascinating book *Africa Counts*, Claudia Zaslavasky documents some of these systems in pre-literate African cultures,[1] and a new branch of study, ethnomathematics, encompasses precisely such attempts.[2] Social factors have indeed deeply influenced the development of mathematics, and this is true in contemporary mathematics, as such commentators as Struik and Restivo have shown.[3]

Writing allows the visual application of some mathematical operations, one at a time, without requiring a person to remember all details. Hence literate cultures tend to produce more elaborate mathematical systems. Further, the acquired knowledge can be transmitted to subsequent generations, leading to a gradual accumulation of knowledge.

But the type of mathematics that develop—their depth and sophistication, and the problems they address—vary from culture to culture as to the first mathematical questions they pose and how deeply those questions are explored. Cross-cultural traffic between regions also influences growth in different ways, depending on the region. In isolated civilizations, or before widespread traffic with other regions, there are particular regional flavors to these mathematical developments. Some areas explore particular problems that may be unknown or simply ignored by others. However, after the sixteenth century, the world tended to become one, scientifically, especially as the discoveries of the Scientific Revolution were accepted across the globe. Largely in the last century, as Western scientific knowledge has been adopted as the only true knowledge to be taught in schools and universities, the unique regional characters of different schools of mathematical thought have been erased.

But post-sixteenth-century mathematical developments in Europe were initially fed by many roots. Contributions came from the Greeks, the Arabs, and (through Arab intermediaries) from many regions afar, principally India (South Asia) and China. But, as with all such transmissions and transfers, every morsel of knowledge is not, indeed cannot be, transmitted. There was a large degree of selection, depending on the recipients' readiness to accept particular intellectual problems as valid, on the transmitters' whims and fancies, and on the ready availability of the material to be transmitted. Much source material may be cognitively hidden for a variety of reasons, and so does not get transmitted. And after the initial transmissions, after the modern enterprise had grown and become formidable, any attempts to search elsewhere would have lost their earlier urgency and importance. In fact, there would even be questions about whether any elements still existed, beyond the bounds of the Western knowledge base, that were worthy of transmission.

Now, mathematicians can be said to play around with their problems, choosing one area left undone by predecessors and then developing it, or starting off on entirely new branches from an original stem—or for that matter, exploring entirely new subject areas. The corollary to this view is that there may be many hidden avenues for further exploration in today's stock of mathematical knowledge. Some could possibly be those very seedbeds that initially fed the modern mathematical enterprise.

There may in fact be elements, previously overlooked, that could be discovered and used. Or there could be elements that were developed further in non-European regions after initial transmission—elements ignored by those who have adopted the now-universal Western package. There could even conceivably be areas and pockets where the original mathematical traditions are still practiced, or were practiced until recently, and which have novel elements that could be incorporated into the current practice. In saying this, I admit that the modern package—given its institutions, large number of practitioners, and systems of developed rigor—is by far the most powerful of all the traditions yet.

But mathematics is not just an accumulation of empirical knowledge. It has some of the characteristics of language; in fact, at least metaphorically it is called *the* language of (at least most) science. It deals in an abstract sense with symbols and their manipulations. Starting from a few symbols, mathematicians add new words, weaving these into larger tapestries of sentences. On the other hand, although there is accumulation, there could be pockets of mathematical language undiscovered elsewhere, as is possible in the very nature of language, which may yield new, unheard-of words and sentences.

The five sections of this chapter are devoted to searching out pockets of untransmitted knowledge and suggesting possible new avenues of exploration. The first section attempts to demonstrate that, in fact, there were differences in approach, subject matter, and results in different cultural areas' approaches to mathematics. To illustrate, I take studies that have been done on three seedbed mathematical traditions: the Greek, the Indian (South Asian), and the Chinese. The second section treats pre-sixteenth-century transmissions to Europe, principally through the intermediaries of Arabs and Muslim scholars, which ultimately led into modern mathematics. Third, I examine elements of post-sixteenthth-century developments that constitute modern mathematics. This section would demonstrate that there are elements of the modern tradition that were known either fully or incipiently earlier in the South Asian traditions before their modern discoveries: not all mathematical knowledge was transmitted when the Arabs ended their role as transmitters. Some were left out of the transmission process. The fourth section takes examples from the post-transmission period in South Asia to indicate that the subjects continued to grow in the region, to make new discoveries that had their parallels in the modern body of knowledge. One can make such observations relatively easily by seeing whether there are parallels to modern developments in the "traditional" South Asian sector.

More difficult is showing that entirely new discoveries were made during the post-transmission period, unknown to the modern enterprise. This task is more difficult than a simple search for parallels, which would itself require a mathematician with some knowledge of old languages like Sanskrit, or a Sanskrit scholar with some knowledge of mathematics, a rare combination. The search for really South Asian developments novel to the modern mathematical arena would require a still rarer combination, a mathematician trained in the latest modern mathematics, with a knowledge of very detailed technical Sanskrit. There is no formal professional preparation for such a class of persons—as was the case when European mathematicians in the Renaissance dug into Greek, Latin, and Arabic sources, or when, in the pre-Renaissance period, Arab mathematicians dug into Sanskrit sources. These concerted scholarly efforts required adequate learning in both mathematics and source languages: Sanskrit, Greek, or Arabic.

Further, to retrieve hitherto hidden, novel mathematical results and have them accepted today as "true" would require going through all the steps of scholarly legitimation, including publication in a technical journal, and extensive discussion. Such an enterprise, again, requires dual citizenship in the two domains. So in the exercise in the fourth section of this chapter (showing that explorations continued in South Asia after the transmission and that novel discoveries are indeed therefore possible) I will take the easy road: that is, by demonstrating parallels between post-sixteenth-century European developments and post-transmission South Asian developments, I will show that mathematical creativity indeed continued in the region although at a much lower rate than did the post-Scientific Revolution march in Europe, with its institutions and other supporting factors.

The fifth section provides current examples (from the last decade) in which scholars with a dual approach have made some significant steps at mining South Asia for new mathematically related knowledge. However, such dual-citizenship exercises, although full of potential, are still in infancy. The illustrative examples I have chosen will vary from those of simple arithmetic, to algebra, to artificial intelligence, to quantum mechanics. The examples I give are those presented in their respective professional arenas, indicating that these explorations are in fact academically legitimate.

THREE SEEDBED MATHEMATICAL TRADITIONS: THE GREEK, THE INDIAN (SOUTH ASIAN), AND THE CHINESE

Mathematical records from the South Asian region, because of its long literary tradition, go back to a very early time—in fact, to the Vedas, the earliest of which is dated circa 2000 BC. In addition, remains of the Indus civilization indicate practical applications of mathematical knowledge in large-scale buildings and in trade.

The early records illustrate an extensive knowledge of geometry. From early Vedic times, the need for the construction of altars resulted in a set of sutras, the *Sulva Sutras*, which codified the construction of altars. Dated to between 800 BC and 500 BC, the *Sulva Sutras* explain the construction of a large number of geometric figures such as squares, rectangles, parallelograms, and trapezia. The Sulvas do not constitute formal mathematical works, but their descriptions cover the material addressed in Euclid's first two books as well as his sixth book. No proofs are given about how the results were obtained, as is often true in many Indian mathematical endeavors, where the result is given and not the form of reasoning.

The Pythagoras theorem is also depicted in the Sulvas in an arithmetical approach unlike the geometric approach of Greek mathematics. It should be mentioned that what has come to be known conventionally as Pythagoras' theorem was also known in several other cultures. Cultures, like the Egyptians, that fed the Greeks had used it for several centuries previously: the *Kahun* papyrus—circa 2000 BC—records the relation $4 \times 2 + 3 \times 2 = 5 \times 2$. The accreditation to Pythagoras in the Greek sources themselves comes very late; the first mention comes from Cicero (circa 50 BC), followed by Diogenes Laertius (second century AD), Athenaeus (circa 300 AD), Heron (third century AD), and Proclus (460 AD). The first five centuries after Pythagoras' death, therefore, have no mention of the alleged discovery. The earliest proof of the theorem in the Greek tradition is no later than Euclid's.[4]

Fractions are also often mentioned in the Sulvas. Irrational numbers are also used in the Sulvas, such as $\sqrt{2}$, $\sqrt{3}$, $\sqrt{1/3}$, $\sqrt{1/8}$. One of the important achievements of Vedic mathematics given in the *Sulva Sutras* was the determination of square roots with a high degree of approximation.[5]

The decimal system of counting goes back in Sanskrit literature to the *Samhitas*, which are part of the Vedic literature. This system continues in the Buddhist and Jain literature (sixth century BC) as well as in epics such as the Ramayana.[6]

The formalization of South Asian culture as one knows it today begins with a shift of the center, from the Indus region in the West to the Gangetic plain in the East. In philosophy, this shift coincided with the period of the Upanishads, the Charvaks and the times of the Jain founder, Mahavira, and the Buddha. It also saw the formalization of many intellectual pursuits, such as the science of language. In a series of stimulating papers, Navjyoti Singh has explored the relationship of the growth of mathematical ideas during this period, compared with that of other intellectual pursuits. He has also compared these ideas with their parallels in the other seedbed cultures—of China and, more importantly because of its hold on modern Western knowledge, of Greece. I will draw on Singh's work here.

There are distinct differences in the mathematical traditions of the three regions of India, China, and Greece. In the first two regions, mathematics took an algebraic turn, while in Greece geometry was emphasized; the Greeks had inherited Babylonian algebra and did not make any additional contributions of significance.[7] Another difference was the method of indirect proof, which was central to the Greek tradition and missing from the other two. The Chinese did not develop a formal logic, only a "dialectical logic," while the formal logics in the other two traditions differed.[8] Singh has enumerated many reasons for these differences in orientation.

In India, linguistics was the first discipline to develop as a science. In the work of the grammarian Panini (fourth century BC), there is a formalization of the science of language not to be seen elsewhere until the eighteenth century. In Greece, on the other hand, the notable development was in mathematics, a characteristic best illustrated by Plato's attitudes to knowledge, as represented in his *Republic* and *Timaeus*. For him, knowledge (in descending order of importance) was organized as follows: arithmetic, plane geometry, solid geometry, astronomy, and harmonics. The Greeks developed a deductive logic, whereas Indians used an algorithmic logic. Singh has shown that this particular direction of Indian mathematical development occurred because there is an affinity, at the foundation of logic, between ancient Indian linguistics and its mathematics.[9]

Zeno and his teacher Parmenides developed the system of reductio ad absurdum in its particular manifestation of the indirect proof and the excluded middle. These logical approaches played a key part in Greek mathematics, but were unimportant in the Indian realm. Zeno's paradox was also a part of this general approach but was never considered seriously in India.[10] On the other hand, a paradox from the realm of languages, the liar's paradox ("What I say is a lie") led to interesting debates in India.[11]

In *Astadhyaya*, Panini developed a structural analysis of Sanskrit, the first such attempt in language. Using a limited number of about 1,700 building blocks—such as nouns, verbs, vowels, consonants, and so on—he built a system using rational rules, out of which one could construct compound words. This is similar to how, in Euclid's elements, a set of axioms and rules allows a large number of results to be deduced. In this sense, it could be said that the algebraic character of ancient Indian mathematics is derived from a tradition of a linguistic character where numbers are represented by words.[12]

In India "operations" in mathematics were more fundamental than "being." In the Greek tradition, numbers were considered rational because they were commensurable with a unit. Those that were not were considered irrational. Because the Indian stress was on operations and not on being, this dichotomy did not arise. There was no rational–irrational dichotomy in Indian mathematics, although there was a dichotomy between exact and inexact numbers.[13] The Indian approach emphasized becoming; the Greek, being.

In India, logic was considered a part of epistemology and was consequently "both formal as well as material, deductive as well as inductive." Generally, logic in India was largely tangential to the growth of mathematics.[14] For example, the Nyaya logic, it has been speculated, grew out of linguistic and medical debates. In the Greek tradition, on the other hand, there are close links between logic, philosophy, and mathematics. There are other, more important differences, such as the Indian use of the principle of mutual exclusion without formulating it to a general law of the excluded middle.[15]

The Greeks developed their crowning achievement—geometry— from an anti-empiricist perspective. Words like *axiom, definition, deduction, proof*, etc., attest to the essentially idealist underpinning of Greek geometry. Arithmetic was reduced to a geometrical form, and incommensurable numbers such as the diagonal of a unit square (square root of 2) were considered "irrational." The Greeks also *used* irrational numbers, as when Pythagoras and others described means of constructing segments of length, $\sqrt{AB}$, $\sqrt{3AB}$, $\sqrt{5AB}$, etc., given a segment *AB*. But the approach is arithmetical and there are no rational approximations of, say, $\sqrt{2}$, $\sqrt{3}$, etc. Further, no problems are discussed that use arithmetical operations on irrational numbers, because there was no arithmetic knowledge of the sort available within the Greek tradition.

Although the Indians first used irrational numbers, a formal theory on the subject is, of course, recent. These formal definitions and theories were propounded by Dedekind, Cantor, and Welerstrass.[16]

This geometric bias limited the arithmetic of the Greeks. In India, on the other hand, numbers came into being on their own. Very early on, their mathematicians conceived of very large numbers, the Greeks only going up to their myriad (10,000). Because the emphasis in the Indian tradition was on operations in mathematics and not what they operated on, negative numbers, zero, and fractions were taken in stride. In the Greek tradition, numbers were believed to have a priori existence. Indians did not idealize geometry, treating it instead as an empirical science; this orientation also gave rise to the rapid development of trigonometry in its modern form.[17] Yet in spite of these differences in cognitive structures, both civilizational areas solved some common problems in mathematics, such as forecasting eclipses.[18]

These different foundational tensions have also structured later developments in mathematics, in the post-Renaissance period. Modern mathematics arose from the fusion of the Greek and Indian traditions, which culminated in the sixteenth-century work of four key persons —Viete (1540–1603), Stevin (1548–1620), Descartes (1596–1650), and Wallis (1616–1703)—who took aspects of the two traditions and developed a utopian history of origins to fit into their particular social epistemology.[19] And the program of science inherent in this synthesis worked itself out in the ensuing centuries.

Given these cultural proclivities, South Asian mathematics developed in a particular direction. Much of the key work developed in the tradition up to about the first millennium AD was transmitted to the West by the Arabs, and I will refer to some results of this mathematics later, in my discussion of onward transmissions. Yet to indicate the particular flavor of South Asian mathematics, it is worth mentioning a few achievements, as seen by European observers in the immediate post–Scientific Revolution period.

Europeans coming to South Asia during the time of the rise of the Scientific Revolution in England could be expected to be dismissive of a civilization which was scientifically backward compared to the new, vigorous Europe. But observers did describe given areas in which the South Asians still excelled and could be admired by Europeans. For example, the local astronomers were capable of predicting eclipses of the sun and the moon with considerable accuracy.[20] Thus Robertson said, in the late eighteenth century, "Many of the elements of their calculations, especially for very remote ages, were verified by an astronomical coincidence with the tables of the modern astronomy of Europe, when improved by the latest and most nice deductions from the theory of gravitation. . ."; "such accuracy is to be expected . . . only from astronomy in its most advanced state, such as it has attained in modern Europe."[21] Laplace (1749–1827) was to declare, "The idea of expressing all quantities in nine figures (or digits) whereby is imparted to them both an absolute value and one by position is so simple that this very simplicity is the reason for one not being sufficiently aware how much admiration it deserves."[22] More recently, in a series of definitive studies, Roger Billard has tested the ancient South Asian astronomical texts and their results, using mathematical statistics and computers. Billard showed not only that regular observations were being made in the region, but that they were precise and fitted well into the observational limits of the time.[23]

WHAT THE ARABS TRANSMITTED TO THE MODERN ERA

Arabic science's foundations were laid down between 750 and 850 AD. During this time, the Arabs first became aware of South Asian astronomy and its mathematics.[24] Afterwards, the Arabs became key transmitters of South Asian knowledge to the West. Some of the details of this transmission have been referred to in chapter 3; what follows is a further elaboration.

The principle of place value has been discovered independently four times. In the second millennium BC the Babylonians used a place-value system on a base of sixty; then, around the beginning of the Christian era, the Chinese used positional principles in their calculations, relying on rod numeral computations, between the third and fifth centuries AD. The Indians used the system that subsequently became universal, and the Maya developed a positional system to the base 20. Only the Indian and the Mayan traditions incorporated zero.[25] The discovery of zero and other developments in South Asia essentially gave rise to the modern arithmetic that we know today.

The mathematical knowledge transmitted to the Arabs included numerals (*Argam-I-hindi*), the rules of computation (*hisab-al-Hind*), and the rule of three (*rashikat*). ("If *p* yields *f*, what will *l* yield?").[26] Algebra and trigonometry were also transmitted. The modern method of multiplication followed, appearing in elementary schools. The result of this borrowing was called *kapata-sandhi*: the rule, given by Sridhara, is to "place the multiplicand below the multiplier and multiply successively in the direct or the reverse order; move the multiplier after each operation."[27] A method of multiplication first stated in Europe in Pacioli's *Suma* as "more fantastic and ingenious than the others," had also traveled from South Asia to the Arab countries. Known in the subcontinent at least by the eighth century, it is described as the method of multiplication in which the numbers stand in the same place, and was called *tasthagunana*. In contradistinction to the approach where "the multiplier moves from one place to another," according to this method, "after setting the multiplier under the multiplicand, multiply unit by unit and note the result underneath. Then multiply unit by ten and ten by unit, add together and set down the result as before; and so on with the rest of the digits. This being done, the line of results is the product."[28] Other multiplication methods were *tastha*, *rupa-vibhanga*, *sthana-vibhanga*, *ista-ganita*, and *gelosia*.[29]

The gelosia method, also known as the method of the quadrilateral, cell, or square, was transmitted by the Arabs to Southern European mathematicians. Latin works on the subject were written by Pacioli (circa 1464 AD) and Tartaglia (circa 1545 AD). The historian of mathematics D.E. Smith says, "it was very likely developed first in India, for it appears in Ganesa's commentary on the *Lilavati* and in other Hindu works. From India, it seems to have moved northward to China, appearing there in an arithmetic work in 1593. It also found its way into the Arab and Persian works, where it was the favorite method for many generations."[30] Modern methods of division appear in the rules laid down by such writers as Mahavira, Sridhara, and Aryabhata II, although they were probably known earlier, as indicated by writings of others as well as internal evidence.[31]

In the Latin and Arabic texts, the Indian rule of three is adopted by the same name. In the sixteenth century, Digges (1572) put it as follows: "Worke by the Rule ensueing. . . . Multiplie the last number by the seconde, and divide the Product by the first number. . . . In the placing of the three numbers, this must be observed, that the first and third be of one Denomination."[32]

Around the middle of the eighth century, the method of extracting the square root was transmitted to the Arab region.[33] A similar approach appeared in Europe at least six centuries later. Smith attributes the later European approach as due to Feurbach (1423–1461), Chuquet (1484), La Roche (1520), Gemma Frisius (1540), and Cataneo (1546).[34]

In 1150, Bhaskara II gave the following definition of algebra, which reflects its importance in the South Asian region. "Analysis (*bija*) is certainly the innate intellect assisted by various symbols (*varna*), which, for the instruction of duller intellects has been expounded by the ancient sages who enlighten mathematicians as the sun radiates the lotus: that has now taken the name algebra (*bijaganita*)."[35] Algebra was known in South Asia itself by various names until its designation was crystallized as bijaganita, at least as far back as Prthudakasvami (circa 860). Brahmagupta, on the other hand, had referred to it as *kuttaka-ganita* (*kuttaka* meaning "pulverizer"; the name reflects the importance this branch of algebra to Indian mathematics), a subdiscipline of algebra which deals with indeterminate equations of the first degree. Another name for algebra was *avyakat-ganita*—the mathematics of the *avyaka-ta*, the unknown.[36] Today, "algebra" immediately evokes the use of symbols to stand for unknown quantities, but this technique was first practiced on the subcontinent. Any unknown term was designated *yavat tavat*, ultimately shortened to the algebraic symbol *ya*. In other works, Sanskrit letters representing abbreviations of names of different colors were used for the unknowns.[37]

Also, elements of Indian astronomy were incorporated into Arabic astronomy: the Indian planetary system; the sine tables; the table of solar declinations; the use of the Indian city Ujjayini as the zero meridian (called "Arin"); the beginning of the South Asian epoch Kaliyuga on February 17–18, 3102 BC, which was designated the "Era of the Flood"; spherical trigonometry; ascensional difference; and parallax calculations. All were used, for example, in solar eclipse calculations.

Relatively rudimentary approaches in the applications of trigonometry are found in Hipparchus (circa 150 BC), Menelaus (circa 100 AD), and Ptolemy (circa 150 AD).[38] But trigonometry changed its character and matured from the time of Aryabhata I (500 AD), when it then began to resemble the modern subject. This version of trigonometry began to be described in Europe (via Arab sources), a detailed description appearing in the 1464 work by Regiomontanus, *De triangulis omni modis*.

The primary Indian influences on Europe, ennabled by Arab emissaries, were: a) the spread of the place-value numerals and associated modes of calculation; b) Indian trigonometry, especially the sine function and its uses, and c) the solutions to different equations, with particular emphasis on indeterminate equations.[39] Yet after acquaintance with Greek authors like Ptolemy and Theon of Alexandria, Arab interest in Indian authors declined. Indian astronomy did, however, continue to be popular in Spain[40] and, through astronomy's impact on navigation, indirectly helped the voyagers of discovery find the Indies.

POST-TRANSMISSION MODERN (EUROPEAN) DEVELOPMENTS AND UNTRANSMITTED SOUTH ASIAN PARALLELS

Ideas passed on from South Asia were, by the very nature of such processes, selective. After (and because of) their transmission, developments in mathematics accelerated in Europe. However, some subsequent European developments in mathematics had also appeared in South Asian texts prior to their appearance in Europe. It is useful to explore these parallels, which could have occurred either because European mathematicians simply borrowed the concepts from translations, or because they built their own systems, unaware of prior South Asian work.

Today's arithmetical methods of getting at square and cube roots appear with Aryabhata I; other authors followed, giving detailed expositions. Aryabhata's rules for square and cube roots are a further confirmation that the decimal place system was entrenched by the time of Aryabhata. These "modern" methods do not appear in Europe until much later, only after Cataneo (1546) and Cataldi (1613).[41]

Leonardo Fibonacci (1202) has been credited with the solution to the problem where, given a side a or b, a right-angled triangle with rational sides is constructed. Similarly, Vieta (1580) has been credited with the solution of the problem where, given the hypotenuse c, one has to construct a right-angled triangle.[42] But both solutions were arrived at using an identical approach by Brahmagupta, born in 598, almost a thousand years earlier. Brahmagupta's *Brahma Sphuta Siddhanta* was the work that introduced the Arabs to Indian astronomy,[43] and historian of science George Sarton called Brahmagupta "one of the greatest scientists of his race and the greatest of his time."[44]

Leonardo Fibonacci (1170–1250) is also associated with the sequence 0, 1, 1, 2, 3, 5, 8, 13, . . . Yet this sequence was well known in South Asia and used in the metrical sciences. Its development is attributed in part to Pingala (200 BC), later being associated with Virahanka (circa 700 AD), Gopala (circa 1135), and Hemachandra (circa 1150)—all of whom lived and worked prior to Fibonacci. Later, Narayana Pandita (1356) established a general relation with Fibonacci's numbers as but a particular case that could be derived from his general one.[45]

Permutations and Combinations

The study of permutations and combinations has a long history in India and is connected with variations in the Vedic metres. Rules for calculation for variations of such metres are given in the work of Pingala (circa 200 BC). Ideas of permutations and combinations were also applied in such spheres as music, medicine, architecture, and astrology.

The Jains developed the subject as Vikalpa, and some simple problems are discussed in the *Bhagabati Sutra* (300 BC), where the number of philosophical positions that can be arrived at by combining a set of basic doctrines taken one, two, three, or more at a time are discussed.[46] This could be considered the first time that philosophical statements were manipulated mechanically, thus foreshadowing predicate calculus or Boolean algebra and symbolic logic. In the latter, statements are manipulated algebraically by means of operators such as "and," "or," etc., and is the foundation on which computers are based. The Jains also used their combinations to arrive at varieties of sensations, given combinations of the five senses.

Indeterminate Analysis

In the work of Aryabhata, for the first time one comes across solutions for what later came to be called indeterminate analysis.[47] The solution in question is for the equation

$$by = ax +/-c$$

This is solved by using the kuttaka, the "pulverizer."

Brahmagupta attempted solutions of indeterminate equations of the second degree. He examined two equations:

$$ax^2 +/-c = y^2$$

and a special case of the general equation

$$ax^2 + 1 = y^2$$

Circa 1000, Jayadeva pointed out that Brahmagupta's approach would yield an infinitely large number of solutions. He then gave a general method for solving equations of this type. This method was later refined by Bhaskaracharya and was known as the *chakravala* or the cyclic method, because the same set of operations was repeated. This approach parallels the later inverse cyclic method in Europe, associated with mathematicians like Pierre de Fermat (1601–1665), Leonhard Euler (1707–1783), Joseph Lagrange (1736–1813), and Evariste Galois (1811–1832). In 1657, Fermat challenged a friend of his, Frenicle de Bessy, to find the solution of the equation

$$54x^2 + 1 = y^2 \text{ for minimum } x \text{ and } y$$

Euler mistakenly called this Pell's equation, after John Pell (1610–1685), not knowing of the Indian antecedents. Hankel described the solution of the Pellian equations as "the finest thing achieved in the theory of numbers before Lagrange."[48] Only in 1767 did Lagrange, using continued fractions, find a complete solution to Pell's equation.[49] Lagrange's method required a long calculation of twenty-one successive convergents of a continued fraction. The earlier Indian approaches involved a trial process but had no reference to continued fractions.[50]

The Jayadeva-Bhaskara approach referred to above requires only a few steps to achieve the same result. The extent of this accomplishment is best stated by a contemporary European commentator, Selenius: "The Chakravala method . . . anticipated the European methods by more than a thousand years, But no European performances in the whole field of algebra at a time much later than Bhaskara's, nay nearly up to our times, equaled the marvelous and ingenuity of *chakravala*."[51]

Infinity

The ancient Greeks' largest number was the myriad, 10^4 (10,000). All numbers, from Harappa times, in South Asia have been to the base ten.[52] But very large base ten numbers appear early on, such as numbers from 10^0 to 10^{12} (the latter known as *parardha*). In the *Ramayana*, the army of the villain Ravana had the following number of soldiers:[53]

$$10^{12} + 10^5 + 36(10^4)$$

while the opposing hero, Rama, had an army comprised of

$$10^{10} + 10^{14} + 10^{20} + 10^{24} + 10^{30} + 10^{34} + 10^{40} + 10^{44} + 10^{52} + 10^{57} + 10^{62} + 5$$
soldiers

These figures, of course, are fanciful and part of a myth. But they indicate that even in late pre-Christian times there were names for powers of 10^{62}. The Buddhist texts also had equally large figures, for example 10^{53}. But the Jains developed a means of dealing with large numbers and had an approach to transfinite mathematics. Their religious system required belief in the concept of large numbers. They had a measure of time, *shirsa prahelika*, that equaled $756 \times 10^{11} \times (8,400,000)^{28}$ days![54] The Jains had numbers with up to 194 digits, and their familiarity with large numbers, fed by their cosmology, led to Jain views on infinity. In the Jain cosmos (*loka*), there are infinite souls and infinite material atoms (*paramanu*). Time has an infinite number of moments. Cosmos, in its spatial extension, has innumerable space points. Further, each of the infinite souls has a spatial extension of innumerable points. The minutest (innumerableth) part of a cosmos therefore has infinite density, possessing an infinite number of infinitesimal material atoms and innumerable souls.[55]

Jains had many other interesting views. They believed that within a body, many souls could reside, just as a number of microorganisms could reside in any given animal. In effect, this meant that in a given monad of space, matter could coexist with souls, allowing for a superposition very much like waves are subject to superposition in the modern descriptions of physics.[56]

Jains developed a number system to capture these discussions of their entities. They were interested in getting at finite determinations of their various infinities, infinitesimals, and innumerableth quantities and describing how these were related to each other.[57] It is useful to contrast the Jains' approach with that of the Greeks. For the Greeks, ideas of order and of a finitude were intimately linked. Cosmos was determinate, finite, and ordered; chaos was indeterminate, infinite, and disorderly. *Aperion*, the Greek word for "infinite," stood for anything that was disorderly and infinitely complex—an arbitrarily crooked line, or an infinitely large entity, for example. Aperion therefore, by its very nature, could not be determined finitely. In contrast, the Jain tradition developed an approach whereby a finite determinateness and an infinitude could both be accommodated.[58] In a pioneering work in the early 1940s, A.N. Singh outlined the Jain theory of transfinite numbers as found in their text *Dhavala*, which dealt with the enumeration of souls. Jain texts have been looked into recently in greater detail, and the ideas of transfinites and infinity found in them have been expanded, principally by Navjyoti Singh.[59]

Before discussing the details of the Jain contribution further, it is useful to define briefly some mathematical terms relating to infinity. "Actual infinity" stands for something which actually consists of an infinite number of component elements. Actual infinity is differentiated from "potential infinity," which denotes entities that are potentially infinite—like the set of natural numbers, in which one can keep counting on and on, but which does not actually have infinite component elements. "Mathematical infinity" represents a number higher than all natural numbers. Because all of these are in a realm beyond finite natural numbers, they are termed "transfinite."[60]

Jains divided all numbers into three groups: numerable, innumerable and infinite. The first category consisted of all numbers from 2 (1 was ignored) to what was considered the highest number, indicated in the example from the *Anuyoga Dwara Sutra* (circa the Christian era): "Consider a trough whose diameter is that of the Earth (100,000 *yojanas*) and whose circumference is 316227 *yojanas*. Fill it up with white mustard seeds counting one after another. Similarly fill up with mustard seeds other troughs of the sizes of the various lands and seas. Still, the highest innumerable number has not being attained." (A yojana is about ten kilometers.) After this highest number N is reached, infinity is given by the following expression:

$$N + 1, N + 2, \ldots \ldots (N + 1)^2 - 1$$
$$(N + 1)^2, (N + 2)^2, \ldots \ldots (N + 1)^4 - 1$$
$$(N + 1)^4, (N + 2)^4, \ldots \ldots (N + 1)^8 - 1, \text{ etc.}$$

In the Jain system, there are five types of infinity: "Infinite in one direction, infinite in two directions, infinite in area, infinite everywhere, and infinite perpetually." Here the idea of infinity was combined with the concept of dimension, so that infinity was defined in one, two, three, and infinite dimensions.[61]

Parallels to these ideas are found in Europe only late in the nineteenth century, especially in the work of Georg Cantor. Until Cantor's work, Europeans (following an Aristotelian injunction) believed that all infinities were equal to each other and therefore the same.[62] Jains had already gotten rid of this idea in their approaches. The Jains' N, the highest innumerable number, finds a parallel in Cantor's concept of aleph-null. Aleph-null is the cardinal number of the set of integers 1, 2, . . . N, which are infinite. It is also considered the first transfinite number. Cantor articulated the idea of a sequence of transfinite numbers and developed an arithmetic to deal with it. The Jains, on their part, also tried to give a whole set of transfinite numbers, aleph-null being the smallest member.

For internal consistency in a world populated with numerous actual infinities, Jains required that there be a rank order in the various groups of actual infinities. For this they used both physical as well as mathematical arguments. Their mathematics of transfinite numbers was therefore fully grounded in the physical basis of actual infinities. They distinguished between two types of transfinites, *asamkyata* and *ananta*—that is, infinities that are, respectively, rigidly and loosely bounded. From this starting point the Jains developed a relatively detailed classification system of transfinite numbers and the mathematical operations to handle them. Their system of mathematical operations on transfinite numbers was established before AD 800, although some of the work has antecedents going back to the third century BC.

Among the strict mathematicians in India, a rough conception of infinity first occurs in Bhaskara, who treats the concept of infinity in his first chapter, in which he presents the concept infinity times a number equal to infinity.[63]

Binomial Theorem

The binomial theorem is an important pillar of contemporary mathematics. Its development was made easy by the discovery that binomial coefficients can be arranged in a triangular array. In Europe this array was known as Pascal's triangle, after Blaise Pascal, whose *Traite du triangle arithmetic* was published in 1665. The triangle, however, had already appeared in the work of several sixteenth-century European mathematicians, such as Stifel, Scheubel, Tartiglia, and Bombelli. In China, still earlier versions of the triangle can be traced in the work of

Chu Shih Chieh (fourteenth century); Yang Hui (thirteenth century), and Chia Hsien (twelfth century).[64]

But an identical array of binomial coefficients, called *meru-prastara*, appeared in a work by Pingala around 200 BC. Recent research has shown how Pingala in fact antedated not only the Pasqual array, the precursor of the theorem, but also the binomial theorem itself. The concrete problem before Pingala was metrical: namely, how, in music, different types of metres (*chandas*) could be produced from one metre of three syllables, four syllables, five syllables, six syllables, and so on. The result was an algebraic rule giving the number of arrangements possible, yielding the binomial expansion.[65]

Motion: Impetus Theory, Gravitation, the First Law

Theories of motion were formalized in Europe between the fifteenth and the seventeenth centuries. Thinkers like Copernicus, Tycho Brahe, Kepler, Galileo, and above all, Newton figure prominently in these efforts. Helped by other lesser figures, they brought together a body of knowledge that helps one understand most everyday motions on earth and beyond. This major task was accomplished by bringing together observational knowledge and combining it with new techniques un-known to the Greek tradition, drawn from subjects like algebra, trigo-nometry, coordinate geometry, and calculus.

Yet some elements discovered by these European pioneers were also known, sometimes in an incipient form, in the South Asian region. So although one should not detract from what was a crowning achieve-ment of the Scientific Revolution, it is useful to record the presence of these elements in another culture, important constituents of the total picture put together by the European pioneers: a theory of impetus; a view of motion as normally occurring in a straight line (the first law of Newton); and relating the forces that make stones fall to a universal force of gravitation that keeps the heavens together (a concept that lead ultimately to Newton's laws of gravitation). These are not all the in-gredients by any means, but without them, Newton's grand synthesis could not have been achieved.

In the Vaisesika texts *Prasatapadabhasya* and *Vyomavati*, as well as in other classical texts, the idea of impetus explains a body's continued motion. When the body is in the first unit of motion, *vega* impetus is transmitted to it; this helps the continued motion. If the moving body encounters another body, the impetus is fully or partially neutralized, and so comes to rest or continues at a lower speed. Motion of bodies like the javelin, arrows, and pestles are described in these terms. It was not before the fourteenth century that an impetus theory appears in the West. Later in Europe, from the time of the sixteenth century, the impetus theory was further elaborated, but had to wait for Galileo to interpret motion in the language of mathematics. Until then, the views of Aristotle prevailed, holding that the person who threw the projectile transferred the impulse to the layer of air next to it. This layer then transferred the impluse to the next, and so forth as the projectile was carried forward. In the sixth century, John Philoponus of Alexandria disagreed with this view and suggested that the projectile got its motive power from the projector himself. The Western impetus theory that was later developed in mathematical terms thus had its conceptual forerunner in the Vaisesika texts.[66]

As for Newton's First Law, the Jain text *Tattvarthasutra* states unequivocally, "All motion takes place following a straight line, *sreni*." It is only when there are counteracting causes that motion veers away from a straight line.[67] This theoretical construct describes motion of material monads (*pudgala*).[68]

"Gravitation" encompasses two broad meanings. One is the force that makes bodies fall to earth, and it is recognized in several ancient cosmologies. The second aspect deals with how planets and other bodies are kept in their paths.[69] In popular scientific mythology, the apple falling on Newton's head brought these two aspects together as a common gravitational concept. However, that the working of normal gravitation (in the sense of the universal pull of a spherical earth) was well understood in South Asia is exemplified by several references to the fact that persons living in polar opposites of the earth are spatially equivalent to each other—that is, any one position on the earth could not be higher or better placed than another. In fact Aryabhata sarcastically referred to the prospect that inhabitants of the South Pole and the North Pole might consider "each other the undermost."[70]

Brahmagupta, one of Aryabhata's successors, elaborated:

> Scholars have declared that the globe of the earth is in the midst of heaven and that the North pole is the home of the *devas*, and the South pole 'below' is the home of their opponents. But this below is without any meaning. Disregarding this we say that the earth on all sides is the same, all people on earth stand upright and all heavy things fall down to the earth by a law of nature, for it is the nature of the earth to attract and to keep things, as it is the nature of water to flow, that of the fire to burn and that of the wind to set in motion. If a thing wants to go beyond the earth let it try. The earth is the only low thing and seeds always return to it in whatever direction you may throw them away, and never rise upwards from the earth.[71]

These views should be contrasted with the views of ancient Platonists and medieval Europeans. The Platonists of Chartres explained the falling of bodies by assuming that bodies of a similar nature tended to come together. This means that a part of a body, removed from its parent, would tend to reunite with it. Thus a stone falls to the earth because it was detached from the earth. On the other hand, fire shoots upwards because it wants to reunite with the fiery sphere at the outer limits of the universe. This theory was borrowed from both Plato (his *Timaeus*) as well as from Aristotle. The latter's view of the cosmos held that the earth was surrounded by a set of concentric spheres, the first three consisting of water, air, and fire. Aryabhata and those who came after him made no recourse to such speculative views.

Aryabhata originally assumed the earth to be stationary, with the planets revolving around it. To explain for kinks in this theory, he also developed a system of epicycles like those of Ptolemy, but with significant differences. One was that he did not have the endless system of circles which the Ptolemic system required; both his eccentric orbits and his epicycles were oval, as opposed to circular. Yet he found this system, too, unsatisfactory, replacing it with a radical new theory where the earth moves and a stationary universe surrounds it. This contradicted common sense, which saw moving heavens. Aryabhata explained relative motion, giving the following example: a person moving forward in a boat sees immobile objects on the shore moving backwards. An argument along these lines justifying the earth's motion was not seen in Europe until the fourteenth century, in the work of Nicole Oresme.[72] Oresme also propounded a theory akin to Newton's laws of motion.

It is one thing to describe the motion of the planets through variations of the epicycles theory. It is another matter to relate the planetary system to gravitation, the second aspect of the gravitation theory. The first type of attraction was of course known in South Asia. But the question of whether gravitation, in the sense of ordering the planetary movements, was known in South Asia is much more problematic. In the eighteenth century, William Jones made the extravagant claim that the whole of Newtonian theory "may be found in the Vedas."[73] This, of course, is preposterous. Yet there are definite indications, in the work of Aryabhata as well as his followers and their Arabic interpreters, that some ideas of universal attraction that operate in the planetary field were indeed known, although in a preliminary form. For example, one of Aryabhata's interpreters, Varahamihira, expanding on Aryabhata's work, says that the earth is kept suspended in the sky "even as a piece of iron is held in the grip of a magnet."[74] Al-Biruni, the Arabic interpreter of South Asian mathematics, elaborated further that "on account of this law of gravitation they [the South Asians] consider heaven too as having a globular shape."[75] This also means that action at a distance was understood as similar to the action of magnets.

Calculus

The idea of infinitesimal calculus first appears in the Greek tradition in the work of scholars like Antiphon (430 BC), Eudoxus (370 BC), and Eudemus (335 BC). In the work of Archimedes (225 BC), one finds the nearest approach to infinitesimal calculus. Apart from these early ideas, the discovery of differential and integral calculus is attributed to Newton (1642–1727) and Leibniz (1646–1716). But there is strong evidence to suggest that Bhaskara II (1150), who lived 500 years before Newton, was a pioneer of some of the principles of differential calculus.

The idea of calculus arose in India as an attempt to solve the following sets of problems: a) the value of π; b) the motion of a planet at any given instant; c) the "position angle" of the ecliptic with any secondary to the equator; d) and the surface area and volume of a sphere. Bhaskara, in his *Siddhanta-siromani*, solves (c) and (d) using the concept of differentiation. Solutions to the problems (a) and (b) required the idea of integration, introduced by Bhaskara in *Siddhanta-siromani*, in Narayana's *Ganitakaumudi*, and in *Yukti bhasa* (circa 1500 AD).[76]

Bhaskara's approach to determining the volume of a sphere considers the sphere to be constituted of a collection of small pyramids, the volume being obtained by integration.[77] In early seventeenth-century Europe, Kepler used a similar means of deriving the volume of the sphere,[78] while modern methods of integration were discovered by Leibniz and Newton.[79]

Bhaskara was perhaps the first to conceive of the differential coefficient and differential calculus. For the purpose of calculating exactly the daily motion of a planet, he thought of the instantaneous (*Tatkalika*) method by taking the length of the day and dividing it into a large number of very small intervals. He then compared the position of the planet at the end of successive intervals. What he termed the *Tatkalika Gati* is the motion of the planet at any given instant. Bhaskara has given the relationship by taking y and y' as the mean anomalies of the planet at the ends of two consecutive moments. He states:

$$\sin y' - \sin y = (y - y') \cos y$$

This is equivalent to

$$\sigma(\sin y) = \cos y \, \sigma \, y$$

Bhaskara states this verbally as, "The product of the cosine of the semi-diameter by the element of the radius gives the difference of the two sines." But it should be noted that this formula

$$\sigma(\sin y) = \cos y \, \sigma \, y$$

was also known to two other earlier commentators, Munjala (AD 932) and Prashastidhara (AD 958).

Bhaskara's differential calculus measured velocity in terms of an "infinitesimal" unit of time, the *truti*. He also knew that the differential dy/dx vanishes when the variable is at a maximum. His approach had elements of the "mean value theorem," which is obtained from Rolle's theorem (1691). Thus he stated in his *Graha ganita, Spasthadhikara,* "Where the planet's motion is an extremum, the motion is stationary; at the commencement and end of retrograde motion, the apparent motion of the planet vanishes." Later mathematicians of the Kerala school, such as Nilakantha (1444–1545) and Acyuta Pisarati (1550–1621) continued this work in South Asian calculus. The former gave an expression for the differential of the inverse sine function, while the latter formulated a rule for obtaining the differential of the ratio of two cosine functions.[80]

However, the idea of a limit of a function was not found in these Indian approaches, being essentially an idea of the modern age. Also, differential calculus was used in India primarily for problems in astronomy and in some problems in mensuration. It was not applied over a wide spectrum, as is modern calculus.

Set Theory

George Cantor (1845–1918) is credited with the invention of the modern theory of sets, but Jain mathematics included a theoretic approach using various types of sets and methods of their manipulation, all of which are discussed in the Jain literature. As the Jains had to deal with a large number of souls and different karmas in their belief system, such set theories became necessary.

These sets consisted, for example, of the Jains' "accomplished souls," "*nigoda* souls" (similar to bacteria and coming to life eighteen times in one breath), "sets of souls in the vegetation world," the "set of all ultimate particles of matter," the "set of all time instants," the "set of all points in empty space," and so on.[81] Souls were classified in various stages of life and in their development toward omniscience. The soul is effected through a complicated system of material particles which, when mapped, give different types of sets in the form of matrices that have a relation to karma. There were means of manipulating different types of sets, and symbolisms for a set theory. Such symbols were arithmetical, algebraical, geometrical, and place-valued.[82]

A later Prakrit-language Jain text of the eleventh century, *Trilokasara*, by Nemicandracarya, describes fourteen different sequences for elaborating number measure. All fourteen sequences are well ordered and parallel the program envisioned by Cantor and later twentieth-century mathematicians like Zermelo, Hadamard, and Hartogs.[83]

POST-TRANSMISSION MODERN WESTERN DEVELOPMENTS AND POST-TRANSMISSION SOUTH ASIAN PARALLELS

If there were novel discoveries in the "pre-transmission period" anticipating the "post-transmission period" in Europe, then it is natural to expect that novel discoveries would continue in South Asia after the transmission. Such discoveries would continue as long as the older institutions for the practice of local mathematics remained. But since the Scientific Revolution's knowledge production in Europe had a strong institutional basis in knowledge "factories" (universities and similar centers), non-European innovations could be expected to be comparatively slower. But documenting fresh discoveries within the South Asian tradition is difficult. Significant developments could lie hidden from the Western-oriented tradition because of the lack of scholar-scientists with dual citizenship in the two knowledge realms referred to earlier. Our only recourse is to a discussion of parallels with modern discoveries, as in the earlier section—but examining those parallel developments in South Asia *after* the golden age of knowledge transmission.

For this purpose, we take only one South Asian region, Kerala, on whose mathematics recent historical research has focused. A survey of Kerala astronomers reveals that over a period of 600 years, from 1237 to 1846, there were over 100 astronomers and mathematicians who between them produced over 700 works. Their work anticipated many results later attributed to their European counterparts, including infinite series for pi, sine and cosine functions, and the development of fast-convergent approximations to them.

In 665, Brahmagupta wrote *Khandakhadyaka*, which describes an interpolation formula through which, once a table of sines is given, the sines of intermediate angles are obtained. The rule he employs is equivalent to what later came to be known as the Newton-Sterling formula: Brahmagupta had initiated a branch of mathematics which later came to be known as interpolation theory.[84] Similarly, a special case of the Newton-Gauss interpolation formula is found in the work of the Kerala mathematician Govindasvami (circa 800–850), who elaborated on the work of Aryabhata and Bhaskara:[85]

$$f(x + nh) = f(x) + n\Delta f(x) + \tfrac{1}{2}n(n-1)\{\Delta f(x) - \Delta f(x-h)\} \text{ for } 30^0 \text{ to } 60^0$$

Madhava of Sangamagramam (1340–1425), perhaps the greatest Kerala mathematician, is noted for the step he took in shifting from the finite approaches and procedures of ancient mathematics to "treat their limit passage to infinity." This shift is the essence of modern classical analysis, and Madhava was very probably the first mathematician to have made major contributions in mathematical analysis. His greatest contributions, however, would be in the infinite series expansions of circular functions and finite series approximations.[86]

Madhava also used a series of numbers to calculate the circumference of a circle, long before Newton performed the same feat.[87] Madhava arrived at infinite series for some trigonometric functions and for π. Madhava enunciated the Gregory and Leibniz's series for the inverse tangent[88] and gave Newton's power series for the sine and cosine,[89] a result later established algebraically by De Moivre (1707–1738) and Euler (1704–1783). Madhava used the series in its finite form before the others;[90] in fact, Leibniz, Newton, and Gregory arrived at these results, using calculus, two hundred and fifty years after Madhava.[91]

Madhava's work also antedates part of the mathematics known as Gregory's series and Taylor's series. One result gives the series for an arc as expressed by its sine-chord and cosine-chord in terms of the radius of the circle and the length of the arc. This formula later can be reduced to the value of the angle θ, given in terms of a series of relations between sines and cosines (Gregory's series, named after James Gregory, 1638–1675).[92]

$$\theta = \frac{\mathrm{Sin}\ \theta}{\mathrm{Cos}\ \theta} - \frac{1}{3}\frac{\mathrm{Sin}^3\theta}{\mathrm{Cos}^3\theta} + \frac{1}{5}\frac{\mathrm{Sin}^5\ \theta}{\mathrm{Cos}^5\theta}\cdots$$

Madhava's approach gave very accurate values for half-sine chords, correct up to eight or nine decimal places.[93] Another series derived by Madhava gives the results for the sin and cosine of $(x + h)$ in terms of the sine and cosine of x, and various multiples of h, and R the radius. The latter is a particular form of the series

$$f(x + u) = f(x) + u \cdot f(x) + \frac{u^2}{2} \cdot f'(x) + \ldots \text{ etc.}$$

which is today known as the Taylor series, after its European discoverer, Brook Taylor (1685–1731).

Parameswara Namaputiri of Vatasseri (1360–1460), another Kerala product, arrived at the results attributed to Lhuilier and Taylor. Lhuilier is said to have discovered a formula for the circumradius of the cyclic quadrilateral in 1782:[94] Parmesvara's formula on the circumradius of a quadrilateral appeared two centuries earlier:[95]

$$R = \frac{1}{4} \frac{\sqrt{(ab + cd)(ac + bd)(ab + bc)}}{(s + a)(s-b)(s-c)(s-d)}$$

Paramesvara also arrived at four-term approximations for the sine function more than two centuries before the Taylor expansion was known in the West.[96] Paramesvara, in his *Siddhantadpika*, also gave a parallel theorem for the mean value theorem of differential calculus.[97]

In his *Sphutanirnaya*, Acyuta Pisarati (1550–1621) enunciated the reduction to the ecliptic that would be presented by Tycho Brahe (1546–1601). Pisarati later produced a simplified formula in his work *Uparagakriyakaram*. Both formulae were reproduced in his work the *Rasigolasphutaniti*.[98]

Series for pi had been worked out by the Aryabhata school, and in AD 499 Aryabhata gave the approximation for pi as 3.1416. In his commentary on the *Aryabhatiya*, Nilakantha Somayaji (1444–1545) stated that pi is an irrational number; that is, the result of operations will always give a remainder. In his words: "The ratio of the circumference to the diameter can never be expressed as the ratio of two integers."[99] Conventionally this result is attributed to Lambert (1671).[100]

A notable achievement, just now coming to the fore through recent translations and their examination by competent mathematicians, is the development of a heliocentric mathematics in Kerala, earlier than the European tradition. This contribution appears in the work of Nilakantha Somasutvan (1444–1545). In his treatise *Tantrasangraha* (1500), Nilakantha significantly revised earlier planetary models of the inner planets Mercury and Venus then prevailing in India. This revision allowed for a significantly improved formulation of the equation for the center of these planets—better than not only the earlier Indian approaches, but also the Islamic and European traditions, until the coming of Kepler a hundred years later.[101]

In later works such as *Golasara*, *Siddhantadarpana*, and the *Aryabhatiyabhasahya*, Nilakantha explains that his computational scheme implies a geometrical model of planetary motion. In the astronomical tradition of India, the stress has been on the obtaining of good computational methods that could, for example, yield the longitudes of the sun, moon, and the planets as well the durations of eclipses. Sophisticated formulae and computational schemes were used for these purposes without emphasizing the geometrical models of planetary motion, unlike the Greek tradition, which emphasized geometrical models.

Sometimes, though, elaborate geometrical models that correspond to the computations are described in the Indian tradition.[102]

In Nilakantha's planetary scheme, the five planets Mercury, Venus, Mars, Jupiter, and Saturn move around the mean sun in eccentric orbits; the sun in turn moves around the earth. Subsequent Kerala astronomers like Jyesthadeva, Acyuta Pisarati, Putmana Somayaji, and others adopted this model of the planetary system. In the European tradition, Tycho Brahe, at the end of the sixteenth century, developed a similar heliocentric model in a reformulation of the Copernican planetary model published in 1542. Yet in Brahe's model, the equation of the center for the interior planets was not formulated properly; the scheme was improved upon by other subsequent European astronomers. Acyuta Pisarati (1550–1621) discusses in detail how to correct for planetary longitudes due to latitudinal effects, using the method of reduction to the ecliptic. In Europe it was, again, Tycho Brahe who formulated this correction.[103]

As is clear from these examples, Kerala mathematicians in the southernmost tip of India developed, over a period of several centuries, mathematical results that anticipated later European discoveries. That these contributions, arrived at by the fifteenth century, did not reach Europe could have been due to the fact that the Kerala school was in the southwest corner of India, beyond the reach of Arab or Persian intermediaries. These innovations also surfaced after the major transmissions westwards had already occurred.[104] Further studies could lead to other discoveries of the Kerala school that not only anticipate later European developments but also yield results new to contemporary discourses.[105]

CONTEMPORARY SEARCHES FOR THE MODERN IN SOUTH ASIAN TRADITIONS

If creativity was possible in various isolated areas after the transmission, then we should be able to find, even today, other scattered pockets of creativity. We should be able to use such resources for contemporary purposes, either as they are, or as foundations on which to build.

The four examples below, from the broad field of mathematically related knowledge, illustrate that the recovery and use of such information is indeed possible. The first example, from the very elementary field of arithmetic, shows that there are still South Asian arithmetic techniques that are more powerful than the generally accepted ones in use today. The second example shows that one could build on mathematical work of a sixth-century South Asian mathematician, indicating once again that old shoots can be built upon. The third is from the computer field and illustrates that early South Asian systems of formal language theory can be used in contemporary artificial intelligence. In the fourth example, quantum theoretical results are derived from a combination of Cantor's theories and Buddhist logic. All of these illustrations are indicative of the potential that lies in civilizational recoveries of knowledge.

Arithmetic Further Developed

The first example is from the work of Ashok Jhunjhunwala, a professor of electrical engineering at the Indian Institute of Technology, Madras. He found the following techniques by looking at how everyday artisans and businessmen in the non-Europeanized sector were dealing with numbers. As a pointer to how similar approaches may be used in other contexts to find cognitively hidden knowledge, it is useful to describe how he made his discoveries. Jhunjhunwala's interest began when he asked a carpenter to make a bookshelf. When the latter started computing the amount of wood required, Jhunjhunwala started his own calculation to crosscheck the carpenter. The carpenter took only about four minutes to do his calculations, whereas he—a trained modern mathematician with a Ph.D. from an American university—took over ten minutes. The carpenter, on the other hand, had only a primary school education. Questioning the carpenter, Jhunjhunwala came across a very simple, but fast, method of calculation.[106] For his own part, Jhunjhunwala had been taught, at age six, a simple technique to check results of addition when he had to help in the family business. He later found, as a university teacher, that by using this method learned in childhood, he was checking tabulations of examination results faster than other examiners. The only other person who checked additions as fast was a person who had also been taught this traditional method. Until recently, such techniques were being used fairly widely in many regions, and Jhunjhunwala emphasizes that the material he has covered is only a small part of the hidden mathematical heritage.

In a recent book, he describes, with examples, eight of these old methods (all of which are faster than conventional methods): means of finding areas; a multiplication method for numbers with some special characteristics called *Nikhilam*; multiplication using the *Urdhva Tiryaka*; squaring using the same method; division using the method; evaluation of powers; square root of numbers; divisibility of numbers; square rooting; factorization; and other methods. Jhunjhunwala also includes methods of catching errors.

To indicate the efficacy of these old methods, I cite below one example from Jhunjhunwala's book, the one involving the carpenter:[107]

Problem: Find the area given by 5ft 1in by 3ft 5in.

a) First multiply the inches part of the breadth and length to get at square inches. Enter this under the inches column as:

$1 \times 5 = 5$

5 ft	1 in
3 ft	5 in
	5 sq in

b) Now multiply the feet part of the breadth and length to get at square feet and write it under the feet column as: $3 \times 5 = 15$

5 ft	1 in
3 ft	5 in
15 sq ft	5 sq in

c) At this stage, cross-multiply the inch part of the breadth with the feet part of the length as well as the inch part of the length with the feet part of the breadth. Now add the two ensuing products: $(5 \times 5) + (3 \times 1) = 28$. Write this between the earlier two results.

5 ft		1 in
3 ft		5 in
15 sq ft	28	5 sq in

d) Divide the result 28 by 12. The resulting quotient 2 is added to the square feet part of the result obtained earlier. Multiply the remainder 4 by 12 and add to the square inch part of the earlier result. The answer is the desired result.

5 ft		1 in
3 ft		5 in
15 sq ft	28	5 sq in

$$2 < \quad \frac{28}{12} \quad > 4 \times 12$$

quotient   remainder

17 sq ft	53 sq in

This description may sound long, because each step is articulated for explanation purposes. In practice, however, it is much swifter than the normal method of calculation, which is of course as follows (and which would be longer than the faster method of the unknown carpenter):

$$
\begin{array}{lll}
5 \text{ ft} & 1 \text{ in} = 61 \text{ in} & 61 \times 41 \\
3 \text{ ft} & 5 \text{ in} = 41 \text{ in} & \overline{\qquad\qquad} \\
& & 2501 \text{ sq in}
\end{array}
$$

2501 sq is equal to

```
144) 2501 (17
     144
     ―――
     1061
     1008
     ―――
      53
```

The answer is 17 sq ft 53 sq in—the same result arrived at by the unorthodox method. The latter method occasionally requires a bit more work than given in the example; Jhunjhunwala explains why and gives the instructions in such cases.

It is not possible, for reasons of space, to give all the types of traditional mathematical calculations covered by Jhunjhunwala and which, one should note, are only a small part of the large tapestry of other traditional solutions that remain undocumented. Yet before bringing this section to a close, it is illustrative to give one more example: the *navasesh* method. Navasesh, *modulo nine*, is the remainder obtained when a number is divided by 9. To check whether a multiplication is correct using the navasesh method, use the following procedure: Check whether 138,529 is the correct answer to the multiplication of 317 by 437. Discover whether the navaseshes square up in the different operations. A navasesh is obtained by taking a number and adding the individual numbers to give their sum. If that result is greater than nine, then the digit of the resulting number should be added. The procedure should be continued until one gets a single digit number.

Based on this method,

the navasesh of the first number, 317, is $3 + 1 + 7 = 11 => 1 + 1 = 2$;

the navasesh of 437 is $4 + 3 + 7 = 14 => 1 + 4 = 5$.

Their product is $2 \times 5 = 10$, whose navasesh is $1 + 0 = 1$.

The navasesh of the result 138,529 is $1 + 3 + 8 + 5 + 2 + 9 = 28, 2 + 8 = 10$,

$1 + 0 = 1$,

which is the same as the earlier navasesh result.

The two results match when the result is correct. Navasesh methods can be applied not only to multiplication, but also, for example, to addition.

Yet another technique described by Jhunjhunwala is straight division, a mental division technique. Jhunjhunwala and Parthasarthy have modified this technique, called *Mischrank*, to carry out large divisions on normal personal computers with significant increase in speed of calculation. He has tried the calculations on a normal PC with various combinations of approaches using the old Indian methods. Below, I give one of his results, to show the speed of the South Asian method.[108]

Technique	Average Execution Time (microseconds)
Normalized straight division	93
Straight division with Mischrank	60.4

Because computers do millions of calculations every day across the world, these applications could mean cost savings in the modern sector. Jhunjhunwala's work has, in the meantime, been published in regular computer journals, where the relative efficacy of the ancient system for computing has been further demonstrated with several examples. The old knowledge has completed a journey, successfully ferried into the current frontier of legitimized knowledge.[109]

The results appear astonishing to those trained in the conventional methods. But as Jhunjhunwala points out, these techniques are based on a deep understanding of the place-value system and some elementary algebra. These methods were worked out long ago—in fact, probably as long ago as the conventional ones that we use now, before they themselves were transmitted to the West. In this light, therefore, the modern conventional system becomes just *one* of a whole set of techniques of doing arithmetic known in the region.[110]

Jhunjhunwala's collection of mathematics at the local level shows how much proliferation there was once the decimal system was understood. Local groups, unfettered by, say, standard school syllabuses as in modern times, would have changed their approaches and discovered new tricks, a process of grass-roots creativity very much like what one observes in different responses to changing agro-climatic conditions across the world and the resultant variations in agricultural practices. Seen in this light, it is clear that Western arithmetic calculations represent only a particular approach, one transmitted through Arabic (and later Latin) translation to become the dominant method. It has been legitimized as the correct (and so the only valid) knowledge transmitted, respectively, from South Asia (into the Arab regions) and from the Arab regions (into Europe). This system allows elementary school children all over the world today to perform arithmetical operations beyond the abilities of the best mathematicians of Greece, Rome, and medieval Europe.[111] But there are other computational approaches that could speed up the system still further.

Extending Bhaskara

The second example comes from Krishnan and his coworkers, who have extended the work of Bhaskara I (AD 600), who developed second-order rational approximations to the sine function. Krishnan's extended and modified result has yielded accurate rational approximations that were computationally efficient. This procedure was then generalized and applied to yield efficient, rational approximations for other functions: e^x, log x, and tan-1 (x). Compared with existing methods for computing problems like that attributed to Chebyshev, the procedure was found satisfactory.[112]

Computer-Based Language Translation

In computing, the mathematical tradition has found fruition as well been expanded upon. At present, researchers are attempting to use the computer for natural language processing through computational linguistics, a subfield of what was earlier called artificial intelligence. In this field, mathematics, systems of logic, philosophy, and psychology meet and are combined into a composite whole. Natural language processing also uses the discipline of linguistics as an important intellectual tool.

In South Asia, linguistics was the first science, as was mathematics in Greece.[113] And there has been a similarity, as noted by Rick Briggs, between a method used for knowledge representation in AI and *shastric* Sanskrit, that is, Sanskrit according to the rules of the fifth-century BC grammarian Panini. It has been noted that the goals of Paninian grammar match generally with those of natural language programming. Paninian grammar aims to understand how the listener obtains the information intended by the speaker through his or her language use. This grammar is especially suited to handle the free word order that is characteristic of South Asian language sentences and it has compared favorably with leading formalisms that have been used recently for translations. Tree adjoining grammar (TAG) is a translation aid for computers which has many similarities with the Sanskrit grammar of Panini. In fact, TAG, with its continuous modification, is beginning to look even more like the charts and other structures of Paninian grammar.[114]

The Paninian approach attempts to develop a theory of human natural language communications capable of answering such questions as: How do speakers use natural language to convey information to the listener? On hearing, how do listeners get at the intended information? The original Paninian concepts, as applied to Sanskrit, have been carried over to similar analyses of existing Indian languages.[115] There have been proposals, such as Briggs', to exploit this similarity as a knowledge representation formalism for natural language processing.[116] An *interlingua* has also been proposed for effective translation. That is, to translate something from English to Japanese, then one would first translate from English to the interlingua, and then to Japanese. An added advantage of the method is that once the sentence is in the interlingua, one can use it for a variety of target languages, in this case, not only for Japanese, but for Hindi or Spanish, for example. Because of the facility of shastric Sanskrit, it has been proposed as such an interlingua. Shastric Sanskrit measures up to the necessary characteristics in that it is language independent, a powerful medium for representing meaning, and is formally laid down.[117]

The goal of natural language processing is the same as that of Panini, namely extracting meaning. This Paninian approach can be used to design an extremely efficient parser. Parsing is the process through which a sentence is analyzed and then given a suitable structure. The Paninian framework has been developed by a group at the Indian Institute of Technology (IIT) Kanpur and turns out to be both efficient and natural in addition to being elegant, and they have described an efficient, very fast parser for the IIT system.[118] Its elegance derives from its ability to handle diverse language phenomena in a unified manner.

Most computational grammars have been designed for languages like English, in which word order is important. Such formulations are then applied almost as an afterthought to handle free word order languages like the South Asian languages, with a resulting drop in processing efficiency. The system developed by the IIT group incorporates two concepts from the Paninian framework, namely the *vibakti* and *karaka*, and uses them to describe languages.

The work at IIT has also led to the interesting concept of *anusaraka*, or language accessor. It bypasses current difficulties in machine translation by separating the language-based analysis of the text part of the problem from knowledge- and inference-based analysis. The machine does the former task, while a human does the latter. *Anusaraka* allows a reader to follow or access text in a language that he or she does not know through a language the reader knows. It attempts the production of a comprehensible output without worrying much about its grammar. The anusaraka system looks at the problem of machine translation in a new way. Instead of seeing the problem as one of pure translation, it reformulates the problems into those of accessing the information in the other language with, possibly, some effort by the reader, who analyses a sentence in the source language and then generates a sentence in the target language.

A basic working model of an anusaraka system has already been demonstrated in translation between Hindi and Kannada. It has a vocabulary of 30,000 root words and has been tested on a variety of Kannada texts—in fact, a good mixture of language samples: popular science, jokes, children's stories, newspaper editorials, etc. The system has been extended also to cover Telegu to Hindi translations.

These efforts are still to be extended to other South Asian languages. Also, suitable man–machine interfaces have to be constructed to provide on-line help when users have difficulty in understanding the text, or if they need to do post-translation editing to make the text more readable.[119] The present research at IIT Kanpur is the outcome of a large number of staff and students working over several years; the depth of that work has already resulted in over thirty papers and theses.[120] By 1995, the system was moving out of the laboratory and into field testing with lay users. The approach is now being applied to translation from and into English.[121] The machine translation efforts at IIT Kanpur are the most comprehensive in India, illustrating that old knowledge can be used for the most modern of technical purposes.

Deriving Quantum Characteristics

Raja Ramanna is considered the person responsible for the Indian atomic explosion over twenty years ago—allegedly for "peaceful purposes"—that brought India into the ranks of countries with nuclear arms capacity. Drawing on insights from Buddhist concepts and logic, he has derived some of the characteristics of elementary particles using the mathematical theories of Cantor (whom we met before, in South Asian parallels in set theory and infinity). Ramanna states that the Buddhist concepts he uses are those of "conditioned reality," *Kshana*, and "nothingness." He does not detail how he used the Buddhist concepts except in a succinct description, given below. But the derivation of his results in the bulk of the paper is all in the conventional Western tradition. We are left with the feeling that he arrived at his conclusions from the Buddhist perspective but translated them into the accepted language of mathematics and physics in current Western discourse.[122]

Basically, he uses Cantor's continuum theory to "derive the systematics of the lifetime and masses of elementary particles." His new perspective, he believes, gives new meaning to the "foundations of quantum theory as Cantor's theory deals with discreteness and continuity, which is the essence of quantum mechanics." He points out that Cantor's construction shows the difference between countable and noncountable infinities, between different types of transfinite numbers. This is a mathematical differentiation between continuous and discontinuous quantities. And as quantum mechanics deals with quantities that are, under given conditions, either discrete or continuous, the Cantor approach can be a starting point for the foundations of quantum mechanics. His point is that the Cantor continuum principle is so general that it can be applied to the study of all problems that deal with discontinuity and continuity.

Ramanna has applied the approach to energy levels in atoms, to the distribution of masses in elementary particles, and to the wave–particle duality in elementary particles, deriving results in the three fields that have been validated by other more conventional methods.

Although his demonstrations are presented through conventional approaches (using Cantor), Ramanna's prior guide has been Buddhist concepts. He notes that he uses the idea of conditioned reality in a mathematical matrix in which different causes of events are dependent on each other. For him, the Buddhist concept of Kshana is the theory of time averages and "nothingness" is the existence of virtual states. Virtual events are those generally predicted by quantum theory.

EXTRACTING HIDDEN MATHEMATICAL KNOWLEDGE

The above examples suggest the possibilities in store for extracting useful methods from the South Asian mathematical tradition. But can we make some broader suggestions and hints, based on epistemological criteria, of what techniques to use? And can we estimate what is there to be extracted?

In discussing how the fusion of the two traditions—the South Asian and the Greek—worked itself out over the centuries, Navjyoti Singh notes that an ideological tension, inherent in the two foundational streams, persisted. This tension, Singh points out, stems from the foundational backgrounds of the received mathematics discipline today, between its empirical, South Asian, algebraic side and its more idealistic, Greek, geometric side.

As part of the original Greek search for self-contained certainties, there was a search for completeness in an axiomatically derived system. But in the 1930s, Gödel would, with his undecidability theorem, prove this search futile.[123] Morris Kline termed this landmark the "loss of certainty" in mathematics. Kline sums up the ultimate effect of Gödel:

> The current predicament of mathematics is that there is not one but many mathematics and that for numerous reasons each fails to satisfy the members of the opposing schools. It is now apparent that the concept of a universally accepted, infallible body of reasoning— the majestic mathematics of 1800, and the pride of man—is a grand illusion. Uncertainty and doubt concerning the future of mathematics have replaced the certainties and complacency of the past. The disagreements about the foundations of the "most certain" science are both surprising and, to put it mildly, disconcerting. The present state of mathematics is a mockery of the hitherto deep-rooted and widely

reputed truth and logical perfection of mathematics. . . . The age of Reason is gone.[124]

In this post-Gödel era, it appears that mathematics is a logically open system. This transformation has created severe strains in the subject. Mathematics is no longer a self-referential subject that folds onto itself: it now has to touch the real world outside. The mathematical world is no longer a Platonic realm of pure ideas. Today, it is like any other subject with which we know the world.[125] The idealist view of mathematics being a logico deductive process is thus breaking down. One should also note that the actual practice of mathematics is different from what it is made out to be by its theoreticians; it has all the characteristics of a pragmatic art, with its all-too-human trials, errors, and frustrations.[126]

This development of a loss of certainty in the dominant metaprogram in mathematics has been paralleled by the growth of a school critical of rationalist mathematics. The intuitionist school was fueled in part by paradoxes in Cantor's set theory. The intuitionists do not view the laws of logic as either eternal or a priori. Logic itself would change according to circumstances. Thus the logic valid in the field of finite objects would not be valid in the infinite realm.[127]

In spite of Gödel's theorem and the intuitionist critique, the formal program of mathematics still flourishes today. Although it is still a minority viewpoint, there has also been a return to the empirical tradition in the mathematics of this century.[128] It is in this direction that the Indian tradition could well contribute. In the Indian tradition, results—whether in medicine, mathematics, or astronomy—were usually stated without giving the prior reasoning. That is, results were generally given in the form of aphorisms which left out detailed reasoning. This approach is found in all disciplines, including grammar, philosophy, and the areas mentioned above. Only the applications of the results are usually given, not how the results were arrived at. A writer on medicine, for example, would give the prescription, not the period of experimentation and trial and error before the final results were obtained.[129]

Yet very little work has been done on the methodology of Indian mathematics. The Indian demonstration of proof, *upapatti*, is found primarily in the commentatorial literature. The upapatti approach does not imply a set of formal deductions from formal axioms, as is the case in the idealistic Western sense. The Indian tradition instead considers mathematics more as a natural science, an area of empirical reality and not of absolute truths. Observation and experiment were acceptable in the upapatti tradition, much as they were in the natural sciences.[130] Studies in these different Indian empirical approaches to mathematics could therefore yield new insights for expanding the subject.

Apart from such broad-brush epistemological approaches, are there specific openings from the South Asian tradition that the present state of mathematics points to? We can speculate on a few areas. It is well recognized, for instance, that there were hardly any significant discussions on the nature of numbers from the time of the Greeks until the time of Frege in the nineteenth century. But there is much discussion on the nature of numbers in the Indian tradition. One such discussion is in the nyaya theory of numbers. Recent Western commentators have given serious thought to the possibility that the nyaya theory may be superior to that of Frege or later developments.[131] This is therefore an area where serious efforts at recovery could be made.

Taking another example, the Jain literature of infinities is still untapped and may in turn yield a lodestone for future developments in mathematics. A modern mathematician, L.C. Jain, having compared these Jain constructions of sets with those of the later set theory of Cantor and others, observes that while they were naive at times, they are full of implications for the later theorems. L.C. Jain hopes that comparisons of the two systems could "help [resolve] the secrets still unexplored."[132]

These examples suggest the potential for mining in mathematics, but the process is just beginning. Histories are only now surfacing, coming to the fore and going beyond the ken of historians to mathematics practitioners, who are realizing that sciences explored earlier could still have a considerable number of undeveloped "stumps" on their knowledge trees. These could be starting points for future developments.

EIGHT

A SEARCH FOR NEW PSYCHOLOGIES

Jack Huber's *Through an Eastern Window* chronicles one of the earliest contemporary discoveries of non-Western psychology, firsthand, by an American academic psychologist. Described as a true voyage of discovery, the book follows Huber's travels to the East in the mid-1960s. He describes how he found that professional psychologists in the East were very imitative of those in the West. They had adopted not only Western scientific methodology, but its subject matter and techniques as well. There was nothing uniquely Eastern in their practice, nothing original. They were taking from the West and offering nothing in return.[1]

Later, in Delhi, Huber saw how pupils of a yoga master were controlling bodily functions to an extent that he could never have dreamed possible. He also read a book by an American admiral on three weeks of mindfulness meditation in Burma. Intrigued, Huber then decided to go to Japan and experience Zen meditation himself, and to go to Burma and explore the Theravada mindful tradition. At the end of his few days in Japan and somewhat longer stay in Burma, he returned to New York. "And," he records, "something had happened to me."[2]

What indeed had happened, as he elaborates further in his charming book, were the effects of a personal exploration by Huber, a Western psychologist, guided by Buddhist practitioners into his own mental world and shown the resulting view from within. These internal views were the effects of meditation. But to quickly refute popular Western misconceptions of the word *meditation*, he observed that it usually has a religious ring to it, but that he had not experienced it as an occult or "spiritual" phenomenon. It was "not religious in any of the usual senses of the word."[3]

These internal observations, "meditations," had transformed him. He contrasted this experience with the Western view, prevalent then (in the early sixties, the heyday of both Freudianism and behaviorism), that mental structures were generally basic and hard. If one was a schizophrenic, one was always so, unless the individual went through a fundamental restructuring by deep psychoanalysis. Patients, it was further felt, had no control over their own destiny, and psychotherapists often acted as if patients knew nothing about themselves.

Huber's meditation instructor, on the other hand, played a different role. He was, at most, a guide who could not produce anything for the meditator. Whatever could be derived was the meditator's responsibility. Meditation, Huber observes, has something of the explorer's spirit: one explores the innermost self.[4]

Huber's was one of many explorations and changes that began to nibble away at the received wisdom of psychology. One other received wisdom was that autonomic control of the involuntary muscles was impossible.[5] But in the late sixties, Neil Miller, applying techniques of conditioning pioneered by Skinner and Pavlov, showed that involuntary muscles could in fact be changed at will. By giving appropriate rewards and punishments, he trained rats to change several involuntary acts by feeding back the results of behavioral changes to the animal. The technique was called biofeedback.[6]

Biofeedback was applied to human volunteers, and it showed humans could also control involuntary muscles. But biofeedback had limited potential. It required a lot of equipment, and it could generally change only one physiological function at a time, depending on what function was being monitored and fed back to the individual.[7]

But for centuries, similar claims regarding control of physiological functions had filtered into the West from the East, emerging from various mind cultivation techniques associated with the broad Indian cultural region. And by the fifties and sixties, scientists and medical personnel in the Western tradition were studying these phenomena using sophisticated instrumentation. In a pioneering study of Indian yogis, B.W. Anand showed significant and dramatic changes in the involuntary physical processes of yoga practitioners. A yogi sealed in a metal box could slow his intake of oxygen, and electroencephalogram measurements showed that there were changes in the electrical activity in the brain. Alpha waves heightened during meditation. These results paralleled those that were being documented by Japanese experimenters on another group, namely Zen Buddhist monks. The results, recorded under controlled scientific conditions, proved that indeed the operations of the presumed involuntary, autonomous system could be changed voluntarily.[8]

Later, American scientists monitored yogis and found some who could control their heartbeats at will, either increasing them to 300 beats per second or lowering them to 30 beats.[9] And Joseph Kamiya was able to train persons to produce alpha rhythms at will, after training. Since then, hundreds have replicated these experiments.[10]

MEDITATION ON THE ELECTRONIC RACK

The most sustained study on the general topic of meditation was conducted at Harvard by Herbert Benson and his colleagues, on the technique called Transcendental Meditation. In this series of studies, financed by the US Public Health Service, the US General Service Foundation, and the Roche Psychiatric Service Institute and done at Harvard University and the Beth Israel Hospital in Boston, Benson describes what he calls the "Relaxation Response."[11]

TM and its progenitor, Maharishi Mahesh Yogi, have a flamboyant and colorful image. In the seventies, Western celebrities like the Beatles took up his techniques. More recently, he has been propagating the rather bizarre view that by mass meditation, one could tap into what he believes is the unified field described by physicists and change human societies. Further, those who follow his techniques are sworn not to divulge the simple technique they learn, giving rise to charges that the whole exercise may be motivated by money. Yet whatever the truth and motivation of Maharishi in some of his activities, TM has spread to many parts of the Western world. Maharishi asked Benson to subject his meditation techniques to scientific study. After an initial no, Benson and his group relented and performed a series of studies in a controlled environment. The technique was now subject to intense study.[12] The results of these and other related studies have been published in the most orthodox of scientific journals, such as *Science, Nature, Scientific American,* and several other more specialized journals. As part of its other, public-relations oriented side, the TM group maintains regular updates of these studies. By the end of the eighties, studies on TM extended to hundreds.

The Benson studies revealed a set of physiological parameters that were effected in TM practitioners. There was a marked drop in oxygen consumption and a decrease in the rate of metabolism. This decreased rate of energy use was different in kind to sleep, hypnosis, or hibernation. There was also an associated change in brain waves, the alpha waves increasing in intensity and frequency. Such changes do not occur during sleep. Another electrical activity found in sleep, namely that associated with rapid eye movements, was also lacking in meditation. There was also a reduction in blood lactate, a chemical associated with anxiety, and drops in blood pressure and cholesterol.[13] Meditation also inhibited secretion of gastric juices, decreased premature heartbeats, and elicited many other beneficial results.[14] Other observed changes were a slowing of breath and heartbeat, lowering or stabilization of blood pressure, a drop in the intake of oxygen, and a decrease in the conductivity of the skin.[15]

Having done these studies, Benson related the marked changes he was observing to what he termed the Relaxation Response, a response in contradistinction to the usual "fight or flight" response. His interpretation of the external correlates of the internally controlled states of the meditators led him to prescribe definite techniques to elicit the relaxation response at will. His book *The Relaxation Response* later became a bestseller. In the book, he provided a simplified package of some TM techniques, giving rise in turn to many of the fatigue-, stress-, and anxiety-overcoming benefits of the original technique. Because of these benefits, TM has been successfully promoted as an antidote to high blood pressure, stroke, hardening of the arteries and heart attack, and a means of increasing alertness, conserving energy, and sleeping better. These techniques have today entered segments of the conventional wisdom in certain health areas. Some large corporations have also incorporated them into their normal routines.[16]

The Relaxation Response, Benson has pointed out, has long existed in several Eastern cultures.[17] He maintained that meditation only requires four simple elements: a quiet environment, an object to dwell on, a passive attitude, and a comfortable position. He described how the response can be simply elicited under these conditions.[18]

Benson subsequently extended his approach, utilizing some of the effects of the Relaxation Response to change ways of thinking and acting and yield what he terms the "Maximum Mind." He found that the Relaxation Response not only helped combat the negative effects of stress, but also, as his subsequent research showed, acted as an opening to a changed life and mind. By eliciting the response, one could encourage the mind to be more plastic. Experiments also showed that as a result, the electrical activity of the left and right sides of the brain became more coordinated. That is, the practice helped create a bridge between the left hemisphere (rational, logical aspects) and the right (creative, intuitive aspects).[19]

In *The Maximum Mind*, Benson describes various techniques that use the response to increase one's physical and mental capacities, citing several controlled research studies. In one doctoral study of elementary school children, those who meditated not only had better academic performance than those in a control group, but also showed, on the basis of standard psychological tests, higher levels of cognitive growth. Other doctoral studies showed that students who perform meditation do business-type problems faster than a control group.[20]

In the early nineties, under scientifically controlled conditions, Benson and his co-workers examined advanced meditation techniques used by Tibetan Buddhist monks. The researchers found that meditators could change their metabolism at will. Resting metabolism could be both raised and lowered, the decreases in metabolism being significant. They also observed marked asymmetry in EEG readings between the hemispheres, in relative alpha and beta activity, and in overall increased beta wave activity.[21]

In a further study of the control of the mind over the body, Benson and his colleagues describe how Tibetan monks high in the Himalayas control their body temperatures. A set of filmed, controlled examinations record some very interesting phenomena. The monks, through mental exercises, raise their body temperatures high enough to dry wet clothes. When they began to meditate, the monks were wrapped in cold, wet sheets. In spite of an indoor temperature of 40 degrees Fahrenheit, the monks did not shiver. Instead, they practiced a form of Buddhist meditation called *gTum-mo* and within three to five minutes, the sheets began to steam. In thirty to forty minutes, the sheets were entirely dry. The experiment was then repeated, for several hours.

In another similar feat, ten monks stood on their haunches on a sharp ledge at 19,000 feet, where the temperature was zero (Fahrenheit). They were clothed only in very light cotton, and any other person would have probably frozen to death. But the monks stood in this position for eight hours, without shivering, practicing a special form of gTum-mo meditation called *Repeu*.[22] Benson surmises that certain practices, including the elicitation of the Relaxation Response, allow the meditators to do these extraordinary things. Although he admits that he does not have all the answers for what he observed among Tibetan monks in the Himalayas, Benson speculates that much potential is still untapped in these techniques.[23]

Benson and his colleagues took a phenomenon claimed by its practioners (whether TM or Tibetan meditation) to be experientially real and which was subject to internal manipulation and control by the practitioners—in fact, to experimentation—and submitted it to study. They then obtained the external correlates of some of the mental activities involved. This experimentation with the inner apparatus is generally alien to the Western tradition and has thus hitherto been the subject of general disbelief. But inner experimentation is common to all South Asian traditions of internal mental culture.

In abstracting his Relaxation Response, Benson did not abstract all of that inner experimentation, which, it should be once again emphasized, is not mystical or spiritual in the usual Western sense, but very much down-to-earth and observable. In fact, Benson makes some gross but understandable errors in his generally sympathetic rendering of the Eastern tradition. For example, he claims that this tradition "has been [based] on pure soul-consciousness, to annihilate the flesh and deny its reality in order to reach absolute freedom." This is a statement that in its simplemindedness would make many people knowledge-able in the area wince.

More recently, Benson's use of the term "Relaxation Response" has been questioned by several writers who are sympathetic to the results of his studies. "Relaxation" connotes "letting go"—to some, an inadequate description. These techniques, since they produce a calming effect, have been called by others "the Calming Response."[24] But the path opened up by Benson and his colleagues, in their serious study of South Asian systems of mental functioning and psychologies, has opened the window for many similar studies in approaches, techniques, and ex-planations other than TM.

MINDFULNESS

Asian psychologies generally posit that most people experience a dis-tortion of perception which, however, can be corrected by techniques such as meditation. This belief agrees with some approaches in Western psychologies, which argue that we are much less aware of our own cognitive processes than we assume. Experimental observations have in fact revealed a large set of such perceptual distortions; Western research now reveals, contrary to past belief, that a large part (if not the biggest part) of our lives is carried out mindlessly. The claims of Asian psycholo-gies that this lack of attention can be reversed by meditation have now been supported by several Western researchers.[25] Mindfulness train-ing is a cornerstone in such efforts.[26]

Nineteenth-century American philosopher and psychologist Wil-liam James was thinking of this concept of mindfulness when he noted that

> The faculty of voluntarily bringing back a wandering attention over and over again is the very root of judgment, character, and will. No one is *compos sui* if he have it not. An education which should improve this faculty would be the education par excellence.[27]

But unaware of non-Western approaches, he added, "It is easier to define this ideal than to give practical directions for bringing it about."

Mindful meditation belongs to the broad Buddhist traditions, both Theravada and Mahayana, including Zen. There are many manuals in the classical literature that outline instructions on how to get into the mindful state and include descriptions of the mechanisms of the non-mindful psychological state as well as the mindful one.

The purpose of mindfulness is to experience and observe what goes on in the mind. Most people are not mindful: the mind wanders. Through meditation, this restlessness is brought under control. There are usually two stages of practice, calming and taming the mind (*Sha-*

matha in Sanskrit, *Samatha* in Pali) and the development of insight (*Vipashyana* in Sanskrit, *Vipassana* in Pali). These techniques instruct one to hold the mind and observe it, and many schools of Buddhism practice Samatha and Vipassana together.

In these techniques, something—usually the breath as it comes and goes over the tip of the nostrils—is a subject of concentration. In the process, one observes that the mind tends to wander, unmindful. The practitioner is instructed to bring the mind back from its wandering and soon realizes that body and mind are uncoordinated. The mind is seized by a constant internal chatter of feelings, thoughts, opinions, theories, daydreaming, fantasies, theories about theories, and so on. It is difficult to keep attention only on the breath. Instead of being just mindful of it, one may even start to *think* about it, which is a very different matter.

Gradually, after practice, one begins to develop awareness, a panoramic perspective. According to one conventional description, "mindfulness" is like the constituent words in a sentence, while "awareness" is the grammar underlying the sentence. In some Theravada traditions, the growth of the panoramic view is discouraged in favor of mindfulness.

But these mindfulness exercises in the Buddhist psychological traditions are to be contrasted with psychologist Wilhelm Wundt's attempt, in the nineteenth century, at an "introspective" approach as a laboratory for the study of the mind. The different "laboratories," namely the different practitioners, that participated in these experiments disagreed about their results. The psychologists of the Buddhist mindful approach would say that the introspection practitioners were only "thinking about their thoughts" and not studying the mind at all. The mindful approach actually cuts through the attitude of introspection. Buddhist practice, in this sense, becomes a tool of study, a bridge between human experience and cognitive science.[28]

The classical texts on the subject describe mindful meditation, *anapana sati*, as also having strong effects on the body. Thus "mindfulness of breathing, however, is more than just tranquilization of emotions; it is a quieting down of all bodily activities," as the authoritative *Encyclopedia of Buddhism* puts it, quoting orthodox texts.[29]

Daniel Brown and his colleagues, reading the classic literature on meditation, noted that there were constant references to increases in perception following this type of meditation. To study these changes, they used two conventional scientific methods: the Rorschach test and the tachistoscope.[30] The researchers recorded significant changes in information-processing and perceptual sensitivity in mindful meditators.

The Rorschach tests revealed important changes, depending on the degree of a meditator's experience and the type of meditation. In those only marginally exposed to meditation, the responses did not differ much from those of nonmeditators.[31] A second group, who had some exposure to insight meditation and had developed some ability at concentration, showed Rorschach results hitherto unknown. Normally, persons read a variety of images into the blots (people, animals, parts of the body, and so on). But the meditators did not see any of these. On the other hand, they saw simple patterns of black and white as they really were and did not organize them into a coherence that was not there. These findings agreed with the classical literature, which stated that meditation helps one focus the mind and reduces the intrusive associations, images, and thoughts that arise.

A third group, called the insight group, were more experienced than the second. They had begun to develop the classical stages of Vipassana, or insight, and showed results that were just the opposite of those demonstrated by the second group. The insight group showed a rich repertoire of associations in the Rorschach test. Normal respondents usually give only one or two responses per card, but these advanced meditators gave ten or more, and could continue if required. Also, in normal subjects, repeated tests days after the original test elicit many of their earlier responses. But in the case of the meditators, fresh responses were largely produced: the subjects were generally open to new associations, thus producing a rich collection of responses.

The fourth group were those who in the classical Theravada tradition are considered to have reached the first stage in the four steps of "awakening" ("enlightenment"). There were several differences in this group's responses to testing. Some viewed the images merely as emanations from their own minds. Interestingly, they could be aware, moment to moment, of the processes by which the stream of consciousness organized forms and images in response to the ink blots.

The last group were those with the most highly developed meditation skills. In the Theravada tradition, they would be those in the final two steps toward enlightenment. Their results were in many ways striking. In one unique feature, the meditators saw both the images associated with the blot, as well as the ink blot itself, as projections of the mind. Nonmeditators, on the other hand, accept the physical reality of the ink blot without question, but some later recognize that they may be projecting images onto it. In another unique response, there was no evidence of the psychological conflicts that are considered by others as the normal part of human existence. This finding corroborates the classical literature, which states that in the final stages of meditation, psychological suffering can be eradicated. In another unique finding, the advanced meditators did not just respond to the individual cards independently. When nonmeditators are presented with ten cards, they do not refer back to the others but respond to each individually. The advanced meditators, however, integrated the ten cards under a common theme.

There were also tests on perceptual sensitivity using the tachistoscope. The tachistoscope flashes images on a screen for brief periods and measures perceptual sensitivity and processing speed, detection thresholds, and discrimination thresholds. The minimum duration of a flash, during which an image can be recognized, is the duration threshold. To measure the discrimination threshold, the interval between two successive flashes is varied and the subject is asked to say whether one or two flashes were seen. At high speed, the two flashes cannot be discriminated. The smallest interval that can be detected between flashes is the discrimination threshold.

Meditators differed from nonmeditators in these tests, demonstrating enduring improvements in visual sensitivity even when not directly engaged in meditation. Meditators also exhibited heightened discrimination ability when they were engaged in deep meditation, but not afterwards. Advanced meditators could describe the moment of the first flash's illumination, its short duration, the instant of its disappearance, the time between the two flashes, the illumination of the next flash, its short duration, and ultimately its vanishing. All these observations were made in the less than tenth of a second between the flashes. These studies indicate that advanced meditators can detect perceptual events that are well below the perceptual thresholds of nonmeditators. These findings also confirm descriptions of perceptual abilities that appear in classical meditation literature.[32]

These tachistoscope results lead us to speculate on the extent to which some of the underlying features of persistence of vision (the direct application of which was Edison's invention of cinematography) were known in these Buddhist Abhidhamma meditation and psychology traditions. *Alatacakra*, the wheel of fire that arises when a fire brand is whirled in the air, is used as a simile in almost all schools of Buddhist thought for a persistence of the transitory. It was considered symbolic of the illusory nature of all things with their restless change.[33] So the phenomenon of persistence of vision, and the correct explanation of it, was known in these traditions. The deeper question is, were the quantitative facts implied by the number of frames per second that consitutes this persistence also known? Intriguingly, it seems possible. Let me digress.

The act of existence, according to the Buddhist view, is only a snapshot in the process of becoming, lasting only the length of one thought; it is the primordial unit of being. "Just as a chariot wheel in rolling, rolls only at one point of the tire, and in resting rests only at one point; in exactly the same way, the [internal] life of a living being lasts only for the period of one thought. As soon as that thought has ceased, the being is said to have ceased."[34] There is also the Buddhist view that thought changes faster than matter, and in the Theravada tradition, a moment of matter consists of sixteen moments of thought.[35] If one were to surmise that this minimal perception of matter requires sixteen elements of thought, then it is not farfetched to imagine that Buddhist practitioners would have known the time interval for the illusion of persistence of vision. In a cinema, thirty-two frames per second creates this illusion, although fewer frames, down to sixteen or ten, would still yield a somewhat jerky persistence of vision. And, if we now remind ourselves of the tachistoscope measurements, where through meditative practice the ability to capture separate elements of flicker have been documented, then it is not so farfetched to hypothesize that by practice, Buddhist observers of the act of perception were able to break down the persistence into those units of perceptions that give the sense of continuity, and count them.

Researchers Boals and Deikman have proffered a cognitive explanation for the general effects of meditation. The cognitive changes associated with meditators result in what Deikman calls "deautomatization" of consciousness. This deautomatization, brought on especially by mindfulness meditation, results in a mode of perceptual organization radically different from the normal one: a process of "cutting away false cognitive certainties." The new mode of perception is, among other qualities, more animated, sensuous, vivid, and syncretic. It results in new experiences beyond the everyday and new perceptual activities hitherto blocked or ignored.[36]

Daniel Goleman similarly observes that meditation results in a flow experience characterized by a merging of action and awareness. It is also accompanied by increased attention to a limited stimulus field blocking out others, a heightened awareness of body states and their function, and clarity regarding cues from the environment and how to respond to them. There is also consequently an optimal fit between the demands of the environment and one's capacity to respond to it. Meditation leads, therefore, to a sharpening of perception and to selective responses to the really important environmental stimuli.[37]

Delmonte also observes selective access to awareness, across various layers of consciousness. He finds that the attentional retraining subsumed under meditation is useful in three ways: augmenting and improving one's system of personal constructs, accessing unconscious material, and creating altered states of consciousness.[38]

Other laboratory research on meditators has shown that they are easily aroused in the event of a perceived threat, but quickly return to normal once the threat is over, with no residual anxiety. Nonmeditators, on the other hand, took longer to be aroused and longer to return to normal. Clearly, the meditators had a physiologically healthier response.[39]

Confirming the classical texts' views that meditation affects the body, a study of monks meditating according to the Theravada Dhammakaya tradition found that serum total protein levels had significantly increased, while serum cortisol levels, systolic pressure, diastolic pressure, and pulse rate had reduced significantly. After meditation, vital capacity, maximal voluntary ventilation, and tidal volume were also significantly lower.[40]

IMAGERY TECHNIQUES

Often meditation and imagery go together in the ancient Indian techniques. Traditionally, visualization techniques have formed a central part in almost all Eastern meditation practices, although visualization is utilized at different stages in different programs. Sometimes it begins the initial meditative state; sometimes, it is used as the primary focus of the later stages. Sometimes it is used to channel "energy" for a particular purpose."[41] Imagery techniques have now been appropriated by many Western practitioners for a variety of psychological goals.[42]

Studies indicate that relaxation training and guided imagery techniques result in significant increases in natural killer-cell activity compared to that in control groups.[43] As part of a UCLA research program, visual or guided imagery has been used to enhance medical treatment by getting patients to mobilize their internal resources of healing.[44]

In the treatment of cancer, visualization techniques have been popularized by the radiation oncologist Carl Simonton. The use of visualization generally in healing has been emphasized in the writings of Mike Samuels. Harvard psychologist, Mary Jasnoski discovered that although relaxation alone increased protection against upper respiratory diseases, if imagery was added on, the effect was enhanced.[45]

Studies in the field of sports medicine have also revealed how in peak performances, the mind relaxes its analytical side and allows the right side to take over, in an almost trancelike flow. An alpha rhythm just before an important stroke in tennis, it appears, is important for success. And what happens in the mind between the strokes is as important as what happens at the moment of striking the ball.[46] Use of meditation-type and imaging techniques among athletes has helped them overcome blocks and nervousness and resulted in more uniform play[47] and has helped many athletes improve their performance.[48]

Visualization is also being used in other therapies including that of autogenic training.[49] In the meantime, anxiety disorder and phobia therapies have recognized that patients' images of danger before and during anxiety attacks can be modified by inducing different images.

Patients are encouraged to bring to the therapy sessions the visual images and daydreams that precede the anxiety attacks. The distorted images are then modified through a variety of image-inducing techniques.[50]

Karen Olness has reduced headaches in children by training them in relaxation imagery exercises, with the result that the children had far fewer migraine headaches than under treatments with conventional medicines.[51] Meditation and guided imagery techniques have also provided additional tools for the dietary counseling of patients requiring nutrituion-related changes in behavior. The sense of inner control, inner healing, and whole-self participation that these mind-body techniques bring about has been a crucial element in enhancing dietary compliance.[52]

The imagery, as well as techniques, of sexuality has been used in the Tantric traditions of both Hinduism and Buddhism for the purpose of their own liberatory programs. Shorn of the liberatory aspects, some of the exercises drawn from these practices have provided exercises to help overcome marital disharmony, including the creation of a ritual before sexual encounters, breathing in a synchronized pattern, continuous eye contact, motionless intercourse, and having the sexual encounter without reaching orgasm.[53]

BEHAVIOR CHANGE IN GENERAL

In addition to these observed phenomena of Asian psychologies, there has recently been much discussion at a more theoretical level between the principles and practices of behavior modifications in the Western tradition and in Buddhism. William L. Mikulas has pointed out that there are many similarities, tracing six commonalities between Buddhism and behavior modification: a focus on the "here and now" and a derived "ahistorical" attitude; emphasis on the perception of reality as it is, without distortion, interpretation, or any metaphysical speculation; the concept of an individualized self being questioned, the person and his or her behavior being seen as separate; accepting change as a central fact, along with the importance of learning; attachment to particular experiences, objects, people, beliefs, and so on, as a reason for malady; and changing what is changeable and then training to let go of the rest. Mikulas believes that pursuing these commonalties should lead to broadening current principles of behavior and behavior modification, and he finds Buddhist literature a storehouse of information for psychology,[54] arguing for greater integration of Buddhist practices in behavior modification packages.

The Theravada tradition of psychology uses meditation and other behavior-changing techniques. The psychological theories of Buddhism, it has been noted, discuss problems of basic drives that motivate behavior, perception and cognition, consciousness, meditation, personal development, and behavior change.[55] There is a strong relationship between theory and practice in Buddhist psychology, especially in the use of meditative techniques and other techniques to change behavior. Modern psychology and Buddhist approaches could therefore interact fruitfully.[56] Padmal de Silva has discussed the parallels between this tradition and modern behavioral techniques such as in thought-stopping and modeling, and in behavior modification for treatment of obesity and stimulus control.[57] He finds resemblances between Western and Buddhist in both concepts and practices.[58]

STRESS TREATMENT, ANXIETY, PANIC, AND PHOBIAS

The Institute of Medicine at the National Academy of Sciences (US) has convened a large number of meetings to explore the effects of stress and emotions on physical health[59] because about 60 percent of outpatient visits to the doctor in the US are concerned with stress or mind/body interactions. Further, one in five primary care visits is related to "major depressive anxiety disorders."[60]

Jon Kabat-Zinn, director of the Stress Reduction Clinic at the University of Massachusetts Medical Center, has used mindfulness meditation to help patients suffering stress related disorders, including chronic pain. Meditation had never been attempted before in a major American medical center, but the results indicated that mainstream Americans did accept such a program with enthusiasm. To Kabat-Zinn, meditation is an "intrapsychic technology" developed over the millennia by traditions that knew much about the mind/body connection; it is an inner science that combines the subjective and the objective. He gets each of his patients to become "the scientist of his or her own body and mind," getting to know themselves. On the basis of randomized trials, these techniques have yielded both physical and psychological symptom reduction that persist over time.[61]

Kabat-Zinn and his colleagues, in a 1992 overview article in the *American Journal of Psychiatry*, notes that there are three major self-regulatory techniques used in the treatment of anxiety, namely meditation, relaxation, and biofeedback. Research had indicated that all three play a role in reducing the physical and psychological aspects in anxiety. A study on the effectiveness of mindfulness meditation showed that such a program can effectively treat such disorders as generalized anxiety disorder, panic disorder, and panic disorder with agoraphobia. The results were maintained over a follow-up period of three years, showing its long term effectiveness. This study on mindful meditation is supplemented by other reports illustrating the effectiveness of TM in these situations.[62] (Relaxation and biofeedback also owe much to meditation and related Eastern techniques, as demonstrated in Robert Benson's work in relaxation therapy and Karen Olness's in biofeedback techniques).

Other stress-related disorders, like anxiety attacks, panic attacks, and phobias, are also being treated successfully with meditation and imagery. A recent text by Aaron Beck and Gary Emery dealing with generalized anxiety disorder and panic disorder, *Anxiety Disorders and Phobias: A Cognitive Perspective* (1985), describes how cognitive changes are made use of in the treatment. Beck is well known for his pioneering work in applying cognitive therapy to the treatment of depression, while his present work is considered to have laid the groundwork for cognitive therapy for phobias and anxiety disorders.[63] Generalized anxiety disorder, he notes, is characterized, among others symptoms, by an inability to relax, tenseness, being frightened, jumpy, and unsteady, feeling weak all over, and feeling terrified. Cognitive therapy for these ailments is based on an educational model, of learning to learn, of restructuring the cognitive field. The techniques used include those from standard meditative practice and imagery modification.[64]

Modifying the affective component is another means of coping with anxiety and phobias. Here too, Beck and Emery describe several techniques. Some are obvious derivatives of standard meditation techniques (although they do not mention this) and include observing the self nonjudgementally and not getting caught in the internal travails and drama of anxiety. This awareness, the authors note, brings the patient back to the present. Self-awareness of this order results in patients' realizing that they "have anxiety" as opposed to "I am anxious." These techniques are taken directly from mindful practice in Theravada Buddhism. The authors also recommend breathing exercises, with the patient experiencing the details internally and consciously. These, again, are directly derived from standard meditation techniques.[65]

As mentioned above, various treatments for anxiety disorders and phobias use image-inducing techniques. Apart from the professional literature, popular self-help books on panic attacks recommend Asian-derived techniques. In Reid Wilson's *Don't Panic: Taking Control of Anxiety Attacks*, meditation occupies an important place. The techniques recommended by Wilson concentrate on a particular sound (as in TM) or heightened awareness (of the Theravada tradition). These techniques have helped patients to face reality, the oft-stated goal of Buddhist meditation.[66] Another popular book on phobias, *Phobias: The Crippling Fears*, also outlines a variety of techniques derived from Eastern approaches, including breathing exercises, meditating on a word, and visualization techniques.[67]

Asthma is another disease that is partially triggered by stress, and popular self-care books on asthma (for example, Geri Harrington's *The Asthma Self-Care Book: How to Take Control of Your Asthma*) have been advocating meditation and meditation-type relaxation techniques for the treatment of asthma.[68]

Meditation has been also used successfully for other stress-related physical disorders and conditions. Psoriasis is triggered by or exacerbated by stress, and physicians have now been urged to supplement the usual psoriasis treatment regimens with stress-reduction strategies involving meditation and other similar approaches.[69] Yoga-derived techniques have informed Lamaze breathing instruction as a powerful aid to childbirth, while athletes are using the slow stretching exercises of yoga to reduce sports injuries.

Meditation has also been used for the treatment of fibromyalgia, a chronic disease characterized by fatigue, sleep disturbance, widespread pain, and resistance to treatment. A variety of evaluative responses—including fatigue level, amount of sleep, global well-being, pain, and "feeling refreshed in the morning"—indicated the effectiveness of meditation.[70] Richard Surwit has also observed that diabetes patients can benefit from meditation-type approaches that result in relaxation.[71] And, trained in meditation practice, patients suffering from drug-resistant epilepsies have shown significant reductions in a variety of indicators, compared to control patients. Positive changes during the observation period of one year included EEG frequency, the clinicoelectrographic picture, and seizure frequency and duration.[72]

Dean Ornish, of Harvard University, recently developed the only system which has been scientifically proved to have reversed heart disease. The program, which does not involve any drugs but uses changes in lifestyle, has been highly publicized and has been featured in TV documentaries and in his bestseller *Reversing Heart Disease*. At the end of the program, clogged arteries have been opened up, bad cholesterol has been reduced and good cholestrol increased, and participants have in general gotten a fresh start on life. Lifestyle changes essential to the Ornish regimen include a change in diet (primarily a shift to vegetarianism), and the use of yoga and meditation exercises. The yoga he advocates is not just stretching, but the yoga of the inner world. Here patients are encouraged to look into themselves and be aware of what goes on inside. Physically and emotionally, they are encouraged to increase their self-awareness. As a result, the effects of stress are reduced and the mind grows more quiet: practitioners experience a natural state of relaxation, joy, and peace. According to the results of Ornish's study, the program integrates patients more more fully with the world rather than drawing them away from it, and helps them perform better.[73]

Some controlled studies have suggested that the practice of meditation by the elderly may in fact extend life through improvements to the cardiovascular system and increase in mental functioning.[74] Yoga-derived techniques have also been promoted as a means of slowing other effects of age, including regimens outlined in such popular books as Raquel Welch's on beauty (Welch's program is primarily a set of yoga techniques).[75]

THE BIOLOGY OF THE INTERNAL SCIENCES

Humans are not made of ethereal substances. They are flesh and blood—biology—and the effects of applied Eastern psychology practices that we have recorded must have a biological basis and should be explainable at the biological level and, further, at the level of chemistry and physics. Evidence at the biological level is now accumulating to confirm the mechanisms of phenomena associated with applied Eastern psychology.

Candace Pert discovered the opiate receptor and several other peptide receptors in the brain and the body, and her work has led to understanding the chemicals that act as messengers between body and mind. It was also discovered that endomorphines and other brain chemicals found throughout the body, including the immune system, form a psychosomatic communication network. Emotions thus bridge the physical and the mental; brain chemicals such as neuropeptides and their receptors are the biochemical correlates of emotion. The brain and the immune system use so many of the same molecules to communicate with each other that one has to think of the brain not only as residing in the head but, instead, as spread all through the body. (This is also the Buddhist view, which treats the mind–body, *nama–rupa*, as one composite.) The mind, under this modern formulation, is in every cell of the body because these brain chemicals appear in so many cells. Emotions, therefore, become very important to the healing process.[76]

The organs of immunity, it is now known, are suffused with nerves. The thymus gland, bone marrow, and the spleen have receptors for neurohormones, neurotransmitters, and neuropeptides. Effects in the brain can therefore change immune function. Research on these themes, begun in the 1970s, has since then matured, and the immune system, it now appears, is intimately connected to the brain, with cells in the immune system responding to signals from the central nervous system. A special place in the brain where the immune function is connected is the hypothalamus, surgical removal of which leads to the suppression of the immune system.[77]

Because of the multiplicity of connections between the brain and the body, it is difficult today to think of the two as separate. David Felton has discovered the nerve fibers that physically link the nervous system with the immune system. If one removed the nerve from an immune system organ—the spleen or the lymph nodes, for example—one virtually stopped all immune responses. Increasingly, notes Felton, these views are now being avidly accepted by once skeptical colleagues.[78] This evidence of the connection between immunity to infectious diseases and brain processes suggests that there is, as it were, a "pharmacy within the brain."[79]

These connections help further explain the interrelation of mind and body, a relation supported by biologically sophisticated research which has shown that a wide range of physiological changes occur with meditation, many of which have clinical applications, including increases in the immune function.[80] At the Ohio State University College of Medicine, Janice Kiecolt-Glaser and Ronald Glaser have shown that older people in retirement homes had a significant increase in their immune responses against viruses and tumors while undergoing meditation treatment. Medical students who pursued similar techniques to reduce the stress of exams had higher levels of helper cells, increasing their immunity to infectious diseases.[81] Those who had the strongest immune responses were those who did the relaxation exercises the most.[82] Other studies indicate that relaxation training and guided imagery techniques result in significant increases in natural killer-cell activity, compared to that of control groups.[83]

After a review of the literature on Zen, Gerhard Fromm has suggested that Zen meditation involves a deliberate development of certain neurophysiological inhibitory mechanisms through psychological techniques. This internal use of core aspects of brain function can yield successful application of Zen techniques and concepts in such varied spheres as martial arts and the creative and performing arts.[84]

Behavioral cognitive therapy actually alters the physical structure of patients' brains. This therapy has resulted in neurological changes similar to those induced by the drug Prozac.[85] Those with obsessive-compulsive disorder given this type of therapy changed their internal neurological wiring. Brain imaging using PET scans, before and after the therapy, have shown these changes very clearly.[86] It is clear that changing the mind also changes the physical aspects of the brain.

LEGITIMIZATION OF PRACTICE

By the mid-nineties, enough scholarly data and clinical experience had been gathered to make some broad statements about the inflow of Eastern methodology into Western psychological practice. After a review of research results, Greg Bogart concluded that meditation is a multidimensional phenomenon that would be useful in many ways in a clinical setting. Meditation brings about states of physiological relaxation that would be useful for a variety of physical symptoms, including stress and anxiety. As importantly, meditation brings about cognitive shifts that can be used for observing and changing behavior and limiting destructive cognitive patterns. The altered states of consciousness developed through meditation may help deeply reorient a variety of a person's attributes, including sense of well being, purpose in life, emotional attitudes, and sense of identity.[87] Greg Bogart, writing in 1991 in the *American Journal of Psychotherapy*, notes that there has been much and growing interest in the possible uses of meditation in psychotherapy. Recognizing the legitimacy of the exercise, regular reviews of the current situation have appeared every eight years or so since the 1970s in professional Western psychological literature.[88]

Taken together, the different studies indicate a variety of effects of meditation: reduced anxiety, empathy, increased perceptual sensitivity, pain tolerance, and self actualization; physiological effects include changes in EEG, reduced blood pressure, changes in hormone levels (such as adrenaline and cortisol) related to stress. Meditation, therefore, has found uses in a list of medical problems, such as anxiety, phobias, asthma, insomnia, cardiac arrhythmia, and chronic pain,[89] as well as addiction, hypertension, and stress. Research has also shown that meditators drift toward stronger mental health, characterized by increased spontaneity, self-actualization, positive personality changes, and decreased depression. Meditation is also associated with a drop in oxygen consumption, cardiac output, blood pressure, body temperature, arterial lactate concentration, respiratory quotient, and arterial gases. There is also an increase in slow alpha brain waves and skin resistance while there is a decrease in beta waves. These physiological changes occur parallel to a state of relaxed wakefulness.[90] Intriguingly, in one set of such studies on geriatric patients by researchers at Harvard, it was observed that the subjects lived longer than a control group.[91] Meditation techniques and their physical correlates may vary according to the particular technique practiced. A comparison of the EEG's of yogis in deep meditation showed that they are oblivious to their external surroundings; in contrast, Zen meditators are strongly attuned to their external environment.[92]

Initially, Western studies of the effects of meditation gave rise to a limited view that meditation techniques were only a relaxation method; this view emerged in large part due to the pioneering work of Benson and his naming the phenomenon a "Relaxation Response." This relaxation model initially made the subject acceptable and accessible to the scientific community. But this limited view did not take into account all the varieties of meditation and seemed, by the early nineties, to have outlived its usefulness. It does not take into account other effects of meditation or the differences among meditation approaches. Further, it seems to equate meditation with other relaxation techniques.[93]

Although the relaxation approach demystified meditation, it did not take into account all the factors that could be involved in the practice. In relaxation itself, there are in fact different modalities. The effects of meditation vary among subjects, and in the same subject at different times, and the relaxation effect does not tell anything about the subjective experience of meditation. There are also variations in meditation approaches, and these variations can create different outcomes. Thus, for example, there are differences between the processes and outcomes of Zen, Vipassana, TM, or kundalini yoga. Further, rhythm, which has been used by the relaxation schools of Benson and others, does not fit all approaches; rhythm can, in fact, create arousal, not relaxation.[94]

It has also been observed that, unlike Western approaches, Asian psychologies ignore the contents of awareness and emphasize its context. Western psychotherapies take for granted the mechanisms that govern mental processes and try to alter them through socially conditioned patterns. Meditation, on the other hand, aims at a more fundamental level, skirting the epiphenomena to focus on the process of conditioning itself.[95]

By the mid-nineties, a process of legitimization that had begun in the seventies was nearing completion. The intrapsychic technologies associated with different Asian psychologies had, in the form of meditation techniques, become legitimate and in some cases virtually commonplace. The variety of approaches in different Asian psychological disciplines were also being realized. Sometimes these techniques have been taken wholesale by Western medical personnel without giving due credit to the original source—the varied and extensive use of "breathing" exercises offers one such example. In the thirties, a Harvard physiologist, Edmund Jacobson, developed a system of relaxation based essentially on yoga techniques. He called this system the "Jacobson Progressive Relaxation" and for the next forty years documented its beneficial uses. He also wrote a popular version of the book, *You Must Relax!*[96]

Western practitioners exposed to the various Eastern traditions of mind culture were also now combining them and making new amalgams unknown to the original traditions. Thus Theravada mindfulness has been combined with yogic practice. *Nova yoga* is one such hybrid, drawing from the works of many psychologists, many schools of psychological thought, and techniques from several yogic traditions, including *hatha yoga, mantra yoga,* and *laya yoga* and has been used as a tension reducer as well as a method of reconditioning negative imagery.[97]

SPREAD AND POPULAR ACCEPTANCE

By 1977, over 400 articles and studies on the psychophysical changes that accompany meditation had already appeared in Western scientific literature.[98] By the late eighties, millions of Americans had tried meditation; it had infiltrated the culture as an important tool in medicine, education, psychology, and personal development. The ranks of practitioners, by this time, included businessmen, professionals, and academics.[99] And, by the early nineties, many techniques of "mind technology" had become, in certain medical areas, nearly mainstream, if not mainstream.

In the West today, therefore, meditation has found a home. According to surveys, more than six million Americans have tried it. Scientific research on meditation began in the sixties, and now a large literature on the subject exists. In addition, there have been translations of the classic texts as well as theoretical debates on the topic. Hundreds of research findings have been published. Almost every week, one researcher has noted, one sees reports in this burgeoning new field,[100] an amount of published work surpassing that done on most other psychotherapies, with the possible exceptions of behavior modification and biofeedback. The published research covers a wide variety of topics, including physiological, psychological, and chemical factors.[101]

In a consensus report released in 1984, the US National Institutes of Health (NIH) recommended meditation (together with changes in diet) above prescription drugs as a first treatment for low-level hypertension. This official recognition was an important milestone in the mainstreaming of meditation. A British follow-up study has shown that four years after the training ended, patients had lower blood pressure.[102] The NIH is also sponsoring research into the use of yoga in some obsessive compulsive disorders and in beating heroin addiction.[103]

The use of many of these techniques is spreading. Several popular journals report matter-of-factly on these approaches as part of normal healthcare coverage. Thus, in October 1992, the *New York Times Magazine* had a special issue on "Good Health," emphasizing mind-and-body medicine. A lead article, "On the Mainstreaming of Alternative Medicine," recorded the transition from a narrow biomedical model to a "biopsychosocial" model of medicine, citing as milestones the acceptance of meditation, yoga, and guided imagery over the course of the previous two decades. Popular news magazines like *U.S. News and World Report* (see August 3, 1993) are now writing of the mental edge and the importance of the brain—and not only physical well being—as the key to peak sports performance. Bill Moyers, in a 1993 TV series on US Public Television and a subsequent book, both titled *Healing and the Mind*, explored the broad theme of the mind's connection to healing. His discussions covered the work of several leaders in the field, including several whose work has been cited here.[104] In mid-1996, *Time* was publishing cover stories summarizing the connections of brain-mind factors to healing.[105] And in the words of a *U.S. News & World Report* (May 16, 1994) health guide for the US, "yoga has to a large extent gone mainstream." An estimated four million Americans were at that time practicing the discipline regularly. Yoga has even become part of the regular post–heart attack regimen in some leading US hospitals. A further indication of this crossover to the mainstream is that yoga is now available as part of Jane Fonda's series of workout tapes.[106]

Another film star, Shirley Maclaine, in her book *Going Within: A Guide for Inner Transformation*, echoes Eastern thought (albeit at a very popular level and often in flowery words).[107] She describes how she could virtually rewire herself internally through Eastern meditation techniques. Her book is unselective in references to chakras, cosmic consciousness, and the like, which a more scientific person would balk at using. Yet her book illustrates not only the efficacy of some of the techniques she has used and which she now advocates for her audience, but also the popularity of the subject.

COMPARISONS OF EASTERN AND WESTERN PSYCHOLOGIES

If Asian-derived mental technologies have found acceptance and are intensifying their inroads, what about Asian mental science and its formal psychologies? One could well accept a cure—like a medical cure used by a forest-dwelling group in the Amazon—without accepting that particular group's explanatory system for the cure.

Roger Walsh, in a review article in the *American Journal of Psychotherapy*, notes that in striking contrast to other disciplines, American mental health professionals have tended to ignore and neglect psychology originating from abroad. Until very recently this bias had also applied to Asian psychologies. Yet in the Eastern classical literature, there are many texts on mental functioning, with numerous detailed discussions. These are couched, Walsh notes, in nonreligious, nonphilosophical, strictly psychological terms and cover such topics as perception, motivation, thought, emotion, conditioning, identity, pathology, and addiction. Without adopting the religious and philosophical baggage, one can assess these purely psychological factors.[108] An illustrative example is Buddhist psychology in the Abhidhamma of the Theravada tradition, which, Goleman notes, describes many psychological states and mental factors—53 in all, including mind moments and objects of awareness. The progression of different meditation states and the higher reaches of meditation are discussed in detail.[109]

With the rising legitimacy of the field over the last few decades, there has also been an increasing literature on comparisons and parallels between Western and Eastern psychologies. For reasons of space, I will limit myself to exploring in relative detail two broad areas of Western psychological theory, namely psychoanalytic theory (that is, Freud his succesors and followers) and humanistic psychology.

Psychoanalysis—Freud and His Followers

In one of the first of such studies, Erich Fromm compared Freud's notion of cleansing the mind's ill health through psychoanalysis with Zen Buddhism. He found that in spite of seeming differences, Zen had much in common with the Freudian approach.[110] Later, Goleman compared the Theravada tradition with the views of Western psychologists such as Jung, Maslow, and Freud,[111] and Coward observed that the mental categories of memory, motivation, and the unconscious are intimately intertwined in the theories of Freud and Jung, as well as in those of Patanjali.[112]

Alan Watts also observed that Eastern approaches were similar to Western psychotherapy in that both are interested in changing the feelings people have about themselves. He found that there was a strong compatibility in therapeutic goals of the two traditions, in such Western approaches as Jung's individuation, Maslow's self-actualization, the creative selfhood of Adler, and the functional autonomy of Allport.[113] Later, Noda argued that there are similarities between the individual psychology of Adler, another estranged disciple of Freud, and Buddhist approaches. The goal of individual-psychological group therapy is to develop the patient's community feeling, and Buddhism has a similar purpose, albeit expressed differently. Noda developed a form of group psychotherapy in which meditation plays the most important role. This meditation emphasized the objects of psychotherapy, such as steadying the emotions, sharpening insights, and looking at one's real self with courage and effecting changes where necessary.[114]

Reich, another psychologist in the same vein as Freud, emphasized sex in the human psyche. His views of combining sexuality and its repression seem at first sight to be far from the austere Theravada tradition, but Burt Kahn has pointed out that there is a school of Theravada meditation, that of the Vipassana tradition of U Ba Khin, which is remarkably similar to Reich in both theory and practice.[115]

Jeffrey Rubin has compared some meditation techniques to the practices of Freud and recommends meditation before a psychoanalysis session.[116] Mark Epstein has also compared Buddhist concepts of evenly suspended attention to the attention paid to the patient by the psychoanalyst. He finds that Freud's proposed optimal attention stance is similar to the Buddhist strategy of equally suspended animation. The Buddhists assert that one could be trained to view changing objects evenly through time, an idea that lends support to Freud's view that a similar state of mind be used by the analyst. Epstein believes this practice would yield an optimal strategy of listening for the psychoanalyst.[117]

John Engler has also compared psychoanalysis and the Buddhism that is practiced in the US and has affirmed that the two could contribute to each other's concepts of psychopathology. Both psychoanalysis and Buddhism aim at achieving optimal inner well-being. Engler has noted that a central element in the Buddhist search for the sense of one's self, namely the appreciation of the illusoriness of continuity and substance, is useful to psychotherapy. He notes that such practices as Vipassana and insight meditation, where attention is paid to the internal passing scene, has much to teach psychoanalysis. Here, an egoless inner state is achieved by rejecting the selection criteria of thought; all internal censorship is stopped and every desire for gratification is thrown away. In the process, the person becomes an acute observer of one's own experiential flow.[118]

Humanistic Psychology

Humanistic psychologists such as Maslow, Frankl, Rogers, May, Allport, and Erikson have many positions in common with each other and with Buddhism. The present writer came across this similarity with Buddhism in the early seventies when researching a Ph.D. in a British university. The discussion below is based partly on the research done at the time, as well as findings in more current research.

Asian approaches, Roger Walsh has noted, establish a hierarchy of motives reminiscent of those of Abraham Maslow and Ken Wilbur. There are also specific techniques to be used in this hierarchy of growth processes.[119] The concept of self-actualization, the pinnacle of growth according to humanist psychology, was developed by Maslow and Rogers. There is, of course, a parallel concept of self-realization in the subcontinental traditions of Vedantic Hinduism, Theravada, and Mahayana Buddhism. The two approaches complement each other and, taken together, provide a larger view of human development.[120]

Broadly seen, Eastern and Western theories of psychological growth have many commonalties, Walsh observed. Both necessitate a growing beyond restrictive perspectives of the self to a larger vision of what it means to be human. The essence of psychological work includes the avoidance of human needs, narcissism, and desensitization, and letting go and openness are elements common to both approaches.[121]

The father of humanistic psychology was the late Abraham Maslow. Discussing the end goal of psychological growth, Maslow states that "our goal is the Eastern one of ego-transcendence and obliteration, of leaving behind self-consciousness and self-observation, of fusion with the world and identification with it, of harmony," and goes on to argue the means of reaching the goal.[122] Clearly, he equates the complicated mechanics of self-actualization with Eastern approaches.[123]

In Maslow's last book, there are increasing references to Buddhist terms. Thus "high Nirvana," "low Nirvana," "Boddhisatva," and "Pratye-kabuddha" appear in Maslow as technical terms, an attempt nearly thirty years ago to infuse Eastern ideas into Western psychology.[124] One should also cite here Maslow's clear assertion that "the new psychology is also in the Eastern tradition."[125]

Viktor Frankl's discussion of the prime motivator of man as a will to meaning also finds very positive echoes in the Indian tradition. To Frankl, the basic concern of man is a will to meaning; self-actualization is an outcome of meaning-fulfillment and self-transcendence.[126] This search for will-to-meaning is, again, the way of the searcher for one's own truth in the South Asian tradition.

Hiroshi Takashima has examined Frankl's therapeutic approach, logotherapy, and believes that this therapy coincides with Buddhist thought. There is, he observes, a three-hundred-year-old Japanese formulation of Buddhist thought that considers human nature as a totality of three dimensions: the body, the psyche, and the "spirit." Buddhists also emphasize the future, not the past, as does logotherapy. Further, both accept that living with disease is part of accepting the unavoidable.[127]

Rollo May, one of the leaders in existential psychotherapy, has observed that "one gets the . . . shock of similarity in Zen Buddhism." This similarity does not derive from a simple identity and similarity in words, but because both approaches are concerned with ontology, the nature of reality.[128]

Goleman, having examined Western psychologies and mainstream models of sanity and good mental health, has noted some remarkable similarities with Eastern approaches. He has noted that Gordon Allport[129] had taken mental health to be realistic perception, self-acceptance, contentment with one's own self, warmth, and compassion. Goleman observes that this is compatibile with Buddhist approaches (except for Allport's emphasis on a strong ego identity).[130]

It has been noted that post–World War II developments in psychoanalysis, in the work of Heinz Hartmann and his concept of "ego strengths" and the "basic virtues" of Erik Erikson, have many "startling" affinities to the Hindu theory of the stages of life (*asrama*). Both the Eastern and contemporary Western versions see human development as stages in life, each stage contributing a specific strength, and all stages interpreted into a whole that seeks the goal of self-realization and transcendence.[131] In his psychology, Erikson[132] has a final stage in the maturing process, characterized by the absence of fear, absence of resentment, and acceptance of life's circumstances. Here, there are strong echoes with Buddhism, except for Erikson's underlying assumption of an ego to be defended.[133]

Similarly, Milton Erickson's psychotherapy has been compared with Zen Buddhist practices in the areas of theory, change, the relationship between therapist and client, and the relationshiop between teacher and student. The two traditions, Goleman notes, "share many philosophical assumptions."[134]

The ideas of Otto Rank, a psychologist well read in philosophy, have strong parallels with Hindu thought. Both characterize life as inherently "painful," although there are four stages through which a person could transcend a life of futility. And at his death, Rank wanted to be cremated rather than buried, choosing South Asian custom over the usual Western practice; the choice is significant as a pointer to his underlying beliefs.[135]

Elbert Russell examined several Eastern and Western means of understanding the mind, including Hindu beliefs, Buddhist meditation, and psychotherapy. After examining categories of consciousness in these traditions, he believes that Eastern and Western approaches can act synergistically to increase human growth.[136] William Merkle, in a Ph.D. thesis written from the interdisciplinary perspectives of psychodynamic psychotherapy and Buddhist meditation and philosophy, found that Eastern and Western systems of thought and practice were complimentary.[137] And Nakamura has pointed out that Edmond von Hartmann derived his philosophy of the "unconscious" from Buddhist inspired thinkers.[138] Many Western researchers, therefore, have seen parallels or complementarity between Western and Hindu-Buddhist psychologies.

Transpersonal psychology, developed from humanist psychology, has almost direct Eastern roots. It arose out of the areas that conventional psychology was not addressing, what Abraham Maslow called "the further reaches of human nature." Maslow identified a series of what he called "peak experiences"; it was later found that several Eastern techniques could induce these experiences at will. And in these traditions, there were in fact whole families of peak experiences.[139] One of the first Western practitioners to examine these Eastern techniques, the Swiss existential psychiatrist Medard Boss, declared that in comparison to Asian approaches, "even the best Western training is not much more than an introductory course."[140]

As summarized here, the encroachments of South Asian ideas of psychology into theory and practice of Western psychology and medicine have been recorded in hundreds of scientific papers and dozens of books. However, one of the earliest to ferry Eastern ideas into psychotherapy was British psychiatrist R.D. Laing, who influenced a significant number of his colleagues in the late sixties. Beginning as a Freudian, through several stages, Laing eventually fashioned a mixture which combined Buddhism with existentialism.[141]

ENCROACHING ON THE MAINSTREAM

To end this chapter, I would like to draw special attention to two books that directly relate South Asian texts to the contemporary Western psychosocial condition. The first explicitly uses Buddhism in its psychology. In the other, a book on psychology, ideas parallel to those of Buddhism are used to describe the modern condition.

Human Minds: An Exploration, by Margaret Donaldson (a professor at Edinburgh University), has been reviewed favorably in both the popular press and the professional journals. The book offers a new theoretical account of the development of the mind from the period of infancy onwards. The framework Donaldson uses provides a single viewpoint from which to look at the relationships between thought and emotion, and a central tenet of the book is that all humans have the possibility of emotional development at every age. Ongoing emotional development is as possible as intellectual development, but because of particular cultural biases, the potential of emotional development has not been realized.[142]

Earlier, Donaldson had studied child development. In her view, human mental development grows through four successive modes of knowing. Infants have only one such mode, the point mode; they are interested only in the here and now. Their perception, thought, emotions, and actions are intertwined. In the next mode, the line mode, the child develops a sense of personal past and future. In the third, or construct, mode the child can think abstractly and reason about events that occur remotely sometime or somewhere. At the highest level, the intellectual or value-transcendent mode, the individual reasons using logic and pure mathematics. When the affective side of the intellectual transcendent mode—the value-sensing mode—is developed, the child is able to value matters not directly connected with his or her personal life and desires. The highest form of this line of development is the type found in certain cultures and approaches such as in Buddhism. Donaldson is of the view that this aspect has been grossly neglected in Western psychology as well as in the broad Western culture. She is particularly interested in techniques and cultures which provide for mode-switching in adults. She takes Buddhism as an example of a system in which this mode-switching is possible. Further, she views Buddhism as a good example of the value-transcendent mode in operation.

Donaldson examines, from a positive perspective, the Buddhist view of emotions and how the mind can be changed to create growth. In her discussion, she uses (among others) four main modes of mental activity, defined by their loci of concern. The point mode has a locus of "here and now," the line mode "there and now"; construct mode does not have a specific place or time, the locus being "somewhere/sometime"; the last mode, the transcendent mode has a locus of "nowhere," that is, not space-time. Through these categories, Donaldson explores how the range of thinking and feeling in humans is experienced and changed. She draws on facets of the growing awareness of babies and children to explore how adults can discover themselves. She describes how persons shift from mode to mode and how Buddhist techniques can be used in such switching. Buddhism, she points out, is a serious attempt at a sustained endeavor of learning how to diminish the role of the line and construct mode in favor of some combination of the "point mode, the value sensing construct mode and the value sensing transcendent mode."[143]

Ellen Langer, a professor of social psychology at Harvard, has written a book whose title, *Mindfulness*, immediately evokes a Buddhist perspective. The theme of the book is how not to be trapped by existing mental categories and automatic behavior. One should outgrow existing mindsets and train oneself to welcome new information and create new categories. Through a series of studies, Langer shows that through mindfulness, individuals can become more creative, the elderly can avoid giving up on life, and executives can become more effective.[144]

Students in her classes, when told of the concept of mindfulness, had drawn the obvious Eastern parallels. Langer, while admitting some "striking" similarities, cautions against making too-tidy comparisons. She believes that behind Eastern teachings of mindfulness are elaborate systems of cosmology that have been developed and refined over time. She, however, hopes that some of the same moral consequences of Eastern mindfulness would also result from exercises such as hers. However, she does not cite any of the numerous well-known articles in psychology which have recently applied Eastern, that is Buddhist, mindfulness. There are also excellent reviews of the results of many such works. These studies have been done without any cosmological baggage. Indeed, contrary to some misconceptions, the original Buddhist approaches do not have any "cosmological baggage." Simply put, they are direct exercises on the mind without any extraneous baggage. In fact, Huber—one of the first Western psychologists to observe such techniques firsthand, in Burma—exulted at mindfully seeing things as they really are (he gives the example of really seeing for the first time the vivid blackness of a crow). The follow-up literature in the technical psychology journals on the perceptual enlargement of mindful exercises that employ these Buddhist techniques are replete with examples of increased perceptual awareness.

Langer, in a telling passage, also evokes mindlessness as "the reverse of entropy, the gradual dissolution or breaking down of an entity or patterns of organization within a closed system."[145] But such statements of everyday reality and the trend toward disorder—that is, of entropy— are the essence of Buddhist discourses. "All compounded things decay" is a central statement in Buddhist rituals, and is therefore one of the earliest statements on the nature of entropy and of an inevitable arrow of time toward dissolution. In fact, these are Buddha's dying words. To counter this inherent decay, Buddhism evokes the mind culture of mindfulness, just as Langer advocates.

Although there are differences in the mindfulness of Buddhism and that of Langer, the many similarities underscore the value of the earlier system to psychology. But if there had been an East-to-West transfer of such concepts earlier, during the previous few decades—or previous few centuries, or for that matter, during the previous two millennia or more— then a book like Langer's could have appeared much earlier. South Asian explorations into the mental realm, it is clear, provide a rich lode to be mined. The examples we have given are only the beginnings of such an enterprise.

SECTION 3
MORE IMAGINATIVE EXPLORATIONS

NINE

TRAVERSING FUTURE TECHNOLOGIES
THROUGH SOME PAST CONCEPTS

A set of "generic technologies" is associated with the history of different human societal arrangements that emerged in the Stone Age and persisted through the advent of agriculture, and into the present. Industrialization began in the late eighteenth and early nineteenth centuries, and technology associated with the steam engine was a key element. Since then, waves of other technologies have swept the industrial system, transforming manufacturing and consumption. The generic technologies that have emerged since the Industrial Revolution are (in the order of appearance) steam, electricity, chemicals and oil-based chemicals, and synthetic materials. Two new contemporary technologies are rapidly maturing at the moment, and are expected to have a much more pervasive impact than all the earlier ones: information technology and biotechnology.

These new "third wave" technologies are more "generic" than any technologies since the Industrial Revolution.[1] Within the next few decades, they and their products are expected to penetrate many niches in the economy, the workplace, and the home. The two technologies would also replace and/or intimately affect life processes and cultural processes, so their impacts could be much deeper and significant than the technological turning points of the paleolithic and neolithic eras.

THE EMERGING TECHNOLOGIES

These two technologies raise fundamental ethical issues which are unique and very problematic. And certain Eastern perspectives appear to be eminently suitable for navigating these issues. Before discussing these "navigational aids" for the future, it is useful to explore some of the key properties of these new technologies.

Information technology and biotechnology are poised to penetrate products and processes as no other technology has done before. The degree of penetration is expected to rise very steeply in the case of information technology, plateauing (if it does) somewhere in the not-too-near future. Information technology substitutes for both products and labor, but in the case of biotechnology, as of now, only a substitution and creation of products is envisaged. But in the future, with the advent of biological-based information technologies, human work would probably be replaced. This substitution would, in turn, increase the penetration of biotechnology, making its profile of uses nearer that of information technology.[2]

The human roles being supplanted through information technology span a wide variety of information-mediated work that has arisen in human societies since paleolithic times. To name a few instances, information technology has affected at least some aspects of hunting and fishing, agriculture, and a variety of roles that developed with the Industrial Revolution. In examining the range of jobs that has arisen since the eighteenth century, one could note that information technology today substitutes for the work of unskilled labor, skilled labor, technicians, and—through expert systems, neural networks, genetic algorithms, and similar artificial intelligence techniques—even creative professionals.

This substitution will have major social repercussions and create dislocations such as unemployment, temporary or permanent. Dealing positively with these technology-induced changes will require major social restructuring, and a full examination of these troublesome issues is, of course, beyond the scope of this book.

This pervasive substitution for and/or enhancement of human information capacities across almost the entire job range is in effect a cloning of human culture. Human culture has given humankind its supremacy over all other animals. Humans learn from each other and hand down, across the generations, useful knowledge. Information technology transmits parts of the culture handed down by humans to both other information artefacts and to humans, in the form of digital records. It also processes that data and therefore transforms and rearranges the culture, making new admixtures that were not there before and, through techniques like genetic algorithms, creating ones that had not been thought of before. Information technology is then a culture-replacing and culture-generating technology, and is very different from its predecessors.

Human culture has existed for roughly 3 million years, ever since hunter-gatherers began transmitting information through the generations. This process increased with the arrival of a speech facility and accelerated further with subsequent human turning points like the Neolithic Revolution (which contributed settled agriculture, an increased division of labor, and a variety of information carriers); and the Scientific and Industrial Revolutions (which saw an explosion of both information as well as information carriers).

But, based on artefacts, the present information revolution supersedes all previous cultural revolutions of information. The silicon chip on which this technology rests allows one to multiply and increase information as never before. Further, the number of such information artefacts and their capacities are themselves increasing exponentially. Over the last few decades, capacities of chips have doubled and costs have halved every two years or so (the so-called Moore's "Law"). Currently this process is occurring in about eighteen months. The result is a proliferation of nonhuman cultural processors.[3]

Very soon, probably within the next five years, such processors (that is, chips) would outnumber the human population, (although still operating at much lower capacities than the human brain potentially could). During 1993, there were about 1.7 billion microcontrollers, the smallest type of chip produced in the world. By the end of the decade, it is surmised that the average Western home would contain about three hundred chips, in everything from entertainment devices to household appliances.[4] Although this degree of penetration will not be repeated immediately in the poorer parts of the world, the particular characteristics of the technology, such as its exponentially dropping price and its ubiquity, will allow it to penetrate the developing world much faster than any other technology has before. Already, the cost of some chips is lower than an agricultural hoe and would soon be lower than the price of a meal in a developing country.

In the cards for the next century is a new set of nonsilicon-based technologies, such as nanotechnology (at the atomic level) and, if some current experimental results are followed up, even at the level of the electron and the light wave.[5] These technologies would yield, probably in the first few decades of the next century, computing capacities that rival the brain's. So the portents are that the penetration of the chip is going to have a much greater impact on the production, processing, and replication of culture than all the discoveries of all previous human epochs, and in a very short time, at that—probably in a matter of decades.

If information technology would have a great impact on humans, then an equally important turning point is in store in the next few decades in biotechnology. Biotechnology, especially in the area of genetic engineering, makes nature's biological heritage plastic. Genes from one organism can potentially be transferred to another; a plant can be made to glow with the fire of a fire-fly. Genes from a sheep and a goat can be made to yield an animal that has characteristics of both. The insect-repelling characteristics of one plant can be transferred to another. A tomato can be engineered to *not* ripen in the normal way. With the transfer of a gene, a sheep can be made to manufacture and yield pharmaceuticals in its milk. All of these feats have been performed, and some of the results—like the nonripening tomato—are even on today's tables. In principle, the genes of one species can be transferred not only between members of the same species, but across the species barrier. We now have, for example, new organisms that cross from microbe to plant to animal. Natural history is ready to be rewritten. Extensively.

The letters of biology's alphabet are also being made explicit. The Human Genome Project hopes to identify the roughly one hundred thousand genes that biologically define a human being, as well as the roughly three billion base pairs in the human genome, within whose strands these one hundred thousand individual genes lie. By 1993, the French had already made a preliminary physical map of the human genome, identifying where the different genes lie in the 46 chromosomes (which are the lengths of the genome "cut" by nature into 46 lengths).[6] The task of identifying the chemical base pairs which code for the genes, and hence the characteristics the estimated hundred thousand genes stand for, are the next steps in the mapping of the genome. The estimate for completion of the map is fifteen years, which would put it in the first decade of the twenty-first century.[7]

Gene maps are concurrently being made for other plants and animals as well. While maps of all extant organisms are not by any means on the cards for the immediate future, one should note that there is considerable overlap among the genetic heritage of different organisms. Thus, humans share roughly 40 percent of their genes with plants and over 99 percent with chimpanzees. Once particular successful responses to the environment had been arrived at in biology's genetic history, they were retained and passed onwards to other organisms. Only when a new environment required novel solutions did evolution produce novel genes. This conservativeness in our genetic heritage means that the total genetic information would probably be only a few orders higher than the human genome, but not of an astronomically higher order. Thus, identifying the major part of the genetic heritage of four billion years of biological history is probably only a few decades away, not millennia or centuries.

With the imminent (in historical terms) decoding of different genomes, the business of rewriting biology now becomes much easier. One could, at least in theory, order at will any characteristic that exists in one organism and transfer it to any other organism. It is as if the entire vocabulary nature has built so assiduously over four billion years were now before us to be re-formed into fresh sentences and new meanings. These words and letters belong not only to other organisms; they are also our own. We can today, literally, be our own biological maker.

In human genetics, interventions have already been made to cure genetic deficiencies. And with a smorgasbord of characteristics to choose from in the future, it is not unreasonable that humans, given the chance, would attempt to reshape themselves to their liking, to "better" themselves.[8] It is a short conceptual distance from using plastic surgery to change a nose, to using gene therapy to change an undesirable gene—or even to insert a desirable gene that would result in that very same shape of the nose. If interventions heighten aspects of intelligence, then drive for that capacity would be even more compelling. Only a fool would not want a son or daughter to be smarter than a parent.

But in this brave new world, genetic characteristics would not be implanted only on individual, impulsive whim. These decisions should be the result of "scientific choice" (albeit of the social-psychological kind). Today in most developed countries, huge data banks exist that categorize populations by age, place of residence, sex, education, purchasing power, and other characteristics. Any advertiser, researcher, or government official can extract various combinations of data from these sources. One obvious use of this information has been niche marketing. According to one's social proclivities, this technology can be used to target individuals for particular products, which are now narrowcast to the individual or group. And the group can potentially be as small as, say, a city block, or if need be—once the data is gathered—a village. All this is possible, thanks to information technology and its ability to store, organize, extract, and shape information.

Now, this capacity to organize information in the entrails of a computer extends also to biological information. In fact, the output of the Human Genome Project is so huge that there is no other way of storing, analyzing, and organizing its data. The project requires not only the relatively simple "bookkeeping" skills that are used in market research but complex pattern recognition AI software, doing nonroutine tasks that can only be described as intelligent.[9] And, like market and other data, this genetic information is stored in silicon form to be extracted at will and, when necessary, displayed and used.

This means that in the future, when a particular gene that codes for a particular characteristic is required, it could be called up from a database and instantly displayed. If this gene is to be inserted into a particular organism, the procedure could also be performed through automatic or near automatic means, whereby the relevant chemical composition of the gene is written in chemical form and "pasted" at an appropriate point in the genome.

In recent production technologies, there has been a move toward computer integrated manufacture (CIM), whereby a computer organizes the total manufacturing process by its control of information.[10] In the case of "genetic manufacture," where the genetic information itself is stored in a computer format, such near-automation of biotechnology production becomes much easier. This means that if one wants the gene responsible for the ripening process in fruit, or for resistance against a particular pest, one could access it through the touch of a keyboard. In the case of humans, the same would apply for the gene for the "correct" shape of a nose. Once identified, the gene could be inserted using, to varying degrees, automation; ultimately, in fact, the process would be fully automated. The system of identification and insertion would then become a seamless blend of information technology and biotechnology. When one takes into account the fact that the identification of genes and their chemical compositions and location in the genome were themselves a product of intensive information technology, it becomes apparent that this blend of technologies extends much further along the chain. But the merger of biotechnology with information technology is ultimately a merger of biological information and information ingrained in information technology, what I will henceforth call "artefactual" information. A combination of two information sources, one drawn from biology and one drawn from artefacts now occurs.

But such mergers occur not only between these two streams, biology and the artefactual. There are also two-way mergers occurring in two other instances, with cultural information as a component. Information technology, at least initially, clones cultural information. For example, databases consist of human cultural information collected by humans and incorporated in computers. The computer reformats the information, creating a new set of cultural information. This transformation process is initially guided by what the human programmer has fed into it, just as a clerk adds up figures in a manner that has been laid down in the rules of arithmetic handed down by teachers. The processed information in the computer is hence a mixture of cultural and artefactual information. The machine bias in this admixture becomes more pronounced when the computer uses techniques whose detailed operations are by definition not known to the human except in the grossest detail and include such "intelligent" techniques as genetic algorithms, fuzzy logic, and neural networks. Their detailed workings are opaque to the individual human user (or even to the programmer), and so their use results in a mixture of cultural and artefactual information.

There is also an admixture occurring in the reverse direction, from artefact to human culture. For example, when a human stares at the output of a computer as I am now doing when I use the word processor, the blips on the screen—processed information spewed out by the computer—enter my brain and change my train of thought. When I interrogate a database, this inflow occurs as fresh information to my brain, fresher than my own regurgitated thoughts spewed out by the word processor. When the database output is mediated by a complicated computer operation, the novelty of the information (to me) increases. When it is transformed through means that by definition I cannot follow, such as the intelligent processes that I have mentioned, this novelty is further increased.

In these actions, the computer acts like a human "other" to me, changing my thoughts and hence, however imperceptibly, my stock of cultural information. When I use a multimedia output this internal change is intensified. And if I use a virtual reality (VR) device—for example, learning to fly in a flight simulator—the feeling of reality in the computer-generated realm is so intense that for most purposes, the mix of the two streams of information—my own cultural information and the artefactual database—become very intimate. A two-way merging of artefactual and cultural information results.

The third merging system involves biological and cultural information. When I order a particular type of genetic intervention, it is a cultural wish, an expression of cultural information. This wish could be for a medicine to relieve a disease or for a shape of a nose for my potential child. This cultural information is then transferred into the biological realm and a particular biological change occurs. The genetic intervention adds a new set of cultural information to the biological, in effect merging the two.

A merger of biological and cultural information occurs in the reverse direction when, through biological intervention, I change my thoughts and feelings, my internal store of culture and its processing machinery. This can occur by ingestion of biological material through a drug that affects my thoughts, reformatting the brain's internal machinery. Or, in a much more intense and longer-lasting fashion, I can change the way I think and feel by genetic means. This prospect is not so fanciful as it appears. Already, Parkinson's disease has been treated by fetal transplants, an intervention at the phenotypal level of the brain's hardware.[11] In the same direction, efforts at growing brain cells and transplanting them are being made in several laboratories in the world.[12] If, on the other hand, the intervention is genetic—for example, to cure a devastating neural disease for which the genetic cause has now been found, (examples are Fragile X syndrome, a form of mental retardation,[13] X-linked adrenoleukodystrophy or ALD, which destroys nerves and leads to cessation of conscious brain functions,[14] and Lou Gehrig's disease, a neurogenerative disease resulting from the death of certain nerves in which patients lose their ability to speak, swallow, move, and even breathe[15]) then the admixture occurs at the genetic level. If, within the next couple of decades, I order an increase in the intelligence endowment of my child through genetic intervention, then the mixture of cultural and biological information is carried further.

In the twenty-first century, as information technology and biotechnology increasingly penetrate existing crannies of culture and biology, these mergings will increase and come to the fore as a major player, if not, in many different ways, *the* major player. The interventions of biotechnology allow one to potentially rewrite the biological realm, four billion years of biological history, into a new format. Information technology, similarly, would rewrite three million years of culture. The paleolithic, the neolithic, and the Industrial Revolutions would pale in comparison to these turning points. It is this deep potential for fundamental transformation and the all-pervasiveness of these two technologies that underlies the importance of these future mergers.

However, some aspects of these combinations have been depicted as the stuff of science fiction and popular horror. From Frankenstein to Terminator 2 and RoboCop, the effects of man's intervention in biological and artefactual nature have been debated and dramatized. And at a more formal, scientific level the prospect of the imminent arrival of the cyborg has haunted our recent scientific imagination. We have even begun to think about whether we should give rights to our creations, our machine hybrids—roughly, one should note, at a time when "we" (that is, the West) have begun to think of giving rights to animals.

But these are not the really serious questions evoked by such couplings. The serious effects will not occur from the temporary arrival of machine-human artefacts, but from other, long-range actions. Current outcomes, at the level of "being," will have less effect than what these couplings will do to histories, in the processes of "becoming" as we unravel new futures. The future is interesting because it is woven as a new pattern, like a fresh carpet from the past through the present onward. The future is not interesting because the schoolboy imaginations of a Jules Verne, a Ray Badbury, or an Arthur C. Clarke have populated it with bizarre creatures, cyborgs, intelligent robots, or "rules" for robots to "follow" à la Asimov (rules that robots will not follow). The serious, "manly" (or for that matter "womanly") questions are of a different order: what are the shapes, textures, and patterns of the the future? These questions are interesting.

MERGED INFORMATION

The combinations just described are not significant simply because they might result in a creature that would last only for its lifetime and then vanish. We are documenting the merging of three *streams* of information. One, the biological, has existed for about four billion years; another, the cultural, for at the most three million years (and in its most effective unfolding, not more than ten thousand years); the last, the artefactual, has a bare half-century of history dating since the advent of computers. Each stream has many common characteristics that will

be profoundly changed by incorporation with the others. I have documented elsewhere these common characteristics and will briefly summarize them here.[16]

A creature's genetic lineage transmits biological information to prog-
eny through a set of unbroken streams that connect the most recent
creature back to the earliest. Similarly, cultural information itself is
transmitted through a chain from human to human, over the course of
generations. This connection links us, in an unbroken lineage, to the
most remote ancestor. Computers, ever since they began operating,
have been storing information to be used in a future set of computers.
This accumulation has also resulted in a lineage, albeit a short one. But
with the proliferation of computing devices, this lineage is expanding
the fastest, proliferating and rapidly creating new types of information. Similarly, cultural information proliferated after its introduction,
producing information much more rapidly than the genetic variety.

Each of the three lineages produces information in interaction with
its environment, changing its internal store of information as a result
of new inputs from outside and then transmitting the processed information onwards. In the case of the artefactual line, the changes can be
very rudimentary, but with new intelligent techniques these lineages
are becoming more sophisticated, adapting to their environments in a
nonroutine way, retaining that adaptation and transmitting it onwards.
So, in each of the lineages, there is a retention of information, a conservativeness, as well as a creation of new information as new environmental challenges are met by the lineage and encoded within it.

Each lineage also has a particularly structured window to its environment, its external world. The approach of evolutionary epistemology is
very useful in describing this fact. Evolutionary epistemology states that
evolution can be described as a process of acquiring cognition, and that
evolution produces cognitive phenomena.[17] Thus, the perceptions of
each biological species are colored by the "historical" lenses that it has
acquired through its genetic evolution. These elements tend to congeal
the experience of that particular species (as well as of the earlier species
from which it branched off). Thus as humans, our historical spectacles
are influenced by the 99 percent of our genes we share with chimpanzees and the 40 percent we share with plants. Biological organisms,
therefore, have different orientations to the external world, different
means of cognizing it. They possess what could be considered different
"world views." Mammals, for example have a different "world view"
from that of birds.[18]

The contents of human minds, cultural products, also have structured windows. Culture varies according to historically formed groupings such as nation, class, profession, and so on. Each such variation is
handed down across generations and constitutes a lineage. The world

views associated with each such sublineage have been studied extensively by social scientists and discussed under the headings of ideologies, forms of social cognition, class consciousness, social constructions of reality, and the more recent work on the social epistemology of science.[19] The world views of a social group change with the environment. Consequently, corresponding to the differing flows of lineages in the cultural realm, there are many differing social cognitive systems.

There are different world views in the artefactual lineage too. The information that is internalized in an artefact is selectively fed from its environment, either through the different sensors with which it interacts with the environment, or through the human hands that feed it, or through the pre-filtering that is done through its programming languages and similar higher-level information operations. As it is internalized by the computer, this selective information yields a set of particular windows to the external environment. If more autonomous techniques process this raw information, the artefacts in the lineage would construct more sophisticated internal representations of the external world.[20] The lineage's world views would thus continuously change, corresponding to the environment that it traverses.

System theorists have attempted to demystify these structured windows, these views from within, these "minds." As systems psychologist O.H. Mowrer puts it, "consciousness is essentially . . . the operation whereby information is continuously received, evaluated and summarized in the form of 'decisions,' 'choices,' 'intention.'"[21] In this light,

> The phenomenon of mind is neither an intrusion into the cosmos from some outside agency, nor the emergence of something out of nothing. Mind is but the internal aspect of the connectivity of systems within the matrix. It is there as a possibility within the undifferentiated continuum, and evolves into more explicit forms as the matrix differentiates into relatively discrete, self-maintaining systems. The mind as knower is continuous with the rest of the universe as known. Hence in this metaphysics, there is no gap between subject and object . . . these terms refer to arbitrarily abstracted entities.[22]

"Mind" (the view from within) and body are intertwined; they coexist. They are inseparable, like two sides of a coin.[23]

The discrete windows within each lineage allow it to "tunnel" through time, each opening its particular epistemological window to the external world. This window of knowing and of becoming continuously creates and unfolds a stream of meaning in a becoming process. If one could observe this flow, traversing history, he or she would witness a stream of structured information, of cognition.

What happens to these characteristics when the three streams merge? The act affects several parameters of a lineage (which I have discussed in detail elsewhere), including conservativeness of memory, creativity, flexibility, reactions to the environment, and rates of evolution. Most importantly, world views change. What this means is that in the coming century, merged evolution processes are going to be very influential in the genetic, cultural, and artefactual fields and will probably be the dominant dynamic affecting us in biology, culture, and artefact.

This extensive intrusion of technology into our biological and cultural beings raises many important questions and concerns about the problems created by these information hybrids. Through these admixtures we are rewriting our very biological and cultural being. Deep ethical questions have emerged regarding how to traverse the future with these new technologies, devolving into a key question: what "grounding" in philosophy, morality, and ethics should we use when we are cutting the biological, cultural, and artefactual ground from under our very being. These questions are much more profound than those surrounding earlier key points in human history. Almost all of these questions relate to the identities to be created by the new interventions.

BIOETHICS OF TRANSPLANTS, REPRODUCTIVE TECHNOLOGY, IMPLANTS, AND PROSTHETICS

At the moment, in genetic engineering's infancy, for example, one sees only the beginnings of such vexed questions. Yet a precursor to their nature is seen in the nongenetic interventions that have already been made in reproductive technology and tissue transplants, with such techniques as the flushing of embryos, in vitro fertilization, sex pre-selection, surrogate motherhood, surrogate embryo transfer, and cloning —all of which have raised a hornets' nest of issues. Some of these interventions have seen confusion and the dramatic reversal of biological and social kinships. For example, when a woman carries, in an act of surrogate motherhood, her own daughter's child, the grandmother and mother are the same person, and the child's identity is multiple— at the same time a child could be son, grandson, and stepfather.[24] The identity of the parent can also be thrown into question. Who, for example, is the parent of a child born through in vitro fertilization from an egg donated by Mrs. A., combined with a sperm from Mr. B., implanted in Mrs. C.'s uterus and given for adoption to Mr. D. and Mrs. E.?

These very complicated social and ethical issues will intensify when genetic characteristics themselves can be moved—in or out—of chromosomes in the future. In such cases, possible parents could be spread over a large number of desirable gene donors, or for that matter, a computerized gene bank. In the extreme hypothetical example, there could be a different donor for a each of a variety of genes: for a nose, for an eye, for a set of teeth, for a particular type of intelligence, to avoid a particular disease, and so on. If one transfers genes from other organisms (the gene for preventing a disease could even come from another species—a transgenic source—which has happened already in the plant field) "parentage" is then a very complicated affair. Ultimately (i.e., in the next few decades) the affair becomes unsolvable within current discourse.

These expansions of biotechnology challenge deeply held cultural assumptions. Recognizing their importance, Western ethicists have called for public debates on these and other issues.[25] But the debates that have already occurred have been conducted within the implicit framework of Western religious definitions of life and ethics.[26] Lurking in the background are certain assumptions, like the existence of a soul. Some of the issues that have been raised by already-existing technology are, for believers in God, only questions of the appropriateness of "playing God," as the title of a book on the topic so aptly puts it.[27]

The Christian church had developed a large and highly detailed set of views on birth, but even within Christianity, such views have changed. Saint Augustine, for example, had held that human life began at quickening, the mother's first sensation of the fetus's movement, which occurs at the fourth or fifth month of pregnancy. Thomas Aquinas, following Aristotle, pronounced that human life begins only at the point when the unborn acquires a soul; males acquired a soul at forty days after conception, females eighty days. Saint Gregory of Nyssa, following Plato, put the beginning of life at conception. However, it was barely a hundred years ago, in the late 1800s, after fertilization was understood, that the Roman Catholic church settled on conception as the beginning of life, giving up the ensoulment concept.

Generally, today's Western bioethics adopts an interdisciplinary perspective, incorporating the views of philosophers, theologians, historians, lawyers, writers, and scientists. Some of the questions and answers are also influenced by the Hippocratic corpus, the taproot from which all Western ethics in medicine has grown. Many medical students take the Hippocratic oath on graduation.[28] Yet in non-Western countries, there has been little debate on these matters, although it is readily admitted by workers in the field that non-Western traditions could well provide different answers to these questions.[29]

There is still another intrusion to human identity: computing, the prostheses for the brain. Advanced information technology, especially that related to artificial intelligence, aims at cloning parts of the mind's behavior. Further, the senses, and thus parts of the "mind," are being extended beyond the boundaries of one's person through electronic means. By pressing a lever, clicking on an icon, or speaking to a device I can affect an event remotely, and that effect can also be fed back to me so that I internalize it. This extension, and the broad cloning of human capacities by computers tends to spread the human identity widely over artefacts. Instead of carrying all my ideas in my head (as I would have done if I lived ten thousand years ago), or on the other hand, only a part, storing other information in written documents (if I lived two thousand years ago), today I can spread my stores of information in all types of computing artefacts. Further, if I had worked in the earlier part of this century, I would have manipulated my information only in my head or in the heads of my fellow humans, except for a few arithmetical calculations that I would have relegated to a slide rule or a calculator. But now, widespread computing artefacts allow me to process my data remotely in machines. To give a trivial example, the form this paragraph or page takes does not require my elaborate intervention; desired spacing, fonts, and so on are determined and processed remotely. I can spread parts of my mental processing over several artefacts.

In virtual reality, the intensity of human-computer interaction reaches an all encompassing threshold of intimacy. Here, the user is immersed in computer-generated visual and other sensory inputs, and he or she manipulates these representations through input-output devices. When virtual reality's intimate output interfaces with autonomous intelligent technologies such as genetic algorithms and neural networks, the computer and human interface will increasingly be like that between humans. The resulting exchange of information will be very intimate. Here, what goes on inside the brain and outside in the artefact is so intermingled that the internal intrusion of prostheses becomes a very important ethical and philosophical problem. Deep problems of identity rear their heads again.

With advances in biotechnology and information technology, therefore, questions of identity and of associated philosophical, social, and ethical modes of navigating the future become central issues. Has this class of philosophical and ethical problems been met and discussed before? If so, would that body of work provide us with some hints for navigating such a future world? There is one such attempt in the field of cultural information to which I shall now turn.

STREAMS OF "INFORMATION" IN BUDDHISM

Culture is information to be stored and processed in the mind. In the minds of humans, culture acts out its role in the form of mental activity, a stream of thoughts. Evolutionary lineages of information are like such extended streams of thoughts. So the flow of our thoughts and our internal life give us a subjective sense of how other lineages behave.

These lineages, in the thought realm, have been the subject of much intense attention in South Asian psychologies and philosophies, especially in Buddhism. Thoughts, as they arise and flow in our minds, have been subjected to a careful observation and analysis that is in many ways parallel—albeit, perhaps notably, only in narrow domains—to the rigor of modern science. As a belief system, Buddhism places a high value on skepticism, accurate observation, and analysis of psychological processes. Its founder declared in the well-known *Kalama Sutta* the supreme need for questioning authority, including his own. In another instance he admonished observers to "come and see" for themselves the truth of observations on the mind.

The validity of *some* Buddhist conclusions on mental processes, including those of the many practitioners who came after the Buddha, has been increasingly corroborated within the last twenty years as Western psychologists and physicians have taken these observations seriously and attempted to validate them (see chapter 5), notably in the area of "meditation," which, shorn of its Western connotations, comprises techniques designed precisely for such internal observations.[30] In this Buddhist literature, a class of questions is raised that directly parallels, some of the questions that occur today in the era of the cyborg and the gene-transplanted hybrid.

I have used the words "lineage" and "stream" to describe the flow of information that connects the present of an information lineage with its past and future. "Stream" and "flow" are the very images often evoked in Buddhist discussions on the flow of mental phenomena as it moves within the mind: that is, culture. In Buddhism, the universe's components are in a state of impermanence, of ceaseless movement; nothing is durable or static.[31] The continuity of life is not realized through an abiding permanent structure, an "I." Buddhism is unique in the philosophies of the world in that it denies the existence of a self or a soul. A belief in a permanent, abiding "me" is radically deconstructed in Buddhism.

Buddhism breaks down the component physical and mental factors that constitute the psychophysical personality into what it calls the five aggregates. They are then analyzed and deconstructed, and determined to be without an unchanging substrate. In the Buddha's own words, "there is no materiality whatever . . . no feeling . . . no perception . . . no formations . . . no consciousness [the five constituent Buddhist aggregates] . . . whatever that is permanent, everlasting, eternal, not inseparable from the idea of change, . . . that will last . . ."[32] And elsewhere, "When neither self nor anything pertaining to self can truly and really be found, this speculative view [of] a permanent, abiding, ever-lasting, unchanging [self] is wholly and completely foolish."[33] A disciple of the Buddha elaborated further that what one calls "I AM" is "neither matter, sensation, perception, mental formations nor consciousness" (the five Buddhist aggregates).[34]

Physical elements change, as do mental phenomena. All are in a state of perpetual becoming; all phenomena are but fleeting strings and chains of events. Because the constituents of an individual change, she or he does not remain the same for two constituent moments.[35] In the Buddhist analysis of identity, there is in fact no individual, only a stream.[36] "Life is a stream (*sota*), an unbroken succession of aggregates. There is no temporal or spatial break or pause in this life continuity."[37] This continuity is achieved not through a soul, but through a stream of becoming.[38]

It should be noted that this analysis is partly arrived at from subjectively observing one's innermost feelings. In fact, one of the objectives of Buddhist mental exercises, meditation, is to observe, experience, and describe this lack of self and of permanence from within one's own streams of thoughts and mental phenomena. From within our own subjectivity, the problem of identity and of an abiding "I" is shown to be false. If this is the real state of "I" from both an external, material point of view, and from an internal, subjective point of view, then what does this entail for our own streams of information, the lineages? All the lineages have the characteristics described here. They are all everchanging streams. Their identity is not in a snapshot existence of being, but a long process of becoming, an unraveling.

From such a perspective, then, questions are raised about identity by the two technologies. The existential angst of being a hybrid, of carrying the genes of plants and animals, can be seen differently. The problem of one's "self" being spread over several artefacts now loses its potential terror. The threat of being a cyborg, of Frankenstein's creature; the concerns of a Jeremy Rifkin, the fundamentalist critic of biotechnology (one of the truest of Christians, much more fundamentalist in his position than those right-wing Christians who protest at abortionist clinics): all can be addressed differently.

Living things, complained Rifkin at one time,

> are no longer perceived as carrots and peas, foxes and hens, but as bundles of information. All living things are drained of their aliveness and turned into abstract messages. Life becomes a code to be deciphered. There is no longer any question of sacredness or inviolability. How could there be when there are no longer any recognizable boundaries to respect?

Further, he continued, "as bioengineering technology winds its way through the many passageways of life, stripping one living thing after another of its identity, replacing the original creations with technologically designed replicas, the world gradually becomes a lonelier place."[39] Buddhism stripped this seeming sacredness and identity over two-and-a-half millennia ago.

A gene does not make a sentient being. Only the stream of a being's existence, an onward-flowing history, constitutes the sentient human or the sentient cyborg. A person does not exist as a unique individual but as a constructed, everchanging flow, a forward-moving lineage. If to this lineage are added new elements, new parts, it is but in the very "normal" nature of such streams. All such streams are constructed from constituents in a constant process, and a person's normal existence consists of such a constructed being. The artificial introduction of elements—say, to the cultural flow, from genes or artefacts—is but another manifestation of the normal construction of such flows. From a realist's perspective, there is no difference.

But such a perspective makes us squeamish, frightened, alarmed, and even disgusted. One would not mind a set of false teeth (even implanted) or a prosthesis for a limb—something like an electronically sophisticated walking stick. But messing up one's interiority, one's subjectivity, evokes an entirely different order of emotions. It is the popular horror of aliens taking over our minds, a fear of one's own consciousness being invaded. After all, the validity of the individual's own subjectively felt oneness is at stake.

But in similar instances, the Buddha himself firmly rejected the views of those who take the thing called the "mind" or "consciousness" to be an unchanging substance. In that case it was better, he argued, for a person to take the physical body as an unchanging "self," rather than thought, mind, or consciousness, because the physical was at least apparently more solid than the mental, which is ephemeral and constantly changing—and so hardly a candidate for permanence.[40] Interiority and consciousness are thus demystified into mundane components. In the ponderous and archaic language of the nineteenth-century European translators of an important Buddhist text: "Were a man to say I shall show the coming, the going, the passing away, the arising, the growth, the increase or development of consciousness apart from body, sensation, perception and volitional formations, he would be speaking about something which does not exist."[41]

This view from within—that is, subjectivity or the feeling of consciousness—is our structured window on the environment, a characteristic of every lineage and an outcome of the history that it has traversed. The window is what is conventionally called "mind." And general system theorists have advised us that such minds are endemic to all systems. Such systems, when observed externally, are physical and material; when viewed from within, they are mind, subjectively experienced.

But experiencing the intrusion of the new technologies that remake us biologically and culturally is disturbing and challenges our sense of self. "This idea that I may not be, I may not have, is frightening to the uninstructed," as the Buddha himself put it.[42] Because the belief in an abiding self is deeply rooted in humans, the contrary position goes "against the current," as the Buddhist texts say on another occasion.[43]

If, then, in the coming future we will inevitably be constructed and reconstructed—from biology, culture, and artefact—what should be our epistemological, philosophical, ethical, and subjectively felt guiding principles? If "we" would then be cyborgs and hybrids, what should the interiority of robots, of constructed hybrids be, as they navigate reality, and tunnel through time in our lineages?

The person is not a "what," but a process. Being is only a snapshot in the process of becoming, lasting only the length of one thought. "Just as a chariot wheel in rolling, rolls only at one point of the tire, and in resting rests only at one point; in exactly the same way, the [internal] life of a living being lasts only for the period of one thought. As soon as that thought has ceased, the being is said to have ceased."[44] There is no stable substratum of the self; the self is just a stream of physical and psychological phenomena that is always perishing. This is the correct view to internalize in the inevitable day of the cyborg. As the fifth-century Sri Lankan classic of higher Buddhist theory *Vissudhi Magga* put it:

> There is no doer but the deed
> There is no experiencer but the experience.
> Constituent parts roll on.
> This is the true and correct view.[45]

One analyzes oneself, knows oneself only to realize that there is no self in the first place. This is not an intellectual knowledge but an internally observed, felt knowledge. And in Buddhism, this elimination of the sense of self sets one free. With the dawning realization that "I am not a thing, but a process," the future becomes open-ended. Buddhism is self-referential: to know oneself is to make oneself, to guide the self that is not there.[46] In the Buddhist analysis, unsatisfactoriness and anxiety become essential to the "I" because these are the "I"'s response to its own groundlessness.[47]

The internal experience of the nonself does not lead to a loss in integration, awareness, or vitality of the mind, that is, of the view from the interiority of the lineages. On the contrary, perception unclouded by false perceptions leads to perceptual clarity. Perceptions of others are enlarged because there is an empathic openness based on a non-judgmental awareness.[48] The fully mentally healthy person, the *arahat*, is expected to enjoy a state of continuous inner delight, attend keenly to all the circumstances of a situation, and respond with skill to every situation.

This is the phenomenology of flow for human thought, and we could extrapolate this perspective to the other two lineages. The Buddhist analysis also suggests a moral compass for the inevitable future of merged knowledge streams; such a perspective includes a profound moral code of altruism, and it is not entirely farfetched to think that these principles could also apply to future scenarios.

But then, what do we make of that "external" baggage of this alien intrusion if it is not "ours," and if in fact "we" do not exist? How do I react to the massive inflow, into my biological and mental interiority, that will occur in the next century? Let me make recourse to a standard exercise in Buddhism dealing with that interiority—"meditation," Buddhist observational practice.

First, one trains oneself to observe one's interiority and recognize its constructed nature, its lack of an essential being. Second, one dispassionately notes also the coming and going away of thoughts, observing them and letting them go.

This is the process of meditative practice. I suggest that in everyday reality, and in the day of the cyborg too, one would indulge in a parallel exercise. We could recognize the constructed nature of our internal and external cyborgs, our own Frankenstein creatures, realizing their real, ephemeral character and using it as our guiding principle to the external world. But at the same time, we can locate where the constructions come from—from this lineage or that, from this sublineage or that, or from a mixture. These, after all, are some of the techniques we all use in analytical thought, incidentally an important branch of Buddhist philosophy (a philosophy which in some renderings, both classical and modern, is called a system of analysis). The analytical faculty is retained and can be used in our new circumstances.

Some Western commentators have alluded to these questions and have come close to the Buddhist position. In his book *Reasons and Persons*,[49] Derek Parfit aroused considerable interest. The *Times Literary Supplement* reviewer considered it "by any standard the most notable contribution to moral philosophy" in the last hundred years.[50] Among the issues raised by Parfitt is the nature of the person. Here, he reaches "reductionist" conclusions on persons and their continuity which, as he acknowledges, are in considerable agreement with the Buddhist position. Parfitt extends this position to a discussion on the nature of rationality and morality in which he denies the importance of self-interested reasoning and of prudence. From the time of Aristotle, it had been a central tenet of Western philosophy that one's best interest lay in prudent assessment. This was not a crude argument for selfishness and greed; the position is best understood by posing it in another way: what is more irrational than consciously to act in a way that frustrates the interests of both oneself and those one loves? An extension of this argument is that morality coincides with self-interest.

Parfitt uses many arguments against this self-interest theory, the most important being his description of personal identity and continuity. He speculates on what would happen in three hypothetical in-stances in which the internal makeup of individuals is changed: an Enterprise-type scan (à la *Star Trek*) and then remote recreation of a human; surgical interventions in the brain that would implant memo-ries; and brain transplants. All raise deep questions regarding personal identity, but what can be believed, through reason, is only a person's physical and psychological connectedness from one instant to another. That is all that can be truly asserted.

But personal identity can be different in both connectedness and continuity. In the extreme, the answer to a question such as "Do I die?" does not have a definite answer. Parfitt applies his discourse on mor-ality and ethics to the notion of past and future selves that could be considered comparable to our present ones. Some of these positions coincide with Buddhist beliefs, as Parfitt readily admits. Thus the self lacks identity, being instead a changing stream of impersonal physical and mental elements; sameness and difference exist within the stream only as matters of degree, not absolutes. The Buddhists would say that in the hypothetical cases of the brain transplants and the other examples of identity change that Parfitt posits, these are "neither the same nor different" (*na ca so na ca anno* in Pali).

Parfitt, however, admits that persons are subjected to experiences and, in turn, act on the world. But to him, this is largely a matter of semantics. "Though persons exist," according to Parfitt, "we could give a complete description of reality without claiming that persons exist."[51] Recognizing that these views, although understable through theory and discussion, may be difficult to experience subjectively, Parfitt says that "Buddha claimed that, though this is very hard, it is possible. I find Buddha's claim to be true."[52] Yet even for Buddhists, the internal grasp of theoretical insight is difficult. In fact, an important meditation de-vice in Buddhism is to acquire this internal view of the changing self precisely by observing the stream of changes in oneself.

Parfitt has faced criticism, of course. In the words of one detractor,

> The Judeo-Christian religions which have shaped Western civilization have all taught regard for human personality, and this understand-ing has been the foundation of our traditional morality. This is not true however, of the doctrines that are now replacing religion. They re-duce individuals to mere manifestations of non-human forces or struc-tures, or to illusions or streams as Buddha taught. . . . Parfitt has taken his cue from such doctrines and produced a profound attack on West-ern civilization.[53]

Another commentary critical of Parfitt reminds us that "the idea that the individual person is of supreme worth is fundamental to the moral, political and religious ideals of our society." For Downie and

Telfer (from whose book *Respect for Persons* this quote is taken),[54] the position characterizes Western thinking. This view of the reality of the person also pervades all contemporary variants of social thinking in the Western tradition, such as socialism.[55] Contrary positions are difficult to swallow.

Another study that evokes some of the same philosophical approaches in charting a future technology is *The Embodied Mind*, by Francisco Varela, Evan Thompson, and Eleanor Rosch (which I discuss in detail in chapter 11). They propose a bridge between the mind as conceived of in science and the mind of everyday experience, through a dialogue between Buddhist meditative practice and cognitive science. The approach was applied to a variety of subjects in neuroscience and cognitive psychology, artificial intelligence, and evolutionary biology.[56] In doing so, they approach (albeit without using our terminology) what we have considered as the three lineages, namely the internal flow of our thoughts (the culture within our minds), the flow of genes (evolutionary biology), and the flow of "artificial thoughts" (artificial intelligence).[57]

Varela and his colleages evoke the flow patterns that one observes internally through Buddhist meditation and find here the key to tackling the other two realms. They tackle the problems of nonself and of everflowing streams, and describe the dynamics of the three lineages. Their discussions are located in specific debates with the research communities in these three areas. They reject the equivalents of the subject–object dichotomy that arises in different forms in all the three lineages (using our terminology).

They consider that the inside and outside of the lineages jointly move forth, "enacted"—in their terminology—by the subject and the external object. A process of coevolution results because the environment is not a given but is enacted and brought into being through a process of coupling. The world is not taken as a given, with the organism representing it or adapting to it. Their Buddhist approach transcends this duality, the outcome being codetermined by both the inside and outside. This position would be the same, whether one's perspective is human–cultural or that based on artificial intelligence. One would act as the environment for the other, and together they would enact an unfolding future.

Although Varela, Thompson, and Rosch's objects of concern differ from ours here, their approach overlaps with some of my key positions, as does Parfitt's. Tackling the problems of the future requires radical reorientations, sometimes even a return to very old—often discomfiting—observations.

TEN

VIRTUAL REALITY: PHILOSOPHY
ON THE NINTENDO

VIRTUAL REALITY AND THE NEW
COMPUTER INTERFACE

An important emerging field that will intimately affect most humans in the coming decades is the intimate data and entertainment interface called "virtual reality" (VR). It has been estimated that in the US, in a lifetime, the average person spends seven years watching TV. To this, if one adds the time working people spend in front of computers, some users in the future could easily spend twenty or more years of their lives inside VR.[1] It will be an inescapable part of life in the coming decades. And this future of the act of living will not be restricted to Western countries.

But VR, which will dominate our lives so intimately, is not just another display. It raises many deep questions, as recent commentators have pointed out. Aside from issues of short-term effects, the questions include philosophical issues, some directly impinging on ontology and epistemology. Because VR is likely to be such a widespread medium, these philosophical questions are not limited to philosophers; they occur to the person in the street. These issues have some direct bearing on a series of parallel questions in South Asian philosophy, a potential field of fruitful interaction. But before exploring this overlap, we have to describe virtual reality, South Asian philosophy, and the philosophical issues raised by the new technology.

VR's foundation is in computer science and computer graphics, especially in the field of interactive computing. Virtual realities, in their usual format, make use of three-dimensional computer-generated graphics into which the user is immersed. The user then manipulates and interacts with these graphics by means of input-output devices.[2] The latter interfacing apparatus includes data gloves, head-mounted tracking devices, goggles, headphones, and wired bodysuits. Through

this setup, the user navigates the artificial or virtual environment—
"cyberspace." (The term comes from William Gibson's celebrated novel
Neuromancer[3] and is used widely to describe the inner space of virtual
reality systems.) Such virtual realities no longer, in effect, just mimic or
represent reality; they *are* a reality.[4]

The rather elementary virtual reality systems now beginning to be
used will grow more sophisticated with developments in the pipeline.
Thus, greater computing power with newer generations of faster computers would increase VR's intimacy. There are also other very radical
approaches on which some British researchers are working, such as
using laser microscanners to write images directly into the retina. These
would, in fact, directly reach the rods and cones of the retina, which
deliver the nerve impulses into the brain. This would be a more direct
form of addressing information than seeing the dots (pixels) normally associated with computer displays and then directing the pixels into
the brain through the eyes. Such a direct system would then have the
same type of visual quality as normal sight.[5]

This sensual intensity can best be illustrated by how direct virtual
perception, applied to help a surgeon, will appear in the not-too-distant
future. Wearing a VR helmet which feeds data from a catheter and its
probe, the surgeon is transported directly into the patient's body, seeing its sights and hearing its sounds. Using the visual and audio information, he or she performs surgery inside the body.[6]

VR, through the use of pneumatic devices, is also expected to have
good tactile systems, better than current systems in "data gloves." Future systems that are being envisaged could also incorporate virtual
taste and smell.[7]

Scientists have recognized that one can understand phenomena
when they are visualized better than when such phenomena are only
conceptualized.[8] Hence the potential uses of VR include, for example,
activities like brain surgery, where a computerized simulation of the
brain is displayed so that the surgeon can practice on the image before
delicate surgery. Another use is in computer aided design (CAD).[9] Generally speaking, VR allows intimate, body-experienced explorations
of hitherto abstract intellectual spaces—without having to go through
reams of data in a conventional database. Through virtual reality, one
"can imagine CAD models that, in effect, come alive. . . . You can enter them at any scale. They could be models of molecules, for example, and you could move about within these molecules with your
whole body to examine their structures."[10] Using VR, wearing a data
glove, a biologist could also get the literal feel of different molecules
and the strengths of the various molecular forces that bind them together. The abstract environment now becomes real in many ways.

Virtual reality systems have already been used in medical systems; in
a virtual orbiting space station; in factory design; in the training of sur-

geons, on a virtual body; in simulated eye surgery; microscopic displays that create microscopic worlds; as aids for the handicapped; in museum displays; and in manipulating and studying chemical reactions.[11] This is only a fragmentary list, indicating the widespread and powerful potential uses of VR. Virtual reality can, in fact, be introduced into almost every human activity, immersing human actors in virtual, artifical worlds. Its impact will be greater than that of literature, film, or theater in the past. VR is in many ways the "first intellectual technology that permits the active use of the body in a search for knowledge."[12]

Further, cyberspace theorist Randal Walser notes that cyberspace gives people the impression that they are bodily transported into new imaginary worlds. Unlike film and theater, VR cyberspace allows the participant not only to view reality but to enter it intimately. The participant has a role—and the power to create the next event. Other media, like film and TV, only show; "cyberspace embodies."[13] But the uses of VR intimately change our views and feel for our environments and reality. VR can thus be defined as giving presence, the sense of being in an environment: not actual surroundings, but the perception of surroundings. Virtual reality exists inside a person's consciousness.[14] VR has, therefore, the possibility of immersing participants in worlds that are almost indistinguishable from the real world. Further, VR environments can merge with real ones.[15]

In VR, one can dip into and explore a self-created artificial world[16] or, in the words of Myron Krueger, an "artificial reality." Krueger developed the idea of an intelligent space which detects, through sensors, various characteristics of the human in the space.[17] This artificial reality allows for a unique blending of feeling and technology.[18]

Through the new technology we can combine a variety of new experiences in a totally unexpected ways.[19] Thus VR allows us to be other creatures, experiencing (at least partially) their worlds. One such artifact allows a person to take on a lobster's body. Giant claws replace hands, and the participant becomes a bipedal lobster.[20] In an ultimate sense, one could perhaps even simulate memories, through the use of computers, to heighten the experience—so that when one takes the form of a worm wiggling through the earth, one is, for all purposes, a worm.

Consequences of the technology become apparent in one of the most preliminary forms of this experience, namely hypertext. Hypertext helps one scan from subject to subject, from association to association, in a free flow of association that fosters an illusion of all-encompassing knowledge and total information.[21] In cyberspace, through the interface one enters a new, relatively independent world which has its own rules and own flavor. The cybernaut—especially when equipped with VR capability—begins as a voyeur but, because of the intimacy of contact, evolves into the fabricated worlds. Information technology fits our minds so closely that it is very difficult to think about: one can travel through it in an infinite direction, but still, in the words of Gibson, it is a cage.[22]

The intensity of close interactions with computers and television displays generally would reach a qualitatively different plateau with this emerging all-encompassing interactive medium. Virtual reality, in the way it feeds data into the brain, has certain parallels to a hallucinogenic world, that is, a biologically felt reality. In fact, Timothy Leary, the LSD guru of the 1960s, became a strong advocate of VR.[23] Hallucinogens are here replaced by *hallucigenres*. Even sex, in one of the ultimate technologies of the system, would become virtual. Slipping into an intimately fitting virtual-reality bodysuit with microscopic actuators and sensors,[24] the partner lies elsewhere in cyberspace and fondles the computer-generated sensory image.

Within the coming decades, it is very probable that virtual realities would shape a large segment of computing. Information consumers, whether at home, school, or workplace, could experience their respective environments through a comprehensive virtual reality—at a virtual workplace—while physically at home.[25] And with the ongoing seamless merger of computing systems with entertainment and communication devices like TV and the telephone, this intimate intrusion of VR will only increase, making up a significant part of the working life of humans.

The boundary between "real" and artificial reality is already vanishing. We have already dislocated part of our senses. A pair of spectacles, a hearing aid, a mobile phone are all portable. Artificial realities would eventually also become portable. One would initially have it in one place, but like air conditioning, after you have it at one place, you will want it everywhere. Physical reality may then become like the forest and the open woods—a nice place to take a trip to. But as for getting things done in the real world, it would hardly be the place.[26]

All these developments are occurring at a very fast pace. The speed of progress in the field is illustrated by the experience of Krueger, who in 1983 wrote a book on artificial reality discussing some ideas of creating the genre. In a 1991 update, *Artificial Reality II*, he states that much of what he had presented in the earlier volume as a vision for the future had been implemented by the time he came to write the second version.[27]

PHILOSOPHY IN VIRTUAL REALITY

Watching someone immersed in VR, say, through headset goggles, is like watching somebody going through an hallucination; we cannot know what is going on in that persons's head.[28] Timothy Leary stated that his first trip in cyberspace reminded him, in an abstract sort of way, of his first psychedelic experience.[29] Further, when two persons, even using a normal electronic screen, play a game while ignoring their immediate physical surroundings, at that moment they take the screen as the real reality. When they pause and come out of the temporary "trance," they apprehend the external reality.[30] With VR the intensity of the trance increases.

According to Michael Helm in *The Metaphysics of Virtual Reality,* the computer interface structures our thoughts and our views of reality.[31] In cyberspace, appearance becomes reality— and the world thus created can be quite arbitrary.[32] In the twentieth century, philosophical questions on the nature of reality have lost much of their earlier import. But VR, Helm observes, may resurrect interest in that philosophy. Virtual reality, he says, becomes the metaphysical engine par excellence.[33]

Krueger adds that a computer can make any imaginable experience available to a person. Artificial and virtual realities now become projections, hallucinogen that could be shared by many people. Virtual reality could therefore become a laboratory where basic questions in philosophy—What is reality? What is perception? Who am I?—can be posed. It can challenge the limits of the imagination, allowing users to dream of things that cannot be experienced, and then experience them.[34]

Cyberspace, because of its virtual worlds, Helm notes, is a metaphysical laboratory that offers a technology for examining what is meant by "reality." Metaphysical questions that are posed in VR relate to questions of presence, simulation as opposed to reality, objectification (first person or third person perspectives), and so on. Questions of ontology, those that describe the difference between real and unreal, appear very early.[35] Such issues have already come up in several guises.

Questions of the construction and nature of reality are presented at the very design level of VR. For example, how should users of the technology present themselves within VR? Should they appear to themselves in the "third person" as just one set of objects among others being displayed? Or on the other hand, should the virtual world be presented through their eyes, from their perspective, as happens to the user in "real" reality? Should there be a link between the persona on the VR set and the user, so that when the persona in VR suffers an injury, a similar injury occurs in the real person?[36]

In VR, a person's sensory input is controlled to create an internal world. However, contemporary neurology and psychology show that in fact, in their normal states, people also live in one or more internal realities, realities generated by neurological and psychological processes. In our normal lives, we build models of reality based on the sense data fed to us, combined with the information-processing capabilities of the brain. We simulate reality.[37] States of consciousness are stabilized versions of these internal virtual realities. So having this parallel, computer-generated VR also has potential in psychotherapy for developing diagnostic, inductive, training, and psychotherapeutic techniques that will both supplement and extend current methods.[38] In fact, in their ability to change the internal states to fit desired goals, VR environments also have potential as ideal vehicles for experimental psychology. The immediacy of VR has therefore already been suggested for use in psychotherapy as a desensitizing device. VR machines may be used, for example, to dissipate the anxiety students feel about examinations.[39] In such cases, various realistic psychological settings are presented to the subject at the stroke of a key.[40] In the Western tradition, having a body creates the "principle behind our separateness from one another and behind our personal presence." In cyberspace, however, this exclusivity is transgressed. One can create one's own particular body in cyberspace unconnected to corporal boundaries. Yet this stand-in self never fully represents the self of the Western perspective. At the cyberspace interface, the body gets deconstructed, as if the spirit migrates from the body; there are images and sensations without a grounded body.[41] A person can talk on a mobile phone; when this experience is augmented so that the communication is intensified—through a virtual-video phone, a person can engage in a heated discussion, arms waving, while on the move—the mind-body problem now takes a new form.[42]

In discussing how to view the body in cyberspace, David Thomas (following Serres), notes that the individual is continuously constructed at an intersection of social spaces and cannot, therefore, be considered to inhabit a particular space. These spaces are the projective (in which we see), the topological (in which we feel), the realm of hearing and communication, and so on. These spaces are fluid, and individual identities are fractured and then rewoven temporarily. But such woven images are eminently cyberspatial and so consonant with the situation of virtual realities.[43]

Old views of the nature of personal identity also vanish in the new electronic world. In cyberspace, one can remake one's identity and take myriad different forms: a male can become a female, a man may be perceived and even touched as a female—or for that matter, as a lobster, and perhaps even eaten as one, if taste buds are virtually excited! Body forms could be prepackaged together with voice and touch and rented out or sold. The multiple personality now becomes commodity fetish. But even in this flighty world, trust, ethics, and risk, the attributes of normal social interaction, would still hold.[44]

In cyberspace, therefore, there is no one identity but multiple versions of the self, blossoming everywhere. Statistical selves, ideal selves, ironic selves, half-alive selves, half-dead selves. So what we have conventionally called reality becomes but a consensus, only a stage in a technique. The twenty-first century will crack that consensus open as these new multiple realities burst forth. We will soon be looking, according to Nicole Stenger, "for the real inside a space of simulation."[45]

Generally, VR raises deep questions touching on some of the core questions in traditional philosophy. Whether there is something out there; whether what we apprehend is all real; whether there is one reality or multiple realities; whether one's perspective is the only one and, if so, whether it is the correct one; whether an individual can have multiple viewpoints on the world and, if so, which is the correct one; what is the nature of the self (is it one or many); does the self exist? VR opens up a host of deep questions and makes them available in concrete form for everyone to explore—even a child.

REALITY QUESTIONS IN SOUTH ASIAN PHILOSOPHY

In a thumbnail sketch, VR theorist Michael Helm gives a brief overview of reality in the Western intellectual trajectory: Beginning with Plato, who denigrated the unmediated physical world and said that his ideal forms were the "really real," Helm traces a path through Aristotle, the religious period of the Dark Ages, and the Renaissance, to modern discussions on the reality questions in modern physics.[46] Generally, the Western debates on what is real include such positions as dualism, empiricism, rationalism, and realism.[47] Plato gave the example of humans in a cave who mistake shadows cast on its wall for the real thing. Using the mind's eye, they are later able to see through the illusion.[48] Cyberspace, it has been said, now offers "Platonism as a working product."[49] But the largest storehouse of permutations and combinations on questions relating to reality has been raised in the South Asian tradition. VR can thus become a laboratory in which the various permutations and combinations of ontology and epistemology discussed in South Asian literature are played out.

Indian thinkers have asked many fundamental questions on the nature of truth and how (and through which means) we know it. How do we know? What is knowledge, and how do we access it? Can we access reality? What is its nature? Is there more than one reality? Do we see it as it really is, or do we superimpose our tangential, manufactured categories on it when we attempt to perceive it? Is reality material or is it something less substantial? Is there a greater reality behind the world of phenomena?[50]

Indian philosophy is divided into many different schools and sub-schools, but there are the six major schools of the orthodox tradition derived from the Vedas: the Mimamsa and the Vedanta, based directly on the Vedas and Samkhya; Yoga; Nyaya and Vaisesika, schools indirectly related to the Vedas; the "unorthodox" traditions of the Charvaks, materialists, Buddhists, and Jains. There are also schools, such as those associated with grammar, that are outside these broad schemes. So when combined with the derivatives and subbranches of these main schools, there are a variety of positions on the nature of reality and how one apprehends it.[51]

These South Asian philosophical positions, in the words of Moore, have an "almost infinite variety of philosophical concepts, methods, and attitudes . . . there are many differing approaches to reality . . . [and] . . . to truth."[52] Indian philosophy is so vast and so varied that it allows for varieties of ontological and epistemological positions. There is also a great emphasis on the inward life, which makes it congenial to the study of the inner intimate realities of VR. Indian philosophy often begins with the internal world and its relationship to the external world, between the seeming reality (the virtual reality inside the mind) and the actual reality. Observing and transforming the inner world is a primary objective of this tradition. To jump within, to observe, to "meditate" are important activities. To the Western outsider, these efforts appear as inaction, but to the observer within the tradition, they are filled with tremendous internal activity and movement of events inside the mind. The direct parallel is to seeing somebody wearing a VR helmet. To the outsider, he or she appears inactive, but inside, the user experiences a world of incessant activity.

To indicate the extent of possible variations, it is useful to traverse a few of the reality models, within the metaphysics of the Indian tradition. For this purpose I am going to make use of an elementary introduction to Western audiences which gives a flavor of the potential and possibilities in the tradition. I begin in the post-Vedic literature, after the philosophical speculations of the Vedic era, which spanned roughly 2000 BC to the seventh century BC. Let me begin with the Upanishad literature, circa the seventh century BC.[53]

Upanishads

In the Upanishads, "ultimate reality" is considered infinite, immeasurable, everlasting and an all pervading unity. Reality has two aspects, cosmic soul and an individual soul. The individual soul, *Atman*, is considered the reality that is tied to Samsara. The cosmic soul, *Brahman*, is the ultimate reality, which pervades the universe. Brahman is "that from which these beings are born, that in which, when born they live, and that into which they enter at death. Brahman is the source of all that exists as well as its supporting system and its end."[54]

The everyday, material, gross world in this formulation has only a limited reality because the sense world is on the lower rung of a series of ascending levels of reality. As the individual rises in levels of experience, he or she becomes cogniscent of Brahman, the higher reality. The subject-object duality prevents individuals from knowing Brahman, which is beyond the ken of the senses and cannot be apprehended through normal epistemological means. Brahman cannot be perceived by the senses because it has no perceptible qualities. It cannot be captured by concepts or language, cannot, therefore, be communicated in the usual methods. It cannot be described except in negative terms: illimitable, indescribable, imperceptible, ineffable. Brahman cannot be known as an object of knowledge because it is also the knowing self. One cannot categorize it or objectify it. Given the limitations of the mind, Brahman can be apprehended only by becoming one with it through immediate, direct, and personal means.[55]

Another important concept in the Upanishads is *Maya*, cosmic illusion. The world is real in the everyday sense, but for those who experience higher levels of consciousness, it is illusory. The world looks illusory to those who have had direct contact with Brahman.[56]

The concept of sheaths is another vital Upanishad teaching. There is a hierarchy of existence, although ultimately all is Brahman. Ultimate reality manifests itself in the form of five sheaths, *Kosa*. These five planes constitute an unbroken series of successively higher levels of existence and are arranged from top to bottom in continuous stages. Together the sheaths constitute a single Reality existing in all their forms and levels in one person. The higher level does not mitigate the lower one.[57] They are like the skins of an onion.

Jainism

For Jainism, reality is many-sided. One could at the most say that a thing *may be*, but not absolutely that it *is*. Jainism believes that in its own sphere each theory may be correct, yet each cannot arrogate to itself the whole truth.[58] In contrast to Western logic, which is dichotomous (that is, either this or that), Jains have a seven-valued logic, as I describe in chapter 11.

In Jainism, reality is divided into two fundamental categories, which are independent and exclusive: soul and matter. The soul experiences pleasure, pain, and knowledge, and is the conscious principle. Matter has no consciousness; it is the object of experience. In Jainsim, reality changes, acquiring new qualities while shedding older ones. Jainism also has an infinite number of souls and matter, reflecting its plural-istic and relativistic nature.[59]

Buddhism

In its long history, Buddhism has been one of the most systematic formulations of reality in the world. Its perspectives have influenced many philosophical positions across Asia. More recently, its process approach has been paralleled by a set of Western philosophers such as Charles Peirce, John Dewey, William James, Alfred North White-head, and Charles Hartshorne.[60]

In Buddhism, no substance is permanent; reality is in a continuous flux. The self is only "a stream of cognitions . . . a series of succes-sive mental and bodily processes which are impermanent."[61] In Bud-dhism, therefore, the universe is continuously changing; imperman-ence and momentariness characterize it. Nothing stable exists behind this constant change. The only reality is that of process and of becom-ing. Identity itself is an illusion, created by the succession of personal events. Individual identity is like a reel of film, which gives the appear-ance of a firm reality.[62] The movie is comprised of snapshots of tempo-rary existence which in the perceiver create the illusion of reality.[63]

In Buddhism, it is assumed that sense reality as well as words and language create illusions. The aim is to go beyond these limitations.[64] The right mindfulness of Buddhism is a direct awareness of things as they really are. Here, one analyzes oneself, knows oneself, only to realize that there is no self in the first place. This understanding is not an intellectual knowledge but an internally observed, felt knowledge. Buddhism is self-referential, so knowing oneself is to make oneself, to guide the self that is not there.[65]

Nyaya Vaisesika

There are other systems of thought and reality in the Indian system. The *Nyaya Vaisesika* affirms that the world is real, in a pluralistic sense, and objects are divided into different categories. Reality itself is dis-tinguished from the perceiving, knowing mind. The *Samkya-yoga*, on the other hand, has an evolutionary view of the universe, dividing reality into two mutually exclusive principles, *Purusa* and *Prakriti*. Prakriti has the characteristics of materiality and change, Purusa of pure conscious-ness. Purusa is in this scheme totally inactive, while Prakriti is full of energy, although nonactive.[66] According to Sankhya, there is no interac-tion between the two principles. One encompasses consciousness and the other the remainder of the universe, including human physical and mental attributes.

Vedanta

The essence of Upanishad thinking has been continued in Vedanta. Vedanta, the attempt to give the later essence of Upanishadic teachings, has several different interpretations of reality, including monism, qualified monism, and duality.[67] Sankara, the chief exponent of Vedanta, for example, holds that because Brahman alone is real, the world, to those who correctly observe, is illusion. The Creator appears as some sort of magician who creates the everyday world as an appearance not to be mistaken for reality. Sankara's description of the relationship between Brahman and the world is that of superimposition. The world is superimposed on Brahman and the perceiver sees this illusionary magic as real. Those who experience this illusionary world instead of Brahman do so in ignorance, through a false perception.[68]

Vedanta philosophy cites three mental states in its model of the mind: waking, dreaming, and deep sleep. Knowledge of all three states was vital in describing the mind completely. In this tapestry are woven ideas which, according to Vedanta, were inseparable from mind stuff.[69] According to Vedanta, these investigations of mind were guided not by whimsy, but by reason.

On the other hand, in Patanjali's yoga system there are five types of mental activities, or *Samskaras*: true cognition, false cognition, imagination, sleep, and memory. When any sort of mental activity ceases, it leaves behind a residual potency. All work of the mind is the product of these preexisting *Samskaras*. The potencies and activities form a causal series, a chain without a beginning. Samskaras have been broadly identified with the category of attitudes used by Western psychologists. Although not identical, they have been shown to be nearly so.[70]

The above very brief description of reality constructions in South Asian philosophy is only meant to give a flavor of the possibilities in the Indian systems. Over the centuries, many variations have been woven into these themes. To take one example, in the various Buddhist philosophical systems a wide tapestry of realities exist, from denial of an external world to belief in varieties of duality. Buddhist schools also describe in great detail the dynamics of the five senses as they interact with the Buddhist sixth sense, the mind, and elicit cognition and the grasp of concepts.[71]

Further, as Gerald Larson and Eliot Deutsch have observed, in India, philosophical positions were taken that anticipate to a large extent many of the positions that have emerged in recent centuries in Europe and America. For example, Sankara's Advaita Vedanta views of maya (that reality is multiple, in contrast to the correct view of an unchanging, unique, quality-less Brahman) are echoed in the work of F.H. Bradley.[72]

Clearly in the Indian traditions there are many possible realities. Worlds—their relation to individuals and how individuals access them —appear in many permutations. It would be possible to create, and then play games in, virtual models of these realities.

KNOWLEDGE OF REALITY IN SOUTH ASIAN SYSTEMS

Given these views of reality, of what is out there (that is, if indeed there is an "out" there), how do we know it? What are our means to knowledge?

The problems of reality have been pictured in both the Indian and Western traditions as oppositions of illusion and reality. In the Indian case, for example, often a rope is said to be mistaken for a snake. Plato evoked the image of prisoners in a cave, mistaking shadows on the wall for the real thing. These illusions can be either full or partial. In the Indian sense, when a person mistakes a rope for a snake, it is a full illusion. But when he mistakes the everyday world for reality it is only a partial illusion. Such illusions include that of rationality, when a person declares that reason alone gives knowledge; this is not necessarily so, reason being only a means to an end.[73]

There are varieties of knowledge problems about reality in the different South Asian schools. In any knowledge situation, there is a subject and an object with opposing characteristics that may suppress each other.[74] In the knowledge relationship there is also a triad comprised of the knower, the known, and the knowledge.[75] Some schools, particularly the Nyaya, hold that in the act of perceiving, the object of perception and the person who perceives "touch" each other.[76]

Indian philosophies also recognize two types of perceptive knowledge, determinate and indeterminate.[77] Indian philosophers differ about unconstructed knowledge, *nirvikalpa* and determinate knowledge, *savikalpa*. They believe that in indeterminate knowledge, one perceives only sensations received through sensory inputs. Determinate knowledge combines this sensory data in a concept shaped by the mind, so this knowledge gives a more complete description of the object.

The Buddhist position is that at the first instance of perception, usually visual perception, one obtains knowledge of bare things. But for Buddhists, what exists in their own right are fleeting moments that form a continuum.[78] In Buddhism, the world continuously changes; it is alive. What is real is the moment in which one lives.[79] Consciousness, according to early Buddhism, emerged from the interaction of the subjective and the objective.[80] *Alatacakra*, the "wheel of fire" illusion created when a firebrand is whirled in the air, is used as a simile in almost all the schools of Buddhist thought. The wheel, a persistence of the transitory, is symbolic of the illusionary nature of things in restless change.[81] For Nyaya philosophers there is no mental projection or construction for determinate knowledge. Determination and knowledge of a thing are but two stages of the act of perception.[82] In the Sankhya school, knowledge arises when matter and consciousness fall into each other's shadow.[83]

The ontological and epistemological positions in South Asia at least partially mirror those of VR, and those parallels have not escaped theorists in the field.

VIRTUAL REALITY AND SOUTH ASIAN PHILOSOPHY

Thus the author of a general nontechnical text on the topic, Howard Rheingold, puts the parallel as follows. The virtual reality

> experience is destined to transform us because it's an external mirror of something that Buddhists have always said, which is that the world we think we see "out there" is an illusion. We build models of the world in our mind, using the data from our sense organs and the information-processing capabilities of our brain—only we are hypnotized from birth to ignore and deny it.[84]

Rheingold's characterization of the orthodox Buddhist position is, however, an oversimplification. Orthodox Buddhism accepts the world; it rejects illusions such as the acceptance of a permanent self. On the other hand, Tracy Cochran notes parallels between VR and visualization techniques in certain callings of Buddhism.[85] The inventor of the data glove used to create tactile sensation in virtual reality puts these parallels in different words: "I think the best way to view virtual reality, especially for people who are interested in Buddhism and their own development, is as a Jungian sandbox. You have to watch how people are using VR and see how what they do with it reflects what they are trying to achieve."[86]

Indian philosophy asks many deep questions about the nature of reality, which is not just accepted as a given: Is there a reality out there? Is the world superimposed on Brahman, and does the perceiver see this illusion and magic as reality? Is what I see all in my head? Do I actually live in a dream? Do other objects outside me exist? If they do, how do I know them? Does the act of observation change these objects, so that the mental spectacles I wear change what I try to see? If I change the spectacles does what I see—that is, reality—change? Is there a grid of concepts, ideas, prior orientations through which one observes the world? If so, can one see the world unconditioned by these lenses? "True reality"—is that a contradiction in terms? For instance, the Buddhist *Madhyamika* philosophy holds that reality is always structured and that the (achievable) goal of life is to see unstructured, unconditioned reality. This is the aim in Zen-type practices of Japan and its precursors in China and South Asia, in which the mind is trained to reveal reality in an unconditioned state.

It should be noted that different South Asian philosophical schools hold their own constructions of reality to be true to the exclusion of others—even the Jains, who subscribe to a pluralistic position. South Asian philosophical debates over nearly three millennia are replete with heated and very sophisticated discussions on these issues. The vast literature on this field runs literally to the hundreds of thousands (if not to the millions) of pages, especially in the classical languages, Pali and Sanskrit. These different South Asian models could provide for rich variations on the approaches to reality.

Virtual reality, in its permutations of reality constructions, provides a mechanism for observing and modeling many different states of reality. Thus, one can have the observer and the observed in different situations and combinations. One can even have the observer observing the world of another observer, through the virtual reality helmet of someone else in the field of perception. Presumably, one could have an almost infinite regress of viewers looking at viewers looking at viewers, and if need be shifting perceptions among them, and so on. The possibilities are endless.

Let me give further examples. Various recent writers have described the modern Western situation of the self. Based on a series of in-depth interviews, Robert Lifton describes a new fluidity of the self. "We are becoming fluid and many sided. Without quite realizing it, we have been evolving a sense of self appropriate to the restlessness and flux of our time. This mode of being differs radically from that of the past, and enables us to engage in continuous exploration and personal experiment." Lifton calls this the "protean self."[87] Echoing this perspective, David Loy has noted that the identity of modern human beings is transitory and open-ended. It continuously undergoes change so that each of us is in a permanent identity crisis. This condition leads to a high degree of nervousness and an increasing feeling of homelessness: modern man has lost his home.[88]

Elster has explored the concept of the multiple self in many versions, especially taking examples from philosophy, psychology, and economics. Some descriptions take the split in the form of the split brain, left and right. Others have said that the human could be best modeled as an individual with several alternative states corresponding to the body's chemical changes. Using the same neurophysiological hardware, different individuals would occupy the body and mind depending on the particular moment.[89] And Gretchen Sliker has also posited a concept of multiple minds and subpersonalities.[90]

But Buddhism deconstructs the fluid self, one of its primary beliefs, in two ways: synchronously, into component physical and mental factors that constitute the psychophysical personality; and diachronically, deconstructing self into "dependent origination" (*pratitya-samutpada*), the linked chain of Buddhist causality. There is no stable substratum to be considered the self in Buddhism. Self is instead a stream of physical and psychological phenomena that is continually perishing.[91]

This breakdown of the self and growth of multiple selves parallels the different realities present in VR. Howard Rheingold illustrates VR's major transformations of the self and identity, pointing out that cyberspace not only offers us new experiences, but also changes how we view ourselves. In cyberspace, the body could change. Choose a new one—the old body is limiting and thus is disposable. One body serves one situation; another, entirely different, one serves elsewhere. Changing one's body at will is bound to have deep psychological impacts. The prerogative changes what one considers oneself to be—the body image. And these new bodies will have problematic ethical, social, and legal implications. If a credit card number is for all purposes an identity as far as one's cyberbank goes, one's different virtual identities could similarly have different legal identities, different from those of the "real" self.[92]

On the VR screen, one models these very different selves in the same manner that today, by clicking on a icon on a computer screen, one shifts among realities presented by various graphic representation. Similarly, VR allows the participant to navigate different selves. Such VR depictions can help bring to life Buddhist descriptions of the varying self and, in turn, those ideas can invigorate different experiments in virtual reality. And, feeding each other, their findings can illuminate the condition of our modern mind.

A fresh look at various South Asian models of reality would be very instructive in designing elaborate virtual reality systems that could traverse many realities and knowledge possibilities. One does not have to believe in the "truth" of these models to use them, in the sense that a virtual reality vision is only a possible model of many depictions of reality as presented to our senses. South Asian writers did not agree with each other's different models of reality. In fact, their rivalries were intense and sometimes bloody: witness the assassination of Aryadeva, the first century AD Sinhala principal disciple and cofounder, with Nagarjuna, of the very radical Madhyamika philosophy.

In these exercises modeling South Asian positions, we can take very much the position of a child playing with Legos, constructing different virtual realities and testing them out in the abstract. Sometimes, such game models would find practical uses. For example, quantum physics has very problematic constructions of reality. A substance can be a particle or a wave; sometimes it can exist and sometimes not. These realities defy common sense. Modeling such a world through VR, in which a researcher could observe subatomic particles through particular constructions of reality, would be instructive.

But VR models need not be restricted to making mathematical abstractions come to life. They can extend to more sensual fields, either for pleasure or for therapy. Thus, in Gibson's description, cyberspace is a powerful place of rapture and eroticism to which the participant submits.[93] In *Neuromancer*, the computer system, an abstract symbolic system, simulates the feel and touch of the lover, creating the ultimate telidonics. Sex is experienced through a simulation. There are obvious echoes here of Tantric philosophy and practice. In Tantra, the form of physical sex is indulged in, but the content of the seeming sexual encounter serves a more abstract purpose.

Recently, there has been much written and researched on states of consciousness in the West. William James wrote in the nineteenth-century,

> Our normal waking consciousness, rational consciousness as we call it, is but one special type of consciousness, whilst all about it, parted from it by the filmiest of screens, there lie potential forms of consciousness entirely different. We may go through life without suspecting their existence, but apply the requisite stimulus, and at a touch they are there in all their completeness, definite types of mentality which probably somewhere have their field of application and adaptation. No account to the universe in its totality can be final which leaves these other forms of consciousness quite disregarded. How to regard them is the question for they are so discontinuous with ordinary consciousness. Yet they determine attitudes though they cannot furnish formulas, and open a region though they fail to give a map. At any rate, they forbid a premature closing of our accounts with reality.[94]

Western philosophy is by and large based on the idea of a single state of consciousness. (However, a different state has entered Western discussions of reality: the dream state.)[95] Asian philosophies, on the other hand, describe multiple states of consciousness.[96] Western philosophizing can thus be considered only one of many possibilities.[97]

A collection of essays on revisioning philosophy notes that much empirical research has been done on altered states of consciousness, and in light of this work, everyday consciousness becomes only one level of awareness. These altered states include those attained through meditation and various Asian internal exercises, and those induced by drugs. Meditators have also claimed that their internal states can be fully appreciated only by others who have had similar experiences. Those who have not meditated find it difficult to appreciate the detailed nuances of the internal experience, which cannot be verbally expressed.

Current research results indicate that depending on these states, insights, memory, understanding, and communication vary. Asian psychologies, which recognize a broad spectrum of consciousness states, provide detailed descriptions of the phenomenology of consciousness, effects on personality, methods of attaining the various consciousness states, and their order of appearance.[98] Asian philosophers have described and expanded on a variety of such internal states. The list is long and includes Buddhist *jhanas* and yogic *samadhis*; consciousness states so strong in equanimity that stimuli cannot disturb the meditator; gentle inner stimuli, such as the faint sounds of *shabd* yoga, or subtle bliss, as in the Vipassana meditation of Buddhism. In states such as those achieved in forms of Zen, the sense of separation between the self and the world dissolves; in other states, all objects and phenomena disappear; in many others, phenomena are viewed as either modifications or expressions of consciousness. These views from the inside, as it were, have been corroborated by different practitioners (see chapter 5).[99]

VR, in allowing different views of reality, also allows us to model these different states of consciousness, and a fruitful interaction between the mechanical and the mental can again result. As mentioned above, in the Upanishads, reality is said to consist of five sheaths. One can model such sheaths in a VR setting, arranging for perception of and from the five different levels. One could play games with the different qualities ascribed to each level and to the whole, and see possible outcomes.[100]

VR raises questions about the nature of the virtual experience. In VR, traditional philosophical questions—What is reality? What is existence? Who am I? How do we know?—become important issues once again. And how do humans change in virtual reality—are they still humans? Philosophy now enters hard engineering. If the earlier searches of philosophy appeared too abstract and even too farfetched, it now seems that philosophy has entered the realm of immediacy, even of play. Scientists are also recognizing that one can better understand phenomena when they are visualized rather than only conceptualized,[101] admitting thereby, indirectly, that thought itself is only one sense, as the Buddhists claim. Generally in South Asian thought, there is much less emphasis on abstraction than on concrete experience and observation; body and mind are intertwined, as are thinking and feeling. Questions relating to the mental construct and its constructor (essential to virtual reality), are regularly dealt with in South Asian discussions, along with discussions of how the mind constructs realities, a perennial problem in Buddhist thought.

In recent decades, there has been much talk of the death of philosophy.[102] It could be enlivened with virtual reality. For the first time, an easily accessible device exists for testing theories and experiencing in a controlled situation the most fundamental of ontological and epistemological questions. And the South Asian experience can expand these discussions.

Mathematics have provided tools for abstracting from the real world and modeling, ultimately yielding logical outcomes from models. These approaches resulted in the mathematized sciences we have known since the sixteenth and seventeenth centuries. A parallel process could work in virtual realities, which are higher-order models of our world than science usually indulges in through mathematics (the world of quantum physics and the like not withstanding). We could model different worlds of reality, play around with them and see their outcomes. We could in fact even recall, speeded up through the computer, detailed verbal arguments on reality that have gone on in the South Asian arena. Conversely, if in our VR reality modeling we come across results that are already well described in the South Asian philosophical literature, then we have the outcomes of these results at the philosophical level. These descriptions would be very "thick" in that they would be the careful considerations of centuries-long heated debate, controversy, and discussion.

ELEVEN

DIGGING DEEPER: EXPLOITING
PHILOSOPHY FOR SCIENCE

Through all his life as a physicist, Einstein has regarded his own work as a search for the general laws
of nature and as closely related to the work of the
philosopher.[1]

Indian philosophy streams back to Europe, and will
produce a fundamental change in our knowledge and
thought.

—SCHOPENHAUER[2]

Kant denigrated metaphysical speculations as idle intellectual chatter unworthy of serious attention, and many twentieth-century philosophers, as varied as Wittgenstein, Heidegger, Carnap, and Ayer, have followed suit and ignored the metaphysical. Questions of reality have lost
their earlier import.[3] Philosophy, therefore, was excised from the sciences, especially since the latter lost its older image as "natural philosophy." Yet, some philosophical questions have reared their heads once
again in science during this century, in the 1920s, when the world of
quantum physics brought new problems regarding the reality that was
observed in the microworld of subatomic particles. Logical schemes like
Boolean algebra and recent variants such as fuzzy logic have also entered the hearts of computer systems. With a resurgence of evolutionary
approaches in biology and in other fields such as computer systems,
psychology, economics, and cultural studies, questions raised by process philosophers such as Whitehead and his followers have assumed a

fresh importance. With the "loss of certainty" in mathematics after Gö-del's theorem of the 1930s, fresh philosophical questions have entered the foundations of mathematics. In cognitive sciences issues of episte-mology, logic, and psychology appear, as does (at times) ontology. Given the plasticity and malleability of biotechnology and information tech-nology, ethical questions with considerable philosophical import have entered these fields. With the developments in virtual reality in the computer field, philosophical questions of ontology and epistemology have now entered the practical world at a near Nintendo familiarity. We will explore these and other issues from a South Asian perspective.

Philosophical questions that touch those of science have arisen in the Western knowledge trajectory, because modern science developed in the Western milieu. So, as a preamble to this chapter, we will have to first briefly demonstrate that South Asian philosophical quests are to be taken seriously. I will accomplish this act of legitimation by pointing out a few of the many serious studies that have come out in the last few decades on comparative East–West philosophy, including work of philosophers like Wittgenstein, Heidegger, and Kant. My treatment is not meant to be exhaustive; this is only a sampling to indicate the rich-ness of this genre.

That a search for Eastern seeds would not be a spurious exercise is seen in the comparative studies literature that has highlighted many similarities between Asian and Western concepts. B. J. Urwick has shown, in considerable detail, that there are many parallels between Plato and Indian philosophy.[4] In this vein, a modern Greek scholar, Vis-silis Vitsaxis, has demonstrated close parallels between Plato and the Upanishads.[5] A strong similarity also exists between Neoplatonism and the Vedanta and yoga systems.[6]

Within the Western tradition, writers have seen parallels between the philosophy of Sankara and the founder of modern thought, Descartes. Thus Scharfstein notes that the celebrated *cogito ergo sum* of Descartes had several Indian parallels and analogues.[7] Scharfstein singles out one philosopher, Sankara (AD 788–820), for discussion and draws signi-ficant parallels between the two. He shows that some of Descartes' ideas may be traced to Augustine and then to Neoplatonic influences. He suggests that cross-fertilization at that early stage between the South Asian and the European traditions, when Neoplatonism was influenced by South Asian thought, could have given rise to the similarities.

Though separated by nearly a millennium, some striking parallels between Fichte and the Vedanta thinker Sankara have been noted, be-tween Fichte's "subjective" idealism and Sankara's "monistic idealism." Both held the position that one could only be sure of consciousness, but they differed on the nature of consciousness and of the Absolute.[8] Kant, Heidegger, and the Upanishad philosophers, Wayne McEvilly observes, all concerned themselves with Being, but were radically different in their essence.[9]

Further comparisons have been made, with Kant and Hegel on the one hand and Nagarjuna on the other. Kant's views were derived partly from Hume, as well as from continental rationalism, although he rejected some of the latter's detailed viewpoints.[10] Yoel Hoffman notes that Nagarjuna and Kant both reject the philosophy of Being with a static reality which is eternal, as well as the philosophy of Becoming with a reality based on incessant change. Further, Hoffman finds points of convergence and divergence between the dialectics of Hegel and Nagarjuna.[11] H.H. Price has also delineated parallels between Buddhism and early twentieth-century thought.[12] Thus, commenting on one of the most important of early Buddhist texts, *The Questions of King Milinda*, he states that it "might almost have been written in Cambridge in the 1920s."

Schopenhauer's "will" is a complicated concept derived from many origins, Western as well as Eastern. His notion of a "blind will" is akin to the idea of *avidya*, or "nescience," in Buddhism.[13] In Schopenhauer's will, there are also traces of the Buddhist concept of the need for continued existence, which Buddhist psychology asserts maintains the wheel of existence.[14] Nietzsche was attracted to Buddhism in his later period because of his relationship with Schopenhauer and Deussen. Freud's central concepts were in turn modelled on Nietzsche's thinking, but Freud did not take up Nietzsche's insights into Buddhism.[15]

According to Coward, there are many similarities between Vedanta and existentialism.[16] The idea that "man and being" belong together is common to both Heidegger and the teachings of Krishna in the *Bhagavad-Gita*.[17] Similarly, the existentialist belief that relationships are complex and go beyond the false dualisms of "I and Thou" is parallel to the Vedanta view that philosophic reflection should not be one-sided, obtained instead through analysis of different dimensions of mental experience.[18]

Fred Hanna has compared the phenomenological approaches of Husserl and Heidegger with writings in Hinduism, Buddhism, and Taoism. He finds similarities and believes the Western and Eastern traditions may have much to offer each other.[19] Martin Heidegger declared of the Zen popularizer in the West, D.T. Suzuki, "If I understand [him] correctly, this is what I have been trying to say in all of my writings."[20] There are also parallels and strong similarities between Husserl's and Sarte's views of the transcendental nature of consciousness and those of Vedanta.[21]

In the case of American philosophers, Dale Riepe has observed that there were stimulating South Asian influences on the Americans William James, Charles A. Moore, Santayana, Emerson, and Irving Babbitt in the areas of epistemology and psychology and in notions of self.[22] William James's concept of the self, according to one observer, "could have been written by a Buddhist," while Whitehead's concept of peace has been described as an American formulation of Nirvana.[23] More recently, Alexander Ma, in a Ph.D. dissertation at Georgetown University, has examined the linguistic categories and rules proposed by Chomsky in his transformational generative grammar (TGG), noting that Chomskian categories and rules have certain similarities to the Buddhist logic of sense perception, inference, and syllogisms.[24]

The above samples indicate that in fact there are many areas of contact between South Asian and Western philosophical discussions. I will now take some specific Indian philosophical traditions and show how there are also possible points of interaction with problems in the sciences, giving examples of where direct results in helping the Western traditions have already been suggested. Regarding the possible uses of South Asian concepts, Helmuth Glasenapp observes that ancient Indian discussions on fundamental issues have many parallels with modern science. Glasenapp identified seven key ideas where Western and Indian beliefs overlap: 1) the enormous age of the universe (discussed in Buddhism and the Puranas); 2) the infinite number of possible worlds apart from our own (in Buddhism, Puranas); 3) worlds that exist even at an atomic level (in yoga, Vasistha); 4) the existence of infinitely small living beings like bacteria (in Jainism); 5) the importance of the subconscious in psychology (in yoga); 6) doctrines of matter in Samkhya and Buddhism similar to those of modern systems; and 7) the idea that the world appearing to our senses is not the most real. South Asia was, in addition, the first tradition to pose questions of what is objectively versus subjectively true.[25] Further, Indian philosophy was not purely theoretical or an idle search, but was always applied for practical ends.[26] Therefore, all insights and explorations made in a field—even those using the most imaginative constructs, far removed from everyday reality—had potential uses. And while most European philosophers claimed their systems were the best or the only true ones, many Indian thinkers took the position that truth manifests itself differently in different minds, allowing for a multiplicity of valid truths.[27]

Indian philosophy also holds a vast store of very complex debates and discussions on epistemology. Metaphysical disputes depend on epistemological views and the latter on theories of consciousness. Thus the problem of realism—idealism, whether objects of knowledge exist outside the act of knowing—is strongly colored by questions on the nature of knowledge and consciousness.[28] Mary Carman Rose, for example, has argued for fruitful investigative interdependence between the contemporary study of the mind and the philosophy of language, methodology, ontology, and metaphysics. In this endeavor, Eastern approaches, she suggests, would provide important inputs.[29]

With this broad background on the legitimacy of East–West philosophical interactions, let us now enter a few science-related areas which interact with the broad domain of philosophy.

DEVELOPMENTS IN LOGIC AND CAUSALITY

Bertrand Russell and Alfred North Whitehead's encyclopedic work *Principia Mathematica* is considered the most comprehensive and systematic treatment of logic today. In its intellectual importance, it is said to equal Einstein's theory of relativity. Yet, as Richard Chi observes, the study has gaps in its treatment of some key concepts. Among the omissions is a systematic treatment of "truth functions." After the publication of the first edition of the *Principia*, newer ideas began coming in to fill this lacuna. In an important article, E.L. Post estimated the total number of truth functions at 16,[30] but being a mathematician, he did not list the full sixteen, which falls to the domain of logic rather than mathematics. In 1921, in his *Tractatus Logico-Philosophicus*, Ludwig Wittgenstein supplied all sixteen truth functions, but being a philosopher, he did not elaborate their logical technicalities. Others, such as Carnap and Lukasiewicz, later extensively elaborated on these functions.

In his second edition, Russell incorporated some of Wittgenstein's ideas, yet he ignored the sixteen truth functions. Unknown to Russell and Post, however, this work had already been done by Buddhist logicians thirteen to fifteen centuries previously. Chi, a philosopher at a US university, points out that Buddhist philosopher Dignaga (circa AD 400–485) anticipated and solved these logical problems left in the air by modern philosophers. Kenneth Inada and Nolan Jacobson remark that recognition of Dignaga's contribution is long overdue.[31]

Dignaga's *Hetucakradamaru* (*The Wheel of Reasons*), a very short outline of a treatise meant for his pupils to memorize, is an extensional study of propositions. The aim of the treatise "is to find out the right kind of propositions which can be used as the major premise of a syllogism to derive a universal affirmative conclusion."[32] Dignaga also deals with classification of propositions, and Chi notes that unlike Western logic's traditional classification of propositions, Dignaga's system is nonarbitrary. Dignaga's classification was not complete, falling six short of sixteen. The full table was, however, completed by his critic and intellectual opponent Uddyotakara, who supplied the remaining propositions. This completed table, Chi points out, is the earliest logical tabulation in history and can be applied to quantificational logic and the logic of classes. This "existentional table" is analogous to the truth table of today's truth functional logic. Although they come from two branches of logic, the tables possess the same structure.

Chi observes that in spite of its reputation, Russell's work was hardly a complete system; Russell himself was aware of some of its inadequacies. Chi places the great work not in the category of Einstein's theory of relativity but of Kepler's pre-Newton work. Kepler provided the background for Newton's work, which in turn, through its inadequacies, made the way for an Einstein. The real equivalent should be "Post's Theory with Dignaga's *Hetucakra* as its avant-guarde." This combination would provide a metalogic, a logic of logic. Different elements in logic, such as the constants, variables, and rules of inference, would then follow systematically. But unfortunately, Chi notes, such a work does not yet exist. Clearly, here lies a large mine of potential knowledge to be developed, with part of its source in the Dignaga tradition. But there are further possible modern uses in South Asian logical schemes.

Aristotelian logic is twofold: *X* is either *A* or not *A*. In contrast, Jains developed a seven-valued logic called the *Sapta-bhangi*: a thing may be (*Syat asti*); a thing may not be (*Syat nasti*); a thing may or may not be (*Syat asti nasti*); a thing may be inexpressible or indescribable (*Syat avaktavyah*); maybe it is not and is inexpressible (*Syat nasti ca avaktavyah*); maybe it is, is not, and is inexpressible (*Syat asti ca nasti ca avaktavyah*).[33]

The Buddhists, on the other hand, developed a four-valued logic called *Chatuskoti*, the fourfold alternatives. F.J. Hoffman, has called it the Buddhist "four fold logic," Staal the "tetra lemma of Buddhism," and Murti the "Buddhist dialectic."[34] The Chatuskoti is expressed in such statements as

    i. The world is finite

    ii. The world is infinite

    iii. The world is both finite and infinite

    iii. The world is neither finite nor infinite[35]

Hoffman has given detailed explanations of the fourfold system in contemporary terms,[36] But a more exhaustive treatment by R.D. Gunaratne[37] shows that Chatuskoti is a "logically consistent, mutually exclusive and together exhaustive set of alternatives."[38] The fourfold system challenges the law of the excluded middle, which is central to Western philosophy. But as Gunaratne points out, Kant also rejects this law where it includes statements like "the world has a beginning in space and time" and "the world has no beginning," etc. He also points out that Hegel did not recognize the law either, because he accepted neither being nor nonbeing, seeking a third way instead.[39]

Further, in the world of quantum physics, there are situations which seem to contradict Aristotelian logic and in which Buddhist logic appears more congenial. In *Science and the Common Understanding*, published in the 1950s, Robert Oppenheimer (the head of the Manhattan Project in the 1940s, which developed the world's first atomic bomb) wrote:

> If we ask, for instance, whether the position of the electron remains the same, we must say "no"; if we ask whether the electron position changes with time, we must say "no"; if we ask whether the electron is at rest we must say "no"; if we ask whether it is in motion, we must say "no." The Buddha has given many answers when interrogated as to the conditions of a man's self after his death, but they are not familiar answers for the tradition of seventeenth and eighteenth century science.[40]

Echoing Oppenheimer, one could suggest that these nodal orientations of Buddhist four-valued logic or Jain seven-valued logic could have modern uses in physics. A rich field of possibility lies waiting to be exploited. One could speculate that, say, Boolean algebra—on which computers are based and which has a substratum of yes/no logic—could be transformed into fourfold or sevenfold logics, with an increased range of possibilities. It could also find uses in such modern computational areas as fuzzy logic, where "degrees of possibility" replace yes/no categories and become "may-be" logic.

Ivo Schneider, of the Institut für Geschichte der Naturwissenschaften (the Institute for the History of Science) of Munich University has raised the possibility of a South Asian contribution in probability comparable to that of its numerals in the development of mathematics. A historian of Greek science, Schneider's interest has been in the development in the Western tradition of a "graduated theory of the probable . . . [which] was the last step before a quantification of the probable occurred, nearly 2,000 years later."[41] These ideas were developed by the Sceptics of the mid-Academy using a nomenclature established by Aristotle.

Consistent in the development of the probability concept in the West up to the seventeenth century was the Sceptics' distinction of different intensities and grades of the probable. In these developments, Arcesilas and Carneades are important names. The former formulated a rule "of the reasonable criterion" to shape decisions in everyday life. Carneades, who followed Arcesilas, had an even more radical construction of the search for truth: life is to be conducted according to probable presentation. His conceptions of probability are as follows. Probability, in the present instance, is used in three senses: that which both is and appears true; that which is really false but appears true; and that which is at once both true and false. Hence the criterion for probability will be the apparently true presentation, which the Academics called the "probable," but sometimes the impression it makes is actually false, so the observer is compelled at times to make use of the presentation which is at once both true and false. But the rare occurrence of this kind—which imitates the truth—should not make us distrust the kind which "as a general rule" reports truly; the fact is that both our judgments and our actions are regulated by the standard of "the general rule."[42]

Carneades then develops different degrees of confirmation, varying from probable, to indubitable, to tested truth. His illustrations, Ivo Schneider notes, are "similar to or identical with examples given by Indian philosophers." The formulation provides for a continuum of possibilities in different intensities of the probable. This formulation is, according to Schneider, "without any exaggeration" the most important step in the development of probabilities before its final quantification in the seventeenth century.[43]

Schneider finds parallels with the Greek approaches of Carneades in the *Syadvada* relational theory of the Jains. Neither subscribes to a dogmatist claim for truth; only a probability is possible. The probability element is in fact engrained in the word "Syadvada," which translates literally as the "may-be" doctrine, implying the possibility of an event. To the Jains, seven categories cover all possibilities of knowledge; they are conditional on the standpoint and the degree of information around a thing. Carneades himself uses these same seven sets of possibilities but concentrates only on three. In the opposite direction, one finds his three categories in the Syadvada.[44] Schneider speculates whether these ideas could have traveled westwards and influenced the Greeks, and then he gives fragments of historical evidence that might support that idea.

The Jains' may-be, probabilistic logic also evokes a comparison with the modern concept, "fuzzy logic," a method invented by Lotfi Zadeh to get away from the traditional yes/no, true/false dichotomies of conventional logic. In its stead, Zadeh introduced the idea of graded membership within defined categories. Thus, instead of saying that A (who weighs 300 pounds) is heavy, B (at 200 pounds) is also heavy, and C (at 100 pounds) is not heavy, the three are considered within a graded membership. Using this approach, one could say that the first statement was 95 percent true, the second 45 percent true, and the last 25 percent true. Fuzzy logic has today become a method of getting computers to deal with shades of grey, "commonsense pictures of an uncertain world."[45] This approach to logical thinking has proved very useful in several applications. Fuzzy logic can help chips, programmed with many overlapping instructions with different grades of importance, to solve a variety of hitherto difficult problems. In many instances, this method has proved cheap, simple, and very effective. Fuzzy logic has become widespread within the last few years in many control devices, like household appliances and elevators.[46] Fruitful interactions could occur between fuzzy logic and the Jains' may-be logic and yield uses in the computational sphere.

Jains also tried to calculate the permutations and combinations of all possible explanations—epistemology on a grand scale. In artificial intelligence, a similar all-encompassing exercise is the attempt to generate solutions by evolutionary means. In genetic algorithms, strings of computer programs are exchanged randomly in the same way that chromosomes exchange genes, so that a solution to a stated problem is arrived at through an exhaustive evolutionary process.[47] Bringing together theories of permutations of explanations from the Jains with genetic algorithms would again be a sensitizing and, hopefully, fruitful exercise in automating modes of thought.

Today, Western logic is generally defined as the "principle of valid inference"[48] or as a "science of necessary inference."[49] It appears as "the symbolic transposition of semantic contents into a mathematical framework."[50] Underlying this approach, functions have a logical system relying on basic operations of algebraic quantification and negation, whereby linguistic formulations are easily transferred into sequences of mathematical symbols.[51] Yet this algebraic logic is not the only one around. Indian and Tibetan traditions, Ter Ellington-Wough has noted, also developed a geometrical logical system.[52] The Western algebraic system uses sequential techniques of quantification and negation; on the other hand, the Indo-Tibetan geometric system "demonstrates configurational relationships of similarity (symmetry) and congruence."[53] Both systems use equivalence, but in the algebraic it is quantitative, while in the geometric it is qualitative. While contemporary Western logic uses constructs of algebraic symbols in algebraic equations, the Indo-Tibetan system presents formulations in *mandalas*, pictorial symbols within geometric constructs.

The geometric logic of the mandala is mathematical and symbolic, its symbols pictorial. The mandala is also multivalued, not single-valued.[54] More than one set of symbolic equivalences can be shown in a diagram and, later, can be interpreted as needed at different levels. Abstract simplicity is abandoned; symbols (plants or animals, human figures, and cultural objects) are chosen because of their "richness and complexity." Combining them creates symbolic composites. If worked out, a grammar of mandalas, Ellington-Wough notes, would give linguistic meanings to the geometric conversions used in drawing a mandala, including "similarity, congruence, concentricity, bisection, quartering, subdivision, inclusion, radiation or projection, tragency, parallelism and perpendicularity."[55] A new vocabulary of logic would then emerge.

In his dissertation Alexander Ma has related the work of Buddhist logician Dharmakirti to the linguistics of Chomsky and differential calculus and to symbolic logic and predicate calculus. Ma shows similarities between Chomskian linguistic categories, the rules of TGG, and the thinking of Dharmakirti. Of the three parts of Dharmakirti's logic, namely sense-perception, inference, and syllogism, Ma concentrates on sense-perception.[56]

Causality has occupied a central position in physics from the time of Newton. With the development of quantum physics and the associated idea of indeterminateness in microphenomena, however, causality has again become problematic. It is clear that causality in the microworld of quantum phenomena is no longer the causality of the everyday world, and these issues became the subject of a major difference between Einstein and the Copenhagen school in the 1930s. The latter's interpretation of quantum physics gave a probability view of causation rejected by Einstein, who remarked that "God does not play dice."[57]

A new view of causality, which emerged in the 1960s and is now gaining influence, does not take a deterministic view of causality in quantum physics. This perspective draws on a probabilistic view of causation, stating that causes increase the probability of their effects, not necessitate them; causes only increase the probability of effects. Apart from applications in the field of indeterminism in quantum physics, this approach has uses in population sciences.[58]

But these efforts have their parallels in Asia. Thus a recent book by David Kalupahana gives a comprehensive account of Buddhist theories of causality. He concentrates on the Buddhist explanation of the operation of both causation and conditionality. Conditionality in Buddhist philosophy, Kalupahana points out, sets the environment for a probabilistic causation, which combined with a number of other causes, gives rise to an effect.[59] The Buddhist discussions on these issues of conditionality cover many facets and go into several technical tomes of source material. An exchange of ideas between one of the oldest systems of conditioned causality and the field of probabilistic causation could give new insights into current debates.

A dense seedbed of fertile ideas and metaphors exists in the South Asian tradition and could provide new conceptual elements for future growth in sciences that use logic, especially those that test knotty questions that transcend Aristotelian logic.

SYSTEMS THINKING

Systems theory is a twentieth-century development that straddles many disciplines and has in turn drawn sustenance in its development from several disciplinary orientations, including biology, engineering, physics, and aspects of philosophy. It sees the whole as greater than its constituents, with properties that can be described. Systems theory also purports to see and explain common phenomena in fields as varied as biology and engineering.

Several South Asian philosophical schemes also assume an interconnected system in their conceptions of the world. These schemes have evoked comparisons with modern systems theory. L.C. Jain, for example, has extracted a fully developed mathematical scheme based on system theoretic elements in Jain mathematics.[60] But as illustration, I will draw extensively on the doctoral work of Joanna Macy, who has connected Buddhist ideas with those of modern systems theory exemplified by a standard text in the field by Irwin Lazslo.

The Buddhist theory of "dependent co-arising," *paticca samuppada*, links Buddhism and systems theory. Buddhists hold the view that many other practices teach love and generosity, but it is the theory of dependent co-arising that distinguishes the Buddhist perspective from others. In both Buddhism and systems theory, everything undergoes fluid, dynamic change, a perpetual flowing—with no experiencer in the stream of experience. Systems theory shares this perspective, as well as others.

Systems theorists try to get at those aspects that do not change in this flux, patterns and regularities. In Buddhism, the emphasis is not on essences of things but on how things function. In systems, the boundary between an organism and its environment is porous, an idea consonant with the Buddhist no-self view. Similarly, systems theory is not interested in the "why" of things, in ultimate causes, but (as in Buddhism) only in the how. Like Buddhism, in systems theory things arise and change through the operation of reciprocal action.

Systems theory is also self-organizing. Laszlo puts forward the position that every system has its subjective dimension. The "mind" is the interiority of things, and matter is the external, observed dimension. Like the inside and outside of a building, one cannot separate the two. The systems solution to the mind-matter problem thus runs counter to the Cartesian dichotomy. Von Bertalanffy, the pioneer of general systems theory, stated this in different words as the structure of a system being the record of its past functions, that is, its partial memory. In such a formulation, present function becomes the fountainhead of future structure. In the Buddhist sense, this is the meaning of the technical word karma (not to be confused with its popular connotations).[61]

There are also parallels in the dynamics of causation. The one-way street of causation from one state to the other was the predominant mode of causation conceived both in non-Buddhistic India as well as in the post-Platonic West. Causality, in the original Buddhist formulation, was not a mere succession of states, but a co-arising: individuals are interconnected, and so thoughts and actions, in the self and in the world, co-arise.[62]

The relationship between the perceiver and the perceived in the Western context had tended to stress one or the other. Empiricists, for example, held the view that the world is the cause of our perceptions, registering itself on neutral and passive sense organs. In contrast, subjective idealists held the polar opposite view that external phenomena are only projections of our mind. Buddhists, however, hold an intermediate position linking both knower and the known, the internal world of the mind and the external world of matter; they are dependently co-arising. Perception, according to the Buddha, results when a sense organ comes into contact with a sense object within its range. These conditions constitute the window through which perception occurs; that window is in turn conditioned by consciousness, and the latter is itself conditioned by the objects perceived. Consciousness, therefore, co-arises with sensory activity. There is no consciousness before or independent of its environment; there is no consciousness in the abstract. Consciousness must always apprehend something, brought into being by its object and conditioned by it.[63]

Consciousness is demystified and deconstructed in Buddhism into something very matter-of-fact, even mundane. Thus: "Were a man to say I shall show the coming, the going, the passing away, the arising, the growth, the increase or development of consciousness apart from body, sensation, perception and volitional formations he would be speaking about something which does not exist."[64] Systems psychologist O.H. Mowrer puts these sentiments as follows: "I will venture the guess that consciousness is essentially a continuous computing device or process. The eternal question is, 'What to do? How to act?' And consciousness, as I conceive it, is the operation whereby information is continuously received, evaluated and summarized in the form of 'decisions,' 'choices,' 'intentions.'"[65] This lack of a quasi-mystical "person" in the view from within is put in the Buddhist texts as follows:

> There is no doer but the deed.
> There is no experiencer but the experience.
> Constituent parts roll on.
> This is the true and correct view.
> Vissudhi Magga[66]

In Buddhism, there is also no belief in reasoning in a vacuum because no eternal a priori truths are unconditioned by the world of the senses. Mental activity generates habits and impulses, which colors the way we cognize the world, through past experiences and associations. Perception becomes, therefore, a strongly interpretative process, with subject and object intertwined. We build our internal worlds, but only on the basis of what is fed to them.

Cognitive systems are from the point of view of systems theory open systems; they export and import matter and information. And they maintain themselves through the operation of feedback processes. Their interactions with the environment are monitored, and the results are then fed back, changing the systems' internal states. What we notice in the world is compared with what we expect from the world in terms of our past experiences. This comparison leads to changes, feedback. Our past conditions our internal codes and the constructs of the mind that structure and filter the data coming from the external world. The Buddhist equivalents of these structuring and filtering codes and constructs are *sankara*, volitional formations or impulses. They have been termed by the Buddhist scholar monk Nanananda Thero as the "ruts and grooves of our mental terrain." Both systems theory and Buddhism recognize the importance of such code-like features in perception; they are the result of experience and are changed by it.[67]

Processes of feedback govern our relationships to the external world. Negative feedback makes the external world, the environment, intelligible to us in terms of our inner constructions and ideas about it. Thus one's codes are projected on the environment such that our assumptions about it are confirmed and so serve our goals. (For example, through such projections an architect has a vision and designs a building.) When there is a mismatch between the reality outside and our constructions of it, feedback steps in. The internal cognitive structure then searches for new ways, for new codes, and so reorganizes itself to match better the reality outside.

General systems theory also assumes that what we know as mind is endemic to all systems. Such systems, when they are observed externally, are physical and material. When viewed from within they are mind, subjectively experienced. Mind and body are intertwined; they coexist and are as inseparable as two sides of a coin.[68] Laszlo puts it as follows:

> The phenomenon of mind is neither an intrusion into the cosmos from some outside agency, nor the emergence of something out of nothing. Mind is but the internal aspect of the connectivity of systems within the matrix. . . . . The mind as knower is continuous with the rest of the universe as known. Hence in this metaphysics there is no gap between subject and object. These terms refer to arbitrarily abstracted entities.[69]

A cell's view of the world is therefore different from that of a higher organism like a human, or from that of a bat (as biologists have pointed out). The view from within, the internal subjectivity, depends on the external interactions, because mind and body, the two sides of the same coin, are interconnected.[70] A cell's "mind events," Laszlo points out, "must feel entirely different from ours, yet they can be mind events nevertheless, i.e. types of sensations correlated with, but different from, physical processes."[71]

In Buddhism, the dynamics of the changing mind-body complex

come from the concept of action, karma. The continuity of our mind-body while it changes is put by the Buddha as follows: "This body, brethren, is not your own, neither is it that of any others. It should be regarded as brought about by actions of the past, by plans, by volitions, by feelings."[72] In another instance, the Buddha puts this centrality of action in these words:

> My action is my possession,
> My action is my inheritance,
> My action is the womb which bears me,
> My action is my refuge.[73]

In the systems approach, too, there is an emphasis on action. The external environment alone cannot change the behavior of a system. It has to interact with the internal state, which is a congealing of a past history. The systems view is that we are not just victims of the past; action appears as choice. As the internal organization increases in complexity, the system becomes less influenced by the environment.[74]

Systems theory was fed by many streams, and it is by no means a complete science like physics and biology. At times an eclecticism governs its approaches, but it presents fertile ground for enrichment by systems-related ideas from South Asia. One area systems theory has drawn upon as well as contributed to is biology, or more precisely, evolutionary theory.

EVOLUTIONARY PROCESSES

The idea of evolution is experiencing a fresh impetus and has reached new areas, with implications not only for biology but in cognition, culture, and computing artefacts. Internationally, many academic groups and fora have arisen over the last decade or so to discuss these issues, and there have been hundreds of articles on evolution, varying from those on anthropology and economics, to biology and AI. There have also been at least a half dozen books on "general evolution" (including one by the present author).[75] The subject is in ferment, judged from the contents of current conferences and journal material. New ideas are emerging, straddling many fields, including those of philosophy and ontology. Practitioners and theorists are foraging across disciplinary boundaries, dredging material from past authors in search of fresh insights. It is here that another fruitful field exists for importing ideas across civilizational boundaries. But before we discuss some of the possibilities, it is useful to recall briefly the history of the Western ideas of evolution.

Evolutionary ideas existed among both pre-Socratic Greek and Indian thinkers. But evolutionary thinking in the Greek tradition was brought to an abrupt end by the anti-evolutionist biases of Plato and Aristotle. Plato viewed the real world as consisting of unchanging forms or archetypes caught solely by thought. Aristotle, on the other hand, viewed the physical world as a hierarchy consisting of kinds of things. For Aristotle the universe did not change and was eternal.[76] Evolutionary thinking was reborn in the West after the Renaissance with advances in natural sciences. The writings of Descartes (1644), Kant (1755), and Laplace (1796) were similar to the later ideas of Lyell (1830) and Wallace and Darwin, after whom begins the modern scientific period of evolutionary theory.[77]

In the Indian tradition, the idea of evolution is found in the Chandogya Upanishad, which states that originally all manifold things existed in an unmanifested condition. Everything sprang from this unmanifested Being, along many paths: fire was produced, then from that water, and from that Earth. From combinations and unions of these elements, everything else was produced.[78]

From the Samkhya point of view, evolution is viewed (in the words of B.N. Seal) as the

> differentiation in the integrated (*samsrsta-viveka*). In other words, the process of evolution consists in the development of the differentiated (*vaisamya*) within the undifferentiated (*samyavastha*), of the determinate (*visesa*) within the incoherent (*ayutasiddha*). The evolutionary series is subject to a definite law which it can not overstep (*parinama-krama-niyma*). The order of succession is not from whole to parts, nor from parts to whole, but ever from a relatively less differentiated, less determinate, more coherent whole. That the process of differentiation evolves out of homogeneity, separate or unrelated parts, which are then integrated into a whole, and that this whole again breaks up by fresh differentiation into isolated factors for a subsequent reintegration and so on *ad infinitum* is a fundamental misconception of the *course of material evolution*. That the antithesis stands over against a thesis, and that the synthesis supervenes and imposes unity *ab extra* on these two independent and mutually hostile moments, is the same radical misconception as regards *the dialectical form* of cosmic development. On the *Sankhya* view, increasing differentiation proceeds *pari passu* with increasing integration within the evolving whole, so that by this twofold process what was an incoherent indeterminate homogeneous whole evolves into a coherent determinate heterogeneous whole.[79]

An important aspect of evolutionary systems is that they are lineages stretching amid change from the past to the present and into the future. There is change, but continuity. The dynamics of these lineages, especially their "semiotics" and "world views," are in some recent formulations coming to the forefront of discussions.[80] There are sometimes direct parallels to these dynamics in some of the philosophical aspects of South Asian religious discourse.

The attention on general evolution has principally centred on biology and culture. These two discourses on long-duration evolutionary chains in modern science—namely of biology and culture—is parallelled by South Asian discussions on the chain of rebirth, *samsara*. One does not have to believe in samsara (I do not) to follow the philosophical discussions, just as one does not have to follow a concept called God (again, I do not) to appreciate the beauty of some of the Western philosophical discussions around the topic. These philosophical discussions on samsara have several interesting parallels with contemporary ones on evolution.

The evolutionary characteristics of the biological lineage are well known. As it courses down through time, this lineage bifurcates into new species that fit better into changing environmental niches. Sometimes there is a sudden disjuncture away from the smooth speciating pattern, due to punctuated equilibria. There is also a phylogenetic ascendancy in the set of lineages, the more complex DNA carrying more information occurring later, there having been more time to acquire a larger battery of responses to the environment. The information is also considered to be in a process of self-organizing, undergoing "autopoiesis," beginning with the self-organization of the earliest molecules from inanimate matter. The encoded information is also a window to the external world and has been seen by some theorists as ecological perspectives or "theories" that change with the environment. Such a gestalt encoded in a genome is seen as a means of structuring the world outside, having its own "subjectivity" on the world, its own "world view" to the environment outside. As the lineages course through time, these "subjectivities" and world views change.

The second information lineage with evolutionary characteristics is that of human culture, which encodes another set of reactions to the external environment apart from the genetic one. It is a means of responding and adapting to the environment, and like genetic information, it is passed down diachronically from generation to generation. The characteristics of these lineages have been described in detail by social scientists as to their adaptation, construction through historical encounters with the environment, transmissions down a lineage, and collective subjectivities in the form of, say, class consciousness or orientation to a profession.

The most rigorous work in the cultural lineages has been in the social studies of one social group, scientists—and their culture. Over the last two decades, the mechanics by which scientific knowledge is constructed though the operation of social factors have been described with an evolutionary epistemology of knowledge. In this formulation, the production of scientific knowledge is influenced by factors in the social environment as well from social factors within the social group doing science. These influences help push science in particular directions, changing its content as well as the trajectory of the lineage. The lineage that results is a tree of knowledge which is parallel to the tree of genetic information, having many characteristics in common with the latter, including speciation (in this case to disciplines and subdisciplines), continuation of a core memory (in the form of the discipline's core knowledge), a process of self-construction (through the continuous social interaction between its members), "subjectivities" attached to each lineage (in the form, say, of a profession's or class's particular view on the world), and sudden disjunctures in cognition in the lineage, in the form of changes in paradigms. These are all common evolutionary characteristics that culture shares with biology (and can be shown to share also with an incipient lineage of information associated with computers).[81]

Some of the central arguments in these contemporary discussions on evolution are derived from *What Is Life?*, by the physicist Schrödinger. In this book, Schrödinger relates life to physical processes, namely, the reduction of entropy. His philosophical outlook here is explicitly South Asian, specifically Vedanta. In the epilogue, "On Determinism and Free Will," Schrödinger states that the individual's free will can be reconciled with the determinism of the world through the Upanishadic equation Atman = Brahman.[82] The influence of Vedantic philosophy on the scientific thought of Schrödinger has been detailed recently by Ranjit Nair.[83]

Some of the most pregnant possibilities for the use of South Asian inputs in evolutionary thought is in the concept of samsara, the flow of life across time. It can be shown that philosophical discussions of samsara reveal many common characteristics with evolutionary lineages. For example, all the discourses on samsara deal with a memory trace handed down in a lineage from the past to the present and on to the future. Although discussions of "lineages" in samsara and in culture and biology deal with different problems posed for different purposes, they have many similarities in the core problems that they address and the answers that they provide, suggesting that at the level of philosophy and at some deeper level, they deal with common issues. It is useful to trace these similarities and then to suggest cross-fertilization possibilities.

"The Samsaric Lineages"

The centrality of a chain of birth and rebirth is common to all the main South Asian traditions—Hindu, Jain, or Buddhist. In all these religious traditions, there is a "beginning-less chain of existences following one another."[84] This "beginning-less" chain is subscribed to even by views holding that souls emerge from the highest deity, Vishnu. Vishnu is not considered a primary beginning point for souls, but rather a periodic emergence from God, where it resides.[85] Similarly, in Buddhism samsara is beginning-less, and a series of I's flows like a stream through a chain of cause and effect, without beginning or end.[86]

The central dynamic sustaining the lineage of rebirths is karma. Many popular readings of karma as strictly predestination—"fate"—are in fact misleading;[87] in its pure philosophical sense, karma only refers to action.[88] Karma is therefore, in this sense, neither predestination nor determinism. An individual is both a slave of the past and the master of the future, bound by fate, yet the instrument of his fate.[89] Karma has a fluid, flexible, and dynamic character whose full working out cannot be apprehended in advance.[90]

Every karmic action results in a residue in a person's memory, and when combined, the residue of these actions forms the essential habit patterns that govern a personality, the way a person reacts or perceives the world. Karmic action with the environment results in a memory that is "deposited" for the future guidance of its lineage. These actions fall into dichotomies such as "good" and "evil," similar to adaptive and maladaptive interactions with the environment in other lineages. Life becomes a continuous accumulation of actions or their results.[91] Karma may be operative through the present life, may be held over from "past lives," and could be stored to be used in "future lives."[92] This "Pool of Karma" (karmasayah) is never empty; it is constantly fed by good and evil thoughts that keep it more or less full all the time.[93]

The "engine," the motive force for the samsaric lineage is therefore karma—action carried out by the individual in the passage of daily living—what in the case of genetic or cultural systems would be called information of the past. The interactions with the environment, of these other two lineages whose results are eventually deposited as information—memory—is carried forward. Karma is the totality of thoughts, good and evil, created in a mind, acting as a kind of psychic memory that carries the body through the cycle of rebirths.[94] In all traditions, karma is latent "rebirth." Rebirth is karma that has become active and manifest.[95] Karma decides the rebirth and the appropriate new environment for the reborn entity.[96] And what is "reborn" in most interpretations is a bundle of attributes, a historical outcome of countless lives.[97] For Buddhists, "Beings are owners of karmas, heirs of karmas, they have karmas as their progenitor, karmas as their kin, karmas as their homing-place."[98] In Buddhism, the karmic transmission process is understood as a subconscious "life stream" (bhavanga-sota) in which are collected life's impressions and experiences.[99]

Yet in the South Asian schemes, not all states are due to previous karma. Certain experiences (*vedayita*) arise from purely physical factors.[100] There would also be occasional "accidents" in which a person may get results not attributed to past actions.[101]

The samsaric lineage carries forward accumulated information—"merit" in Buddhism—that is the result of its karmic actions. This information affects the direction of the lineage, but in the nonpopular formulation, it has also a novelty-creating indeterminateness to it. The present is a result of the past, but the trajectory of the future, although restrained by the past, is still open. This openness and creativity has parallels to what has been termed autopoiesis and self-organization in the biological and social fields.[102]

The vehicle for this memory in most South Asian traditions is a material or nonmaterial soul, and in the Buddhist tradition, a "no-soul." These South Asian descriptions of what is transmitted have many parallels to descriptions of the transfer of biological and cultural information. As illustration let us take Richard Dawkins' view (in his evocative book *The Selfish Gene*) of the body simply as a carrier for the gene:

> [Genes] swarm . . . safe inside gigantic lumbering robots, manipulating the outside world by remote control. . . . Their preservation is the sole reason for our existence. They skip like chamois, free and untrammelled down the generations, temporarily brought together in throw-away survival machines, immortal coils shuffling off an endless succession of mortal ones as they forge toward their separate eternities.[103]

Clearly the above description has many parallels to the view of the Hindu "soul" in relation to the body. Thus: "As a person casts off worn-out garments and puts on others that are new, so does the incarnate soul cast off worn-out bodies and enter into others that are new" (*Bhagavad-Gita*).[104] The Buddhists do not have a genome-like structure, like those in the Atman theories, which physically carries the information. Yet here, too, "Man quits his mortal frame, when joy in life is past / Even as a snake is wont, its worn out slough to cast" (*Uraga Jataka*).[105]

In Buddhism, life as a vital principle is denied, and change and process is considered the essence of life. Life is not a property, qualification, or condition but rather an action and reaction to its environment. Life is a process constantly evolving from a set of conditions and reforming depending on other conditions.[106] The Buddhist continuity across births is not through a soul or a genome-like structure but through a stream of becoming (*bhava sota*).[107] In this sense the Buddhist view of the no-soul is nearer to the autopoietic views of recent theorists in biology—of how the future is conditioned by the past but is created through a process of self-construction, a process of becoming.[108]

In both the biological and cultural spheres, there is an evolutionary process with phylogenesis, a tendency to build up more information with the passage of time. The samsaric journey, too, has an evolutionary direction, with greater memory being accumulated along the way. The 'I' going through the samsaric lineage, being the result of past actions, carries with it certain attitudes and so particular means of dealing with the environment to which it is born. This is a "world view," a subjectivity from the lineage's point of view, as it were, parallelling similar subjectivities that have been described in the two other lineages.

While the samsaric lineage has the characteristics of action with its environment, the creation of information, memory retention, phylogenesis, and associated subjectivity in common with the other lineages there are differences, too. The lineage does not break up into branches, and no speciation occurs as in biology and culture; the individual entity does not, with the passage of time, divide into sublineages, as in speciation in biology.

All three lineages—genetic, cultural, and samsaric—have a historical continuity, coursing through time and conserving elements of their memory when they interact with their environment. Information accumulated from earlier encounters with the environment is connected with present information,[109] and each information lineage goes through processes similar to what has been termed autopoiesis, the process of self-organization and creation of novelty.[110] As the lineage moves forward, it creates novel arrangements, and there is a phylogenetic ascendancy which results, with the passage of time, in more complex information systems. Associated with the information in the lineage is a world view, its own "egocentricity" (to use the word derived from Morin).[111] Such an egocentricity is normally built up historically in gradual steps within a given lineage.

Now, could there be any areas that have been explored in the samsaric lineages which could be used to throw light on the common characteristics of others, like the genetic or the cultural? More specifically, are there any insights, hints, and metaphors that arise from South Asian discourses on their lineages that could be transferred to the other domains? There could well be, given the long philosophical obsession in South Asian traditions with this class of problems. Modern science has always foraged the past Western philosophical tradition for fresh ideas, as in Dalton's return to the Greeks' concepts of atoms or in contemporary theoretical physicists' return to core intellectual questions in the Greek tradition. The Greeks, however, did not develop a sophisticated tradition of evolutionary thought, both Plato and Aristotle having an anti-evolutionist bias.

The various philosophical explorations of the nature of samsaric lineages in the South Asian tradition have been made with different assumptions, which then were driven to their logical conclusions. These provide a seedbed of ideas on the general nature of lineages as well as a set of formalizations on time-directional processes. But lineages are process systems and recent theorists of general evolution have turned to process philosophy for guidance.

Process Philosophies and Evolution

Buddhism as an older process philosophy is intimately tied to concepts of time and temporality: not only are we born and destined to die, but we live our lives as a series of experiential moments.[112] Whitehead resurrected process philosophy in the West this century, in Heraclitus' view that all things flow.[113] According to Robert Neville, Buddhism, as a process philosophy with the longest unbroken tradition in the world, has much to offer modern process philosophers. Whitehead's work becomes therefore the first fully developed metaphysical system in which Buddhism finds general sympathy. His views of process consist of momentary occasions of experience. Buddhism, on its part, brings together the internal flow of phenomena and the external environment in which it operates, as well as the interconnectedness of all phenomena, *patichcha samuppada*.[114]

The contemporary process philosopher Charles Hartshorne (who with Peirce is one of the philosophers influencing the new evolutionary thinking) finds many commonalities between Whitehead and Buddhism:

> Ancient Buddhism and Whitehead agree, against Aristotle, and also against Bergson, but in nearer agreement with Plato's Timaeus, that concrete reality consists of momentary actualities that successively become, this succession being what we call change.

> If we are looking for concrete definite unitary wholes of reality, we should recognize that the individual now is always a new such whole. The Buddhists, whom Peirce admired, saw this.[115]

In evolution, things flow as a process, have an external and an internal dimension, and are interconnected. Recent discussions in general evolution, including the semiotics of information, have seen interactions with philosophical positions like those of Whitehead. There are discussions exploring a variety of positions in process philosophy at the cutting edge of general evolutionary theory today.[116] Some of the philosophical positions evoked in evolutionary discussions are those of Whitehead, Peirce, and Mach, figures who have been directly or indirectly linked to Buddhism. It is clear that recourse to direct Buddhist philosophical discussions could help in some of the issues facing evolutionary theory.

Philosophical concepts are somewhat like pure mathematics; they can sometimes be transferred into the applied field as explanatory devices in the sciences. Logical schemes have been transferred, for example, to the field of computer science and artificial intelligence. Similarly, elements in South Asian philosophy relating to evolution can be a storehouse for potential new developments in evolutionary theory. One example that comes into mind is Buddhist observational practice on one flow system—the thought realm, the carrier of culture—and discussions on the degrees of freedom and determinateness in our thought flows.[117] How much we are restrained by the past, and how open is the future are central problems for all lineages.

Another application of Buddhist process philosophy in evolution, especially with relation to ethics, occurs when in the foreseeable future, direct intervention in the evolutionary process occurs through genetic engineering. A whole host of philosophical, ethical, and scientific questions are raised as a result (see chapter 9, above).

I end this chapter on the uses of South Asian philosophy for modern science by mentioning a recent book that has used Buddhist approaches to solve a common set of problems in evolution, cognitive sciences, and AI.

COGNITIVE SCIENCE, ARTIFICIAL INTELLIGENCE

Cognitive science is a new multidisciplinary matrix that covers a variety of subjects, such as neuroscience, linguistics, cognitive psychology, AI, and aspects of philosophy.[118] There is no tight distinction between philosophy and cognitive science; philosophers like Descartes, Locke, Leibniz, Hume, Kant, and Husserl were, in this sense, proto–cognitive scientists. In the words of Jerry Fodor, "In intellectual history everything happens twice, first as philosophy then as cognitive science."[119] In an important recent book on the subject, *Embodied Mind*, Francisco Varela, Evan Thompson, and Eleanor Rosch note that cognitive science is, however, Janus-faced: one face is turned toward nature and so sees cognitive processes as behavior, while the other face is turned toward the human world and so has a prelidiction to see the cognition as experience.[120] Their approach continues a tradition set by the French philosopher Merleau-Ponty, in which Western scientific practice now requires that one sees one's body as both a physical structure and lived experiential structure, having both biological and phenomenological characteristics, an outer and an inner existence.[121]

Cognitive science has realized over the years that the cognizing subject is not a unified entity. This realization has many parallels with the Buddhist view of a nonunified self in its meditative practices and psychology. Varela and colleagues propose a bridge between the mind as discussed in science and the mind of experience, by means of a dialogue between Buddhist meditative practice and cognitive science. They suggest taking a bold step in terms of philosophical traditions. Having summarized some of the leading Western philosophical positions and finding that they no longer have a foundational value with respect to other cultural activities, such as art or science, Varela, Thompson, and Rosch suggest dipping into other philosophical traditions—specifically, Buddhism and its views of the nonself.

The Buddhist no-self doctrine, they point out, helps one understand the process of the fragmentation of the self that is depicted in both cognitivism and connectionism. The Buddhist nondualistic philosophical position in between the subject and the object ("middle way") has resonance with some recent ideas in cognitive science and with the perspective of Merleau-Ponty.[122] The authors clear the usual popular misconceptions in the West about meditation and point out the observational character of Buddhist mindfulness/awareness practice. The Buddhist experiences of observing the mind are, they note, in the tradition of scientific observation and can lead to discoveries about the behavior and nature of the mind. Their position is that if cognitive science is to incorporate human experience, then it must have a means of exploring this dimension—a means already provided by the Buddhist practice.[123]

When one is mindful, one has an embodied, open-ended reflection. This reflection is not just on experience; it becomes a type of experience itself, cutting through habitual thought processes and preconceptions. The Cartesian dichotomy of mind and body is now found to be only a limited point of view, the perspective of a disassociated mind and body. The wandering mind, turning away from the body, is a limited experience that can be changed. In fact, mind and body can be brought together, and then mindfulness can result. Descartes' split was the outcome of a specific experience of unmindful, disembodied reflection. Reflection, even from a theoretical perspective, need not be disembodied.

"Cognitivism," so-called by Varela and his colleagues to describe the hitherto dominant tendency in cognitive science, assumes that the brain can be seen as a device that manipulates symbols. In this tendency computers provide for a model of how the brain works, and so cognition can be discussed as computations of physical representations in the brain. A technological embodiment of the cognitivist approach is in artificial intelligence, the search for human solutions in special software. An important result of this approach is the realization that there can be cognition without recourse to an entity called the self.[124]

But persons have an experience of the self, what the authors call "the I of the [mental] storm." The idea of a nonself is central to all forms of

Buddhism. The individual has an experience of a self, but Buddhist analysis and observation leads one to dismiss the self as something abiding and real. There is a feeling of self in everyday reality, but this is illusory. In particular, the Buddhist *Abhidharma* texts on psychological analysis examine the nature of self in great detail and show its non-substantiality. The position of *Embodied Mind* is that this perspective, once adopted and internalized by science, can be transformative to the AI discipline.[125]

It has dawned gradually on AI practitioners that the more difficult thing was to program not the activities of, say, scientists but those of a baby. It was apparent that a more fundamental and difficult type of intelligence was involved in the mental world of the baby as it learned to make sense of the world. The expert's and child's roles have to be reversed in the scale of performance, allowing for a new paradigm in looking at the cognitive world, in which the brain becomes the seat of metaphors and ideas of cognitive science. Some of these newer approaches have a definite parallel, the authors note, with the central Buddhist idea of no-self.

Although there is no ultimate permanent self, there is coherence in our lives. We act and feel as if we had a self in a process of "dependent co-arising" (*pratityasamautpada* in Sanskrit), a Buddhist concept that expresses how transitory yet emergent properties arise. The Buddhist category of psychological causality, karma, is central to understanding how this process of co-dependent origination occurs. Under the type of analysis done by the Buddhist *Abhidharma* approach, there is a description of how direct experience emerges without recourse to the idea of a self. In this it has obvious parallels with the emergent properties of "societies of mind," an AI approach put forward by Marvin Minsky. The authors find that Minsky also has no recourse to an abiding self except as a society of changing ideas. This position is very close to the Buddhist view, but although cognitive science says that there is no self, it is "virtually forced" (in the words of Minsky) to maintain the concept. But the Buddhist tradition does not require that one maintains this fiction. It also offers freedom of action different from the idea of freedom as randomness usually resorted to by computer specialists.[126]

Today the world of representation in the cognitivist approach is undergoing major changes. Developments in cognitive research are requiring that one shift away from the idea of the world as independent and extrinsic, to the view that the world is inseparable from the structures that get involved in self-modification. The world is not represented in self-organizing systems; it is "enacted" as a domain that is inseparable from the structure that is embodied by the cognitive system. Varela, Thompson, and Rosch find an exposition of this approach in the theory and practice of the Buddhist "middle way," which denies the project of foundationalism, namely that not only is there an abiding perceiving subject, but that the world itself is pre-given and ready-made for one to discover and experience.[127]

The Buddhist approach incorporates the everyday world of common sense. Common sense, which has been considered largely some sort of folk experience, is now coming into the forefront, especially in Continental philosophy. The approach has pointed out how one's knowledge depends on being in a world inseparable from one's body, language, and history—in other words, one's embodiment. In this formulation common sense becomes our personal and social history. In such a situation, mind and world, the knower and the known are in a process of dependent co-origination. This enaction and the bringing forth of different worlds can thereby also be seen in the structural couplings that different animals have with the world. As a consequence, birds, fishes, insects and primates have enacted or brought forth different means of perceiving the world—seen, for example, in their different worlds of color. Different histories have given rise to a rich diversity, different forms adapting to varying ecological niches.[128]

Varela and his co-authors now move their attention to a problem in another discipline with a parallel set of problems: adaptation in evolutionary biology. The adaptationist program in its neo-Darwinist formulation has come under criticism recently. They point out that what cognitivist theory is to cognitive science, neo-Darwinism is to evolutionary theory. The neo-Darwinian approach emphasizes natural selection as the main factor in biological evolution. It admits that there are other factors that influence evolution but plays down their importance. The authors' criticism takes into account these other factors, such as linkages of genes with each other, the influence of factors other than genetic in the development of an embryo, the random drift of genes in the genetic composition of organisms, the fact that organisms sometimes do not change for considerable periods of time even though their environments change, and that the individual is not the only level of selection.

As a result of this type of criticism, evolutionary biologists have begun to move away from an emphasis on fitness alone to a larger, and as yet incompletely formulated, theoretical perspective. *Embodied Mind* provides a point of view on this emerging new perspective on evolution from a Buddhist-inspired viewpoint. The authors find that there is an exact parallel here to their discussions on the current impasse in some of the cognitive science positions that work on the representative position. Rather crudely and awkwardly put, representationism in cognitive science is the homologue of adaptationism in evolutionary theory. Their suggested approach to evolution would still include selection but in a lesser role, operating to discard what is not in keeping with reproduction and survival. Yet selection is not a constant guide to fine-tuning fitness. Natural selection becomes only a guarantor that what ensues from the intermingling of genetic variety in a population satisfies only two constraints, that of reproduction and survival. This perspective shifts interest to the large variety and diversity found in biology. The evolutionary process under these circumstances is thus only satisfying and not optimizing. This formulation is also compatible with a recent description from a post-Darwinian perspective, that evolution allows one to put together parts and items in complex collections.

In the conventional view it is assumed that the environment exists prior to the organism, to which the latter fits in. This is of course not so. The organism is not by some mysterious means parachuted into the environment. The environment itself is changing. As a leading theoretician, Richard Lewontin, puts it:

> The organism and the environment are not actually separately determined. The environment is not a structure imposed on living beings from the outside but is in fact a creation of those beings. The environment is not an autonomous process but a reflection of the biology of the species. Just as there is no organism without an environment, so there is no environment without an organism.[129]

Living beings and the environment are linked in a process of co-determination or mutual specification. In this light, environmental features are not simply external features that have to be internalized by the organism, as both the viewpoints of representationism and adaptation would have it. Environmental features are themselves results of a long history of co-determination. In this light, the organism becomes, as Lewontin puts it, both the subject and object of evolution.[130] The processes of co-evolution result in the environment not being pregiven but being enacted and brought into being through a process of coupling.

Varela and colleagues now point out that taking the world to be pregiven and the organism to be representing or adapting could be considered a dualistic approach. Buddhism, however, transcends this duality in its co-determination and middle way perspective, which is also found in the newer AI approaches—in cognitive technology as opposed to cognitive science—and in some recent robotic research. The authors highlight the attempt of Rodney Brooks to develop "Intelligence without Representation" in robotics. Instead of representing the world in the form of an internal program in the computer, Brooks' approach is to use "the world as its own model" and develop a class of autonomous robots that will coexist in the world with humans. In his approach to this problem Brooks does not break down the system into functions, as is the usual case. He instead breaks it down to its component activities. This approach has yielded several robots (which he called "Creatures")[131] and would allow (Brooks believes), in the relatively short term of a few years, for the arrival of sufficiently intelligent Creatures. Brooks' work is also an embodiment of the enactive mode outlined by the authors: the Creature brings forth its world as it develops.[132]

The authors of *Embodied Mind* have fruitfully applied what they have called the enactive approach, derived from Buddhist practice, to several areas of science. They believe that the inner directions of research in several fields, such as cognitive psychology, neuroscience, linguistics, artificial intelligence, evolutionary theory, and immunology are driving toward such an enactive orientation. The result is a breakdown of the strict subject/object dichotomy. The enactive program is therefore no longer to be dismissed as the quirk of a few eccentrics: cognition is embodied action, and is thus tied to lived histories. Further, these lived histories are the result of evolution as a process of co-determination, incorporating what the authors have called natural drift.[133]

These researchers applied a Buddhist-inspired perspective to three different realms—cognitive science, evolution, and AI robotics—because all three realms deal with a similar class of problems. They are involved in problems that deal with the interaction of an outside and an inside world and the lack of a substantive, permanent base from which to view these interactions. The authors also ruminate on further scientific developments from this Buddhist perspective. They believe Buddhism, especially the perspective from a nonself point of view, can shed some light on the collapse of foundations in several disciplines. They claim that "at the very least, the journey of Buddhism to the West provides some of the resources we need to pursue consistently our own cultural and scientific premises to the point that we no longer need and desire foundations and so can take up the further tasks of building and dwelling in worlds without ground."[134]

This approach to the living world has been echoed by other recent writers. Andrew McLaughlin has observed that the existing image of nature in science, derived partly from Descartes and Bacon, is "neutral," devoid of value and meaning, and creates problems in the current world. He has therefore suggested that the view of nature as an interconnected network, which is the view of both ecology and Buddhism, be adopted. This interpretation gives a wider framework and a sufficient foundation to both practical and ethical aspects of human perceptions of, and relations with, nature.[135]

In a similar vein, Linda Olds has examined a variety of non-Western and contemporary ontological metaphors of interrelatedness. She finds that positions as varied as feminist theology, Buddhism, Taoism, the process philosophies of Whitehead and Hartshorne, and the systems theory and philosophical perspectives of Laszlo and Bateson express, in similar fashion, aspects of the modern condition and its relation to the physical world. These similarities provide important entry points for an ontology and ethics of interrelatedness.[136]

This rather speculative chapter has explored areas where the rich philosophical traditions of South Asia could be mined for modern science. Here, as in earlier chapters, it is clear that science is intertwined with the philosophical condition of humans.

TWELVE

TOWARD A NEW MILLENNIUM

The previous six chapters have given concrete examples of how South Asian knowledge has been or could be harnessed in some specific areas. But these suggestions do not exhaust the possibilities; they only suggest the potential. The examples have come from my own familiarity with some of the subject areas, as well as from what has been already attempted. So if this mining approach is followed diligently, what lies in store for the future?

I have mentioned that there are two broad means of acquiring mined knowledge: by splicing crucial nodal points of South Asian knowledge directly into the contemporary knowledge structure, and—unexamined to this point—the uses of metaphor. Having devoted several chapters to mining of the first kind, I will devote somewhat more space here to how to tap into the rich lode of metaphors in the South Asian tradition. Let me begin by delineating in some detail how metaphors, "the pregnant mother to science," have been used hitherto, and from these lessons I will suggest broad methods to tap South Asian systems.

THE TRANSDISCIPLINARY CONSTRUCTION OF SCIENCE

Where existing theory does not match reality, disciplines open up. At these crucial times, new views and new metaphors cross disciplines, and new theoretical perspectives are formulated. How such imported metaphors function has been the subject of some recent work.

Using Metaphors

It has been noted that Aristotle believed that metaphors could not convey knowledge. But from the Renaissance onwards, as the heritage of Aristotle has been dismantled, this view has been increasingly rejected. Vico, Nietzsche, Coleridge, and Croce all saw the importance of metaphors in understanding facts. Metaphors, it seemed, were as efficient in communicating ideas in the natural sciences as they were in poetry.[1] Several decades ago, Pepper demonstrated that theories in science originate in metaphors.[2] His root metaphor theory influenced other metaphilosophers like Rorty and Nozick.[3]

Two principal aspects shape discussion of the use of metaphors. In the ellipsis or comparison theory, metaphors are actually similes, while the interactions or juxtaposition theory holds that metaphors, apart from any simile-like characteristics, also generate new meanings.[4] There are many examples of explicit or implicit metaphors in science, varying from metaphors related to gender to those borrowed from other disciplines, including metaphors of the mind used in the physical sciences.

Feminists have relied on two basic arguments in their view of science, namely that males and masculine metaphors dominate science.[5] Feminists have shown from historical studies in language and analyses of text, as well as empirical data, that the disciplines covered by logical-mathematical intelligence are gendered. Sue Curry Jansen shows that artificial intelligence and its models and disciplinary assumptions suggest masculinization, which results in a "phantom objectivity" in the knowledge produced by the discipline, knowledge which ignores different constructions of reality.[6]

Metaphors from biological evolution influenced several other disciplines, including paleontology, comparative anatomy, and ontogeny.[7] Clerk Maxwell compared Faraday's tubes of forces in his descriptions of electromagnetism to muscles: the tubes could shorten or widen like muscles.[8]

Haddne holds that the mathematical, mechanistic world view of early modern science mapped analogously the relations of commodity production and exchange. Commercial reckoners mediated between these external social relations and the content of science. They gave a fresh meaning to ancient mathematical concepts and helped establish the view that all sensually infruitable events can be explained in terms of the notion of qualitatively similar bodies.[9]

Study of archival material shows that Darwin was aware in detail of the ideas and language of his cultural circle, all of which was crucial in the presentation of his work. His methodology and metaphysics were, for example, deeply influenced by the views of the realist school of philosophy of Scotland.[10] It is within these and other strands that he created his many metaphors and insights,[11] and where technical terms were lacking, he presented his views in metaphors.[12]

From its infancy, the computer has been associated with metaphors drawn from labor and labor organizations. An important figure in this area was Charles Babbage, noted for contributions both in the design of the "well run" factory and the computer. In the former, his *On the Economy of Machinery and Manufactures* (1835) was seminal. Here he took the ideas of Adam Smith on the division of labor and applied them to organizing factory production. Later, in his pioneering attempts to build a computer through his "Difference Engine" and "Analytical Engine," he linked this division of labor in the factory to computation. From that time, the computer has been linked to a particular division of labor.[13]

Metaphors from psychology have influenced computer design and vice versa. Fred Van-Besein, reviewing the 1985 *PSYCHON* yearbook, which is a report on 63 projects in psychology, finds that 50 percent of the metaphors present in the texts are drawn from computer technology.[14] But many of these computer terms are actually a re-borrowing, because the computer discipline originally borrowed many metaphors from psychology.[15] Models derived from the mind can have a powerful impact on a host of disciplines, not only in computers but in a large number of other disciplinary areas, as is illustrated by the case of association of ideas.[16] Although facets of the association of ideas were seen in writers prior to Locke, including ancients such as Plato and Aristotle and more recent thinkers like Hobbes, it was David Hartley (1749), influenced by Locke, who developed the central concepts. Explicitly drawing upon the ideas of Newton, Locke, and Gay, Hartley developed a systematic psycho-physiology combining the mind's association of ideas with the nervous system's corpuscular vibrations.[17] Through a learning theory, Hartley consequently postulated a mechanism for the concepts of adaptation and utility. His scheme was limited to individual experience, but by extending it beyond an individual lifetime, it offered a useful mechanism for ideas of evolution and progress.[18]

Utilitarians like Joseph Priestley, Jeremy Bentham, James Mill, J.S. Mill, and William Paley later drew upon Hartley's views to develop their own various social, psychological, and economic theories.[19] Condorcet drew upon a continental brand of associationism for his theories of continuous and inevitable social progress. Similarly, William Godwin drew on Hartley's ideas for his theories of progress. Adam Smith, for his part, drew on the same source for his descriptions of wealth and economic equilibrium. On the other hand, Malthus, drawing on Adam Smith and thus also indirectly on Hartley, arrived at a pessimistic vision of the human future, a contrast to Godwin and Condorcet.[20] Drawing on the continental tradition of association of ideas, Lamarck developed his theory of the inheritance of acquired characteristics in evolution. The English parallel to Lamarck was the evolutionary theories of Erasmus Darwin, Charles Darwin's grandfather. He also drew heavily on Hartley's ideas to theorize the transmission of acquired characteristics and evolution. The evolutionary views of Charles Darwin did not draw explicitly on the associationist perspective but were influenced strongly by the associationist-inspired Malthus.[21] Associationist ideas continued to influence psychology directly and indirectly, as witnessed in the work of William McDougall, C.S. Sherrington, I.P. Pavlov, Freud, Wundt, William James, and George H. Mead.[22]

The power of a dominant metaphor to infiltrate many disciplines, enriching and rearranging their contents, is illustrated by the example of associationism, but the rich store of metaphors on the mind in South Asian traditions presents a treasure trove of further possibilities. But if this be so, are there any discussions of how to capture metaphors and use them? There are.

Metaphors, it has been observed, help both formulate and communicate insights in science.[23] They help "insight capture"[24] and possess definite cognitive roles as they cross disciplinary boundaries. Because they spur the imagination, they touch a deeper level of conceptual understanding than do paradigms.[25] Metaphors are also condensed thought forms which act as heuristics to nudge the imagination away from existing patterns.[26] As Julie Thompson Klein has noted in an important book on interdisciplinarity, "metaphors" can be a variety of things—illustrative devices, and paradigms or didactive devices that lead to new models. They could also have heuristic properties, or they could create new meanings. Metaphors from another discipline could sensitize scholars to fresh questions in their own field. Metaphors provide new insights and act as probes. If research is incomplete, a borrowed metaphor could help in induction.[27]

Metaphors can therefore be used in a variety of ways to nudge new insights. A useful method, according to Anthony Judge, is to "re-read" existing conceptual patterns as metaphors, thus mining the past for "fossilized knowledge."[28] In this manner, the Western scientific tradition regularly mined its intellectual past, as in the Renaissance, when it selectively rediscovered some of its Greek roots. This could also be the means of mining other knowledge traditions and their vast knowledge reservoirs for potential new concepts.[29]

Judge has offered some guidelines for this re-reading of metaphors.[30] First, we have to identify source patterns—in this case the large reservoir of non-Western concepts and experimental results. Next, we must recognize the target domains—current scientific disciplines and their present disciplinary impasses, along with problems in existing paradigms. Then we have to see how far the imported patterns fit the problems of the target domain. The final step would be to evaluate the utility of the insights yielded by the match.

Judge has suggested a "pattern database" that would help in this systematic search for metaphors. Such a database would be analogous to current databases of intellectual property, such as patents, where keywords (for example) are used to access patterns. With a suitable interface, the user of such a database would be able to browse associatively through metaphors, determining how isomorphic and matched are the patterns revealed, based on the searcher's needs. The pattern database would include material from a variety of sources, including, according to Judge, sciences (mathematics, natural, and social), humanities and arts (fiction, drama, music, and paintings), mythology, theology, philosophy, folklore, aphorisms, and traditional wisdom stories. A database of metaphors could discover previously unknown patterns in the work at hand, a function parallel to the use of mathematics to solve a problem in natural science.

Such a metaphor database, in allowing constant comparison of patterns with the material at hand, broadly fits into the technique of discovering new grounded theory in social science, as described by Barney Glaser and Anselm Strauss.[31] Variables in social science are numerous, and discovering patterns is therefore difficult. Glaser and Strauss describe a method of seeing patterns by constantly comparing the research material; constant comparison evokes patterns in the researcher's mind. In a similar manner, going through the metaphor database and comparing it with the raw data at hand would elicit new insights for current impasses in science.

In recent science, these inflows of metaphors have come primarily from the European heritage. However, a vast soup of empirical knowledge, metaphors, and theoretical constructs also exists in the non-European world, varying (as noted in this volume) from sophisticated debates on the nature of ontology and epistemology, to discussions in psychology, the nature of mind, mathematics, and medicine. An infusion from this source would help enlarge our scientific horizons.

Disciplinary lineages, in the modern sense, are a recent, post–nineteenth-century phenomenon. But lineages of culture have existed from the beginnings of humanity. In simpler societies lineages consisted of knowledge handed down from parent to offspring. In civilizational entities, with a greater division of labor and more sophisticated problems, lineages took a form reminiscent of lineages in modern disciplines. Sometimes, they were for all purposes like disciplinary lineages, or more narrowly a disciplinary program that followed a common research program, having similar problems and a similar set of approaches.

In the South Asian case, these lineages take the form that stretches from teacher to pupil, down the line: the *guru-shishya-parampara* (the teacher-pupil generational lineage), which has enriched all callings, including specialized activities in mathematics and astronomy. One finds these lineages in all the traditions, whether in Vedic schools such as Sulbakaras, Jyotiskaras, the Kusumpura school, the schools of Asamkadesa or Ujjain, and Jain or Buddhist schools. Varahamihira alone gives the names of twenty scholars that had come before him in a practice equivalent to modern citation practice in contemporary scientific papers. One builds up, through a chain of authorities, a frame of legitimacy for one's problem, methodology, and approach.[32]

Because such lineages search for what they believe is valid knowledge, some split up to factions leading to sublineages. Sometimes, the jealousy guarding these lineage boundaries can be so intense that there is no discourse between them—often metaphorically, sometimes literally. Explorations within such lineages could be a rich source of metaphors (as well as of directly transferable, nonmetaphorical information) for enlarging current lineages in the science disciplines.

Apart from lineages in subjects dealing directly with science, there are lineages and vast storehouses in the humanities and philosophical traditions. South Asia is one of the largest repositories of literature in the world, and apart from works in the two classical languages of Sanskrit and Pali, there are large collections in the regional languages. Part of this humanistic literature consists of a large number of allusory texts, perhaps the greatest source for metaphors. In poetics, in the *Kavya Shastra*, allusions and metaphors are studied and recorded in great detail. If one wishes to let the imagination soar in controlled freefall, this (and related texts) is a literature to jump into.

Rarely does the act of scientific creation consciously transfer metaphors from one realm to another. The process takes place naturally, subconsciously, although clearly the approach can be cultivated. Unconsciously, Copernicus transfers the metaphor of a circle intuitively into the movements of the heavens. Later, he checks the metaphors against his figures. Once the geometrical pattern is established and the gestalt switch made, Kepler can come along and refine the broad gestalt into an ellipse. To be able to play the metaphor-searching game unconsciously, one would have to be immersed in South Asian culture; a person immersed in the culture is not restrained to think through that culture, and taps the storehouse of metaphors without conscious effort. But, unfortunately, South Asians doing science have themselves been socialized completely into the Western frame. They have—to use the title of an earlier book of mine—*Crippled Minds*.[33] They tend not to think in a free-floating, unconscious manner; they are always inclined to look over their shoulders about what their mentors in the West say, or would say.

But those South Asians who can think associatively—unconsciously—on their own, can come up with major breakthroughs. The classic example is the Indian mathematician Ramanujan, "discovered" by the British mathematician Hardy and brought to England earlier in this century. Today Ramanujan is considered one of the greatest mathematicians the world has known, as a recent American book (Robert Kani-

gel's *The Man Who Knew Infinity: A Life of the Genius Ramanujan,* 1991)
readily testifies to.[34] Mathematicians are still poring over his notebooks
to dredge for nuggets. But Ramanujan did not follow an imitative Western method. A man fully rooted in his peculiar traditions, he believed
that his results came up on the tongue of his god Narasinha.[35] Yet, even
without unconsciously tapping the metaphorical database of a civilization, one can consciously train oneself to the practice. An outsider to a
culture, can also train her- or himself to be empathetic to a culture and
tap its roots.

The metaphorical uses of non-Western sources are virtually limitless
in their potential. Only the human imagination can limit its possible
uses: metaphors are an added tool in the arsenal of indirect approaches
to mining.

Some Potentials

But do we have any inkling of the potential that exists in the other
allusory approach—philosophy? Here, one could get some hints from
South Asia itself. The Jains discussed permutations and combinations
of theories.[36] One could presumably make calculations similar to
theirs, based on the soup of concepts from South Asia that could yield
possible solutions and theories for given problems. This mechanical,
exhaustive approach would be roughly similar to the mechanical approach to knowledge found in the computer technique called genetic
algorithms, in which solutions to problems are randomly selected and
then tested mechanically to fit into the criteria for a solution. Aspects
of knowledge creation, then, are partly automated. In a similar manner,
a Jain exercise could yield an exhaustive list of theoretical positions
to then be selectively mined.

The South Asian region has also been the site of extensive discussions
on, possibly, all the variations in ontology and epistemology that the
human imagination can think of (see chapter 11). But such philosophical questions seemed to have been solved "forever" in the seventeenth
century by Descartes' subject and object. Yet today, this seeming ontological and epistemological certainty has broken down in many areas—
particularly in quantum physics, where the relationships between the
observer and the observed, between subject and object have spilled
beyond the Cartesian dichotomy. As we demonstrated in chapter 11, it
also appears to have broken down in aspects of biology and cognitive
sciences. Consequently some of the deeper philosophical questions
are raised afresh and once again shifted to the arena of science. It is
also possible that similar philosophical questions lie lurking in other
disciplinary areas too.

Now that the certainties of a Descartes and a Bacon, constructed at the beginning of the scientific endeavor, are breaking down, there could well be much deeper questions in store regarding the relationship of humans with nature, including those related to understanding nature. Here, fresh insights drawn from other traditions could well be useful. An examination of ontological and epistemological questions drawn from the wide South Asian canvas could well give hints or directions for the pursuit of different areas of scientific interest. Can we formalize some of these ontological and epistemological positions?

All the variations in the chemical elements were set to order in Mendeleev's periodic table in the nineteenth century. It accounted for the chemical elements then known to exist, and through gaps in the table predicted unknown ones. Similarly, it would not be difficult to have an equivalent "periodic" table of all the positions in ontology and epistemology. The philosophical tapestry of South Asia could at least provide an initial foundation for such a table. It provides a table not only of almost all possible "realities," but also of almost all the possibilities of apprehending and knowing them. An examination of such a table would help one locate existing positions in different scientific disciplines, like the periodic table helped locate different chemical elements in different slots. Gaps, when applied to existing fields of science, could hint at further exploration, just like gaps in the periodic table pointed searchers to new chemical elements. A further examination could yield hints and guidelines for implications in the different scientific fields, even suggesting areas for new research.

Further Estimates

The South Asian Indologists, it had been observed, have only been searching for "occasional scraps of contemporary relevance from the remains of a civilization that for them, is perhaps as dead and as alien as it is for the West."[37] But in this book, I have been recording examples and techniques for making these civilizational stores relevant by splicing them to the modern knowledge enterprise. Can we have any measure of the potential that exists in splicing-in techniques?

The chapter on Ayurveda includes potential lists in medicine from a variety of perspectives. Let us consider one more subject: mathematics. The works of the Kerala mathematicians mentioned earlier (see chapter 7) are mines of mathematical ideas and applications that have hardly been touched. The few correlations with more recent Western discoveries show the potential of this mine.[38] Only a very small number of their texts have been published as yet. Of these, only a few have been studied by professional mathematicians. A detailed study would undoubtedly produce, if past results from this tradition are a guide, further useful outcomes.[39]

But the Kerala tradition is only a part of a much large tradition. One US estimate of manuscripts in *Jyothisastra* (roughly, the areas of mathematics and astronomy) cites 100,000,[40] yet the recently published *Sourcebook of Indian Astronomy* lists only 285 works.[41] Clearly, a large number of manuscripts still await further exploration, and a still larger number remain to be imaginatively accessed and built upon.

This is only a crude indication of the potential in one subject area, mathematics. During the last hundred years, probably two thousand *catalogues* and lists of known Indian manuscripts written in the various languages of the subcontinent have been compiled. Each of these catalogues lists about two hundred manuscripts, so we are talking about approximately 400,000 manuscripts. Others have estimated that all Indian manuscripts available today, in all languages, amount to some 500 million (fifty crores). But hardly any of these are being read today.[42] This figure, which incorporates our previous estimates for some of the specific fields, gives an idea of the huge potential in store.

Mining for Whom?

The examples of, and suggestions for, mining raise the question, "Mining for whom?" There is no clear answer. Mining could be used as a giant vacuum pump that will suck up knowledge for monopolistic multinationals. In the conventional debates related to this issue, one usually assumes the example of a tribal group in the rain forest, whose knowledge is extracted, perhaps the genes of its medical remedy taken, patented, put in a product, and then sold back to them. Clearly, such a scenario is not the most ethically sound and has been the recent subject of bitter acrimony. But these tribal groups, in the late twentieth century, are often the last carriers of such knowledge. Once exposed to the outside world, their children are leaving traditional pursuits, so unless something is done very fast, this knowledge would be lost forever. Under such a possibility, the issues are not so clear cut. If one were arguing from the perspective of a loss of knowledge, one could say, "Let the bad wolf in"—as a lesser of two evils.

The world today, however, is not the clear cut center-periphery colonial model of the fifties. The world's economic axis is changing to other civilizational areas, specifically to Asia. South Asia, our subject of exploration in this book, may not be the fastest growing in the world today, but in historical terms, it is growing rapidly. Observers expect that it will grow much faster in the years to come, and given its huge population, it will be part of the reason that the economic axis of the world will tilt strongly toward Asia in the next few decades. This will mean that in a globalized world, the future multinationals will no longer cluster in the West. Already, several of the key players are Asian.

But having your own multinationals will not necessarily answer the equity question in science. Presumably, they might bring in more civilizational elements and enlarge the catchment area for science. But the multinational world, in the globalization game bent on maximizing profits, could yet distribute science in the old Western way.

The emerging multinationals are also not necessarily centered in just one nation; multiple ownerships form as shares are traded globally. This is no longer the time of the first "multinationals." The early colonials —the British East India Company and its Dutch equivalent—had monopoly control, from their country, over those they traded with. Secondly, there is a tendency, with the growth of such institutions as mutual funds, for multinationals to be broader-based in ownership.

The emerging paradigm of production wedded to cheap computing power is also not the Fordist mass production and consumption one. It is a paradigm of limited production runs fitted to niches in the market. The world of production may be dominated in the future by multinationals, but they will increasingly tailor their products to narrower and narrower markets. For several years now, marketers in the US have tracked consumption patterns and targeted mailings based on very specific criteria. For example, my buying particular books and going to particular types of conferences has ensured that I am on the mailing list of many of the publishers who produce books that match my interests. This cheap targeting—by junk mailers, at times—is a prelude to global niche marketing in the future. Under such niche marketing, interests of narrower and narrower segments can be catered to in the search for profits. The practice will take more time to reach across the world, but the trend in a globalized world is in that direction.

Science is produced not only in the private sector. The key discoveries and the most disinterested science are still expected to come from universities and similar public institutions. Here the reach of commercial interests is less powerful and, where it exists, more indirect. If mining approaches are adopted in these public institutions, they will reflect the particular internal dynamics of such institutions.

So the answer to the question "Mining for whom?" is not simple. A mining exercise in science will not solve the problems in society at large. The mined knowledge will be added to existing social structures, and the act of mining in other civilizations will open up vistas to and from these cultures. Some would have produced their science in a predominantly patriarchal lineage, because all classical civilizations have been patriarchal. Some of these classical lineages would have more female input than others; some less. The act of enlarging will, however, tap these patriarchal lineages more frequently than women's contributions, for the simple reason that patriarchal contributions are dominant.

The search for a larger science has, therefore, many built-in contradictions. Completely ideologically satisfactory solutions are not possible, like attempts at starting new lineages. We have already pointed out that this would be like starting an entirely new evolutionary tree in biology, a project no longer possible in any lineage. Modern science exists and it works. The feminist philosopher working on her manuscript uses the same computer and software program that I use. Such software, as we have already pointed out, has grown up within a strong patriarchal trajectory, yet it exists, and the only realistic option is to use it.

Onto this knowledge system selected civilizational knowledge, knowledge from women's cultural lineages, and ethno-knowledge can be grafted. The act of getting a vital science system that is fed by all the historical experiences of all segments of the population is a long, drawn-out enterprise. Grafting on knowledge from outside is more like biology, of dealing with givens and maximizing what is at hand, as far as possible.

The worlds of ideas were brought together and packaged in Europe in a fresh vigorous amalgam, and post-fifteenth-century developments —the Renaissance, the Scientific Revolution, the Enlightenment, and the great discoveries in the nineteenth and twentieth centuries—have been the outcome. Now, a new packaging is possible. A new historical moment has opened up in which the world is increasingly being interconnected in a pervasive spread of globalization. As a result, no single node can in the future act as a clearinghouse for all the world's knowledge. The role of a dominant center over a periphery is eroding as near-instant communication begins to spread information and knowledge creation across a wide network. Globalization is opening up centers of dominance away from one central place to a shifting web (and the moving denizens of the Internet). Academic groups are spreading across the world. Further, new historical research is coming up across the world, bringing out hidden nuggets.

Inputs into science from different civilizational sources could occur because of this globalization. Already, mainstream R&D is getting relocated for considerations of cost in some cheaper Third World countries like India. Another science-related technological activity, software production, has been relocated for similar reasons, and India's software production today rivals that of many developed countries. Globalization is thus opening up to different cultures, and cultural colonization is moving away from the classic mould, to niche marketing and niche production, as has already happened in several industries across the world. Globalization will suck up cultural elements from the different global cultural systems and "remarket" them to global niches. As a result, there would be greater openness for knowledge than in the first cultural colonization in the mercantile period, or the second in the Industrial Age.

The potential for enlarging the existing knowledge system by tapping into civilizational stores is vast. The authors of *Embodied Mind* state, "It is our contention that the rediscovery of Asian philosophy, particularly of the Buddhist tradition, is a second renaissance in the cultural history of the West, with the potential to be equally important as the rediscovery of Greek thought in the European renaissance."[43] How far this may be true is for the future to decide. However, a strategic alliance of feminist approaches, ethno-knowledge, and regional civilizations' knowledge is probably in the making. Through their combined efforts the results of millennia of human enquiry which have been lost from the Euro-patriarchal view could be resurrected. A true global science, taking into account multicultural politics of facts, would result.[44] Undoubtedly the knowledge system would become less chauvinistic and less parochial, while maintaining the rigor developed in the last few centuries.

For whom do we mine civilizational knowledge? For us all.

Notes

ONE / INTRODUCTION

1. Brown, Malcolm W., "Scientists deplore flight from reason," *New York Times,* June 6, 1995.

2. Price, D.J. de Solla, *Little Science, Big Science* (London, Macmillan, 1963).

3. Science Foundation Course Team, *Science and Society,* Science Foundation Course Units 33–34 (The Open University Press, 1971).

4. Dizardt, Wilson P., *The Coming Information Age* (New York and London, Longman, 1982).

5. Quoted in Horton, Allan, "Electronic Access to Information: Its Impact on Scholarship and Research," *Interdisciplinary Science Reviews,* 1983, Vol. 8, No. 1, p. 67.

6. Ibid.

7. Langley, P.M., "Rediscovering physics with Bacon 3," *Proceedings of the International Joint Conference on Artificial Intelligence* 6, 1977, 505–507.

8. Waldrop, M. Mitchell, "Learning to drink from a fire hose," *Science,* 11 May 1990, 248:674–675.

9. Ibid.

10. Feyerabend, P., *Against Method: Outline of an Anarchist Theory of Knowledge* (London, New Left Books, 1975).

TWO / THE TRAJECTORIES OF CIVILIZATIONAL KNOWLEDGE

1. Goonatilake, Susantha, "Colonies and Scientific Expansion (and Contraction)," in *Review: Journal of the Fernand Braudel Center,* 1982, Vol. V, No. 3; and Goonatilake, Susantha, *Aborted Discovery: Science and Creativity in the Third World* (Zed Press, London, 1984).

2. Barnes, Barry, *T.S. Kuhn and Social Science* (London and Basingstoke, Macmillan Press, 1982), p. 90.

3. Schrödinger, E.C., *Science and Theory and Man* (Dover, New York, 1957), pp. 87–88.

4. Ibid, p. 88.

5. Bernal, Martin, *Black Athena: The Afroasiatic Roots of Classical Civilization,* Vol. I, *The Fabrication of Ancient Greece 1785–1985* (Free Association Books, London, 1987).

6. Bernal, J.D., *Science in History* (Hawthorn Books, 1956).

7. Ibid, p. 2.

8. Needham, Joseph, "The Emergence and Institutionalization of Modern Science," *Sociology of Science—Selected Readings,* Barry Barnes (ed.) (Harmondsworth, Penguin, 1972).

9. Hessen, Boris, "The Social and Economic Roots of Newton's Principia," *International Congress of the History of Science and Technology,* London, 1930.

10. Forman, Paul, "Weimar Culture, Causality and Quantum Theory 1918–27: Adaptation by German Physicists and Mathematicians to a Hostile Intellectual Environment," *Historical Studies in the Physical Sciences (iii),* (Philadelphia: University of Pennsylvania Press, 1971).

11. Lederman, Leonard L., "Science and Technology Policies and Priorities: A Comparative Analysis," *Science,* 237, 4819, 4 September 1987, pp. 1125–1133.

12. Yearly, Steven, "The Social Construction of National Scientific Profiles: A case study of the Irish Republic," *Social Science Information,* 26, 1, March 1987, pp. 191–210.

13. Shinn, Terry, "Progress and paradoxes in French science and technology 1900–1930," *Social Science Information,* 28, 4 December 1989, pp. 659–683.

14. Crow, Michael, "Technology Development in Japan and the United States: Lessons from the High-Temperature Superconductivity Race," *Science and Public Policy,* 16, 6 December 1989, pp. 322–344.

15. Luukkonen, Terttu, "Publication Structures and Accumulators Advantages," *Scientometrics,* 19, 3–4, September 1990, pp. 167–184.

16. Lancaster, F.W.; Lee, Sun-Yoon Kim; and Diluvio, Catalina, "Does Place of Publication Influence Citation Behavior," *Scientometrics,* 19, 3–4, September 1990, pp. 239–244.

17. Gaillard, Jacques, "La Science du tiers-monde entre deux mondes: science nationales ou science internationale? Quelques reflexions et implications politiques."

18. Mackenzie, D., and Barnes, S.B., "Scientific Judgement: The Biometry-Mendelism Controversy," in *Natural Order,* Barnes and Shapin (eds.), 1979.

19. Farley, J., *The Spontaneous Generation Controversy from Descartes to Oparin* (Baltimore, Johns Hopkins University Press, 1977).

20. Mackenzie and Barnes.

21. Scharfstein, Ben Ami, et al., *Philosophy East/Philosophy West: A Critical Comparison of Indian, Chinese, Islamic and European Philosophy* (Oxford, Basil Blackwell, 1978).

22. Rose, Hilary, and Rose, Steven, *The Radicalization of Science* (London, Macmillian, 1976).

23. Collins, H.M., "The Seven Sexes: A Study in the Sociology of a Phenomenon, or the Republication of Experiment in Physics," *Sociology,* Vol. 9, 1975.

24. Gilbert, G.N., "The Development of Science and Scientific Knowledge: The Case of Radar Meteor Research," in O. Lemaino (ed.), *Perspectives on the Emergence of Scientific Discipline* (The Hague and Paris, Mouton, 1976).

25. Pinch, T.J., "What Does a Proof Do If It Does Not Prove," in E. Men-

delsohn, P. Weingart, and R. Whitley (eds.), *The Social Production of Scientific Knowledge* (Dordrecht, Reidel, 1977).

26. Pickering, Andrew, "Editing and Epistemology: Three Accounts of the Discovery of the Weak Neutral Current," *Knowledge and Society: Studies in the Sociology of Culture Past and Present*, 8, 1989, pp. 217–232.

27. Ibid.

28. Wynne, B., "C.G. Barkla and the J. Phenomenon: A Case Study of the Treatment of Deviance in Physics," *Social Studies of Science*, Vol. 6, 1976.

29. Latour, B. and Woolgar, S., *Laboratory Life: the Social Construction of Scientific Facts* (London, Sage, 1979).

30. Fleck, L., *Entstehung und Entwicklung einer Wissenschaftliche Tatsache*, (English translation F. Bradley and T.J. Trenn), published as *Genesis and Development of a Scientific Fact* (University of Chicago Press, 1979).

31. Barnes, Barry, *T.S. Khun and Social Science* (London and Basingstoke, Macmillan, 1982), p. 10.

32. Restivo, Sal Episteme, *Mathematics in Society and History* (Boston and London, Dordecht, 1992).

33. Horowitz, Irving Louis, "Limits of Standardization in Scholarly Journals," *Scholarly Publishing*, 18, 2, January 1987, pp. 125–130.

34. Hallberg, Margareta, "Science and Feminism, Vetenskop och feminism," *Sociologisk—Forskning*, 27, 3, 1990, pp. 27–34.

35. Jansen, Sue Curry, "Is Science a Man? New Feminist Epistemologies and Reconstruction of Knowledge," *Theory and Society*, 19, 2, April 1990, pp. 235–246.

36. Witt, Patricia L.; Barnerle, Cynthia; Derouen, Diane; Kamel, Freja; Kelleher, Patricia; McCarthy, Monica; Namenworth, Marion; Sabatini, Linda; Voytovich, Marta, "The October 29th Group: Defining a Feminist Science," *Women's Studies International Forum*, 12, 3, 1989, pp. 253–259.

37. Keller, Evelyn Fox, *Gender and Science* (New Haven, Yale University Press, 1985).

38. Jansen, Sue Curry, "The Ghost in the Machine: Artificial Intelligence and Gendered Thought Patterns," *Resources for Feminist Research/Documentation sur la Recherche Feministe*, 17, 4 December 1988, pp. 4–7.

39. Haraway, Donna, "Situated Knowledges: The Science Question in Feminism and the Privilege of Partial Perspective," *Feminist Studies*, 14 (Fall 1988), pp. 575–599; Harding, Sandra, *The Science Question in Feminism* (Ithaca, NY, Cornell University Press, 1986); and Harding, Sandra, *Whose Science? Whose Knowledge?* (Ithaca, NY, Cornell University Press, 1991).

40. Shiva, Vandana, *Staying Alive: Women, Ecology and Development* (London and New Jersey, Zed Books, 1988); Mies, Maria, and Shiva, Vandana, *Ecofeminism* (Halifax, Nova Scotia, Fernwood Publications; London and New Jersey, Zed Books, 1993).

41. Alcoff, Linda, and Potter, Elizabeth (eds.), *Feminist Epistemologies* (New York and London, Routledge, 1993).

42. Bourdieu, Pierre, "The Peculiar History of Scientific Reason," *Sociological Forum*, 6, 1 March 1991, pp. 3–26.

43. Harding, Sandra (ed.), *The "Racial" Economy of Science: Toward a Democratic Future* (Bloomington: Indiana University Press, 1993).

44. Pagel, Walter, "The Spectre of Von Helmont," in Teich and Young (eds.),

Perspectives in the History of Science (Boston, Reidel, 1973), p. 104.

45. Heisenberg, W., *Physics and Philosophy* (London, Allen & Unwin, 1963), p. 7.

46. Wiener, Philip P., *Dictionary of the History of Ideas* (New York, Charles Scribner's Sons, 1973), p. 89.

47. Schrödinger, E.C., *Science and Theory and Man* (New York, Dover, 1957), p. 92.

48. Ibid.

49. Ibid.

50. Evenari, M.; Shanan, L.; Tadmor, N., *The Negev, Challenge of a Dessert* (Cambridge, MA, Harvard University Press, 1982).

51. Gould, Stephen Jay, "The Politics of Evolution," *Psychohistory—Review,* 11, 2–3, Spring 1983, pp. 15–35.

52. Ibid., p. 348.

53. Manier, Edward, "'External factors' and 'Ideology' in the Earliest Drafts of Darwin's Theory," *Social Studies of Science,* 17, 4 November 1987, pp. 581–609.

54. Ibid., pp. 373–376.

55. Feuer, L.S., "Teleological Principles in Science," *Inquiry: An Interdisciplinary Journal of Philosophy and the Social Sciences,* 21, 4, Winter 1978, Universitets Forlaget, Norway.

56. Holton, Gerald, *Thematic Origins of Scientific Thought: Kepler to Einstein* (Cambridge, MA, Harvard University Press, 1973).

57. Ibid, p. 29.

58. Turner, Ronny E., "Language and Knowledge: Metaphor as the Mother of Knowledge," *California Sociologist,* 10, 1, Winter 1987, pp. 44–61.

59. Ibid.

60. Rothbart, Daniel, "The Semantics of Metaphor and the Structure of Science," *Philosophy of Science,* 51, 4 December 1984, pp. 595–615.

61. Klein, Julie Thompson, *Interdisciplinarity: History, Theory and Practice* (Detroit, Wayne State University Press, 1990), p. 93.

THREE / THE BACKGROUND TO CROSSFLOWS

1. Auboyer, J., and Goepper, R., *The Oriental World* (New York, McGraw Hill, 1967), p. 11.

2. Auboyer, J., and Goepper, R., *The Oriental World* (New York, McGraw Hill, 1967).

3. Rapson, Edward James, *Ancient India* (Cambridge University Press, 1914), pp. 87–88.

4. Jaggi, O.P., *Indian System of Medicine,* Vol. 4 of *History of Science and Technology in India* (Delhi, Atma Ram & Sons, 1973), p. 207.

5. Halbfass, Wilhelm, *India and Europe: An Essay in Understanding* (Albany, SUNY Press, 1988), p. 10.

6. Chowdhury, Amiya Kumar Roy, *Man, Malady and Medicine—History of Indian Medicine* (Calcutta, Das Gupta & Co. Ltd, 1988), p. 67.

7. Chowdhury, p. 67.

8. Jaggi, p. 207.

9. Jaggi, p. 207.

10. Chowdhury, p. 67.

11. Rawlinson, H.G., "Early Contacts Between India and Europe," in *Cultural History of India,* A.L. Basham (ed.) (Oxford, Clarendon Press, 1975), p. 427.

12. Ibid., pp. 427–428.

13. Ibid., p. 427.

14. Ibid., p. 428.

15. Urwick, B.J., *The Message of Plato* (London, Allen & Unwin, 1920).

16. Vitsaxis, Vissilis, *Plato and the Upanishads* (New Delhi, Arnold-Heinemann, 1977).

17. Bose, D.M.; Sen, S.N.; and Subarayappa, B.V., *A Concise History of Science in India* (New Delhi, Indian National Science Academy, 1971), p. 573.

18. Chowdhury, p. 67.

19. Bose et al., p. 582.

20. Hamilton, W., "History of Medicine, Surgery and Anatomy—Healing Art of Hindustan," 1831, p. 43, quoted in Jaggi, p. 212.

21. Jaggi, p. 213.

22. Ibid., p. 208.

23. Basham, A.L., *The Wonder That Was India* (New York, Grove Press, 1953), p. 497.

24. Ibid.

25. Halbfass, p. 12.

26. Jaggi, p. 209.

27. Weber, A., *The History of Indian Literature* (London, Theodore Zachariae, 1876), p. 246.

28. Dodwell, S. (ed.), *Cambridge History of India* (Cambridge, 1922), pp. 419–420.

29. Halbfass, p. 17.

30. Paranavitana, S., *Sinhalayo* (Colombo, Lake House Investment Ltd., 1969).

31. McEvedy, Colin, and McEvedy, Sarah, *The Classical World* (London, Hart Davis, 1973), p. 47.

32. Nakamura, Hajime, "Buddhism," *Dictionary of the History of Ideas* (New York, Phillip B. Wienertedt, Scribner's, 1973), pp. 254–257.

33. Halbfass, p. 16.

34. von Glasenapp, Helmuth, "Indian and Western Metaphysics," *Philosophy East and West,* 111, 3, October 1953.

35. Rawlinson, p. 436.

36. Goleman, Daniel, *The Meditative Mind,* (New York, G.P. Putnam & Sons, 1988), p. 151.

37. Rawlinson, p. 436.

38. Goleman, p. 151.

39. Halbfass, p. 18.

40. Ibid., p. 3.

41. Ibid., p. 436.

42. Ibid., p. 436.

43. Ibid., p. 227.

44. Ibid., p. 8.

45. Ibid., p. 19.

46. Sen, S.N., "Influence of Indian science on other culture areas," *Indian Journal of the History of Science,* 5, 2, 1970, pp. 332–346.

47. Nakamura, pp. 254–257.

48. Datt, B.B., and Singh, A.N., *History of Hindu Mathematics,* A source book, Part 1 (Lahore, Motilal Banarsidas, 1935), p. 96.

49. Abdi, Wazir Hasan, "Interaction of West Asian and Central Asian Science," in Kuppuram G. and Kumudamani K. (eds.), *History of Science and Technology in India,* Vol. 4, "Science" (Delhi, Sundeep Prakashan, 1990), p. 340.

50. Jaggi, O.M., *History of Science and Technology in India,* Vol. 6, *Indian Astronomy and Mathematics* (Delhi, Atma Ram, 1986), p. 49.

51. Joseph, George Gheverghese, *The Crest of the Peacock: Non-European Roots of Mathematics* (London, Penguin, 1990), p. 306.

52. Ibid., pp. 10–11, 306.

53. Basham, 1953, p. 496.

54. Lach, Donald F., *Asia in the Making of Europe,* Vol. 2 (University of Chicago Press, 1977), p. 399.

55. Singer, S., *A Short History of Science* (Oxford, 1943), pp. 148, 162.

56. Bose et al., p. 135.

57. Ibid., p. 211.

58. Lach, p. 408.

59. See Lach.

60. Panikkar, K. M., *Asia and Western Dominance* (London, George Allen & Unwin, 1953), p. 28.

61. Worcester, G.R.D., *Sail and Sweep in China,* HMSO, London, 1960, p. 5.

62. Lach, p. 419; Panikkar, p. 28.

63. Lach, p. 483.

64. Boxer, C.R., *Two Pioneers of Tropical Medicine: Garcia D'Orta and Nicholas Honardes* (London, Hispanic and Luso-Brazilian Councils, 1963); Lach, p. 434.

65. Ibid., pp. 509–518.

66. Ibid., pp. 64–69.

67. Basham, p. 500.

68. Ibid., p. 65.

69. Alvares, Claude, *Homo Faber: Technology and Culture in India, China and the West 1500–1972* (Bombay, Allied Publishers Private, Ltd., 1979).

70. Ibid., p. 58.

71. Juleff, Gill, "An ancient wind-powered iron smelting technology in Sri Lanka," *Nature,* Vol. 379, 4 January 1996, pp. 60–63.

72. Ibid., p. 62.

73. Basham, p. 5.

74. Wiener, Philip P. (ed.), *Dictionary of the History of Ideas* (New York, Charles Scribner's Sons), p. 67.

75. Ibid.

76. Bernal, J.D., *Science in History,* p. 333

77. Inada, Kenneth K., and Jacobson, Nolan B. (eds.), *Buddhism and American Thinkers* (Albany, SUNY Press, 1984), p. vii.

78. Kline, Morris, *Mathematics in Western Culture* (New York, Oxford University Press, 1965), p. 239–240.

79. Datta, Bibhutibhusan, and Singh, Awadhesh Narayan, "Use of permutations and combinations in India," *Indian Journal of History of Science,* 27, 3, 1992.

80. Bernal, J.D., p. 343.

81. Jacobson, Nolan Pliny, "The possibility of Oriental influence in Hume's philosophy," *Philosophy East and West*, XIX, 1, January 1969, pp. 17–38.

82. Poussin, La Vallee, *The Way to Nirvana* (Cambridge University Press, 1917), pp. 38–39.

83. Hume, David, *A Treatise of Human Nature*, ed. Selby-Bigge (Oxford, Clarendon Press, 1896), II.ii.5 (363).

84. Jacobson, p. 18.

85. Ibid.

86. Ibid., p. 26.

87. Halbfass, pp. 57–59.

88. Mukhopadhyay, R.K., "History of Science and Two metamorphoses of Mind," *Occasional Paper 9*, Project of History of Indian Science, Philosophy and Culture, New Delhi, 1991.

89. Nakamura, p. 255.

90. Faber, M.D., "Back to a crossroad: Nietzsche, Freud and the East," *New Ideas in Psychology*, 6, 1, 1988, pp. 25–45.

91. Nakamura, p. 255.

92. Quoted in Jackson, Carl T., *The Oriental Religions and American Thought: Nineteenth Century Explorations* (Westport, CT, Greenwood Press, 1981), p. 47.

93. Jackson, *Oriental Religions and American Thought.*

94. Riepe, Dale, "The Indian Influence in American Philosophy: Emerson to Moore," *Philosophy East and West*, XVII, 14, January-October 1967, pp. 124–137.

95. Henderson, Linda Darlymple, *The Fourth Dimension and Non-Euclidean Geometry in Modern Art* (New Jersey, Princeton University Press, 1983), p. 32; Ouspensky, P.D., *The Fourth Dimension.*

96. Jammer, Max, *Concepts of Space: The History of Space in Physics* (Cambridge, MA, Harvard University Press, 1954), p. 180.

97. Einstein, Albert, *The Meaning of Relativity* (New Jersey, Princeton University Press, 1946), p. 99.

98. Miller, Arthur I., *Imagery in Scientific Thought: Creating 20th Century Physics* (Cambridge, MA, MIT Press, 1987), p. 39.

99. Graves, John Cowperthwaite, *The Conceptual Foundations of Contemporary Relativity Theory* (Cambridge, MA, MIT Press, 1971).

100. Einstein, pp. 103, 107.

101. Born, Max, *Einstein's Theory of Relativity* (New York, Dover, 1965), pp. 3, 313.

102. Frank, Phillip, "Einstein's Philosophy of Science," in Coley, Noel G., and Hall, Vance M.D., *Darwin to Einstein: Primary Sources on Science and Belief* (Harlow, Essex, Open University Press, 1980), p. 149.

103. Jackson, Carl T., "The Meeting of East and West: The Case of Paul Carus," *The Journal of the History of Ideas*, XXIX, 1, January-March 1968, pp. 74–75.

104. "Mach and Buddhism," in Blackburn, John T., *Ernst Mach, His Work, Life and Influence* (University of California Press, 1972), pp. 286–355.

105. Ibid., p. 287–288.

106. "Mach and Buddhism," p. 289.

107. Ibid.

108. Bass, L., "Schrödinger: A Philosopher in Planck's Chair," *British Journal of Philosophy of Science*, 43, 1992, 111.

109. Hoffman, Yoel, "The Possibility of Knowledge: Kant and Nagarjuna," in Ben-Ami Scharfstein et al. (eds.), *Philosophy East/Philosophy West, A Critical Comparison of Indian, Chinese, Islamic and European Philosophy* (Oxford, Basil Blackwell, 1978), p. 27.

110. Moore, W., *Schrödinger: Life and Thought* (Cambridge University Press, 1968), p. 114.

111. Nair, Ramjit, "The Science of Schrödinger and the Vision of the Vedanta," paper presented at Erwin Schrödinger Symposium, Sorbonne, Paris 1992, and "Indo-Austrian Workshop," University of Delhi, 1994, and forthcoming as a chapter in D. Tiemersma (ed.), *Between Science and Culture*, Kluwer, 1997.

112. Inada and Jacobson, p. vii.

113. Bose et al., p. 573.

114. Basham, p. 497.

115. Bose et al., p. 467.

116. Lach, p. 399.

117. Bose et al., pp. 334–338.

118. Alvares, p. 63.

119. Wiener, p. 67.

120. Maslow, A.H., *Towards a Psychology of Being* (New York, Van Nostrand, 1968).

121. Jacobson, p. 17.

122. Hoffman, p. 27.

FOUR / TRANSFORMATIONS

1. Westaway, F.W., *Quest: 3000 Years of Science* (London and Glasgow, Blackie and Son, 1936), p. 177.

2. Jeans, James, *The Growth of Physical Sciences* (Cambridge University Press, 1947), pp.170–175.

3. Westaway, pp. 177–183.

4. Ibid.

5. Gay, Peter, *The Enlightenment: An Interpretation*, Vol. 1, *The Rise of Modern Paganism* (New York, Alfred Knopf, 1967), p. 94.

6. Bose et al., pp. 140–141.

7. Sen, S.N., "History of Science in Relation to Philosophy and Culture in Indian Civilization," *Occasional Paper 6*, Project of History of Indian Science, Philosophy and Culture, New Delhi, 1993, p. 17.

8. Scharfstein, Ben-Ami, *Philosophy East/Philosophy West: A Critical Comparison of Indian, Chinese, Islamic and European Philosophy* (Oxford, Basil Blackwell, 1978), p. 30.

9. Ibid.

10. Ibid.

11. Inada, Kenneth K., and Jacobson, Nolan B. (eds.), *Buddhism and American Thinkers* (Albany, SUNY Press, 1984), p. vii.

12. See chapter 7, "Mathematics," for details.

13. Basham, A.L., *The Wonder That Was India* (New York, Grove Press, Inc., 1953), p. 497.

14. Bose, D.M.; Sen, S.N.; and Subbarayappa, B.V., *A Concise History of Science in India*, Indian National Science Academy, New Delhi, 1971, p. 590.

15. Bose et al., p. 467.

16. Sen, pp. 20–23.

17. A popular rendering of these parallels is in Fritsjoff Capra's *Tao of Physics* (London, Fontana/Collins, 1976).

18. Hiley, B.J., and Peat, David F., *Quantum Implications: Essays in Honour of David Bohm* (London and New York, Routledge & Kegan Paul, 1991); Krishnamurti, J., and Bohm, David, *Future of Humanity: A Conversation by J. Krishnamurti and David Bohm* (Madras, Krishnamurti Foundation, 1987).

19. Debus, Allen G., "Alchemy," *The Dictionary of History of Ideas*, Vol. 1 (New York, Charles Scribner's Sons, 1973), pp. 27–34.

20. Jeans, pp. 152–154.

21. Bose et al., p 334–335.

22. Debus, pp. 27–34.

23. Bose et al., pp. 318, 589.

24. Bhatia, S.L., "Renaissance and the Evolution of Medicine," *Indian Journal of the History of Medicine*, 18, 1973, pp. 1–11.

25. Singh, L.M.; Thakal, K.K.; and Deshpande, P.J., "Susruta's contributions to the fundamentals of surgery," *Indian Journal of the History of Science*, 5, 1, 1970.

26. Sastry, V.V.S., "Indian knowledge of blood circulation," *Bulletin of the Indian Institute of the History of Medicine*, 5, 2, April 1975, pp. 57–65.

27. Ibid.

28. Christophersen, Axel, "Big chiefs and Buddhas in the heart of the Swedish homeland," in *Thirteen Studies on Helgo*, The Museum of National Antiquities, Stockholm, 1988.

29. See chapter 9.

30. Alvarez, Claude, *Homo Faber: Technology and Culture in India, China and the West 1500–1972* (Bombay, Allied Publishers Private, Ltd., 1979), p. 58.

31. Alvarez, p. 62.

32. Dharmapal, P., *Indian Science and Technology in the Eighteenth Century* (New Delhi, Impex, 1971).

33. Dharmapal, p. li.

34. Quoted in Alvares, p. 64.

35. Pearson, L.C., *Principles of Agronomy*, 1967, p. 79, quoted in Madras Group, "Indian Agriculture at the Turn of the Century," *Readings from the PPST Bulletin* (Madras, PPST Foundation, n.d.), p. 56.

36. Quoted from Dharmapal in Madras Group, "Indian Agriculture at the Turn of the Century," p. 57.

37. Madras Group, pp. 61–62.

38. Mollison, J., *A Text Book on Indian Agriculture*, Vol. 2 of 3 vols. (Bombay Agricultural Department, Government of India, 1901), p. 6.

39. Bose et al., p. 353.

40. Madras Group, pp. 61–62.

41. Ibid.

42. Voelcker, John Augustus, *The Report on the Improvement of Indian Agriculture*, 1893, Reprinted 1990 (New Delhi, Agricola Publishers), p. 232–233.

43. Mollison, Vol. 1, pp. 43–45.

44. Howard, Albert, *An Agricultural Testament*, 1940, quoted in Madras Group, "Indian Agriculture at the Turn of the Century," p. 92.

45. Quoted in Jacobson, p. 17.

46. Sen, pp. 20–23.

47. Ibid.

48. Jayatillake, K.N., "Buddhism and the Scientific Revolution," in K.N. Jayatilake (ed.), *Buddhism and Science* (Kandy, Sri Lanka, Buddhist Publications Society, 1980), pp. 1–12.

49. Bose et al., p. 80–81.

50. *Encyclopedia of Buddhism,* published by the Government of Ceylon, Colombo, 1979, Vol. V.

51. Basham, p. 320.

52. Smart, N., *Doctrine and Argument in Indian Philosophy* (London, Harvester Press, 1964).

53. Descartes, René, *Discourse de la methode*, Part V, 2, quoted in *Dictionary of the History of Ideas*, Vol. 1, p. 242.

54. Goleman, Daniel, *The Meditative Mind* (New York, G.P. Putnam & Sons, 1988), p. xxii.

55. *Buddhist Encyclopedia*, Vol. 1, p. 186.

56. Ibid.

57. See chapter 9 for details.

58. Goonatilake, *Aborted Discovery,* pp 160–163.

59. Vide chapter 8.

60. Nisbet, Robert, "The Myth of the Renaissance," *Comparative Studies in Society and History,* Cambridge, 1973.

61. Rahman, A., *Bibliography of Source Material on History of Science and Technology in Medieval India—An Introduction*, Council of Scientific and Industrial Research, New Delhi, 1975.

62. Bose et al., p. 592.

63. Joseph, George Gheverghese, *The Crest of the Peacock: Non-European Roots of Mathematics* (London, Penguin, 1990), p. 215.

64. Kaye, G.R., *Indian Mathematics* (Calcutta and Simla, 1915), quoted in Joseph, p. 216.

65. Joseph, p. 215.

66. Basham, p. 164.

FIVE / INDIGENOUS KNOWLEDGE

1. Schrödinger, Erwin C., *Science Theory and Man* (New York, Dover, 1957), pp. 86–88.

2. Price, D.J. de Solla, *Little Science, Big Science* (London, McMillan, 1963).

3. Pfeiffer, John E., *The Emergence of Man* (New York, Harper & Row, 1969), p. 135.

4. Simpson, G.G, *Principles of Animal Taxonomy* (New York, Columbia University Press, 1961).

5. Speck, F.G., "Reptile Lore of the Northern Indians," *Journal of American Folklore*, 30, 141, Boston and New York, 1923, p. 273.

6. Smith, Bowen E., *Return to Laughter* (New York, Natural History Press, 1954), p. 19.

7. Smith, A.H, "The Culture of Kabira, Southern Byuku Islands," *Proceedings of the American Philosophical Society*, 104, 2, Philadelphia, 1960.

8. Handy, E.S.; Craighill and Pukui; Kawena, M., "The Polynesian Family System in Ka'u Hawaii," Parts VI, VII, VIII, *Journal of the Polynesian Society*, 62 and 64, Wellington, N.Z., 1953.

9. Conklin, H.C., *The Relation of Hanunoo Culture to the Plant World*, Dissertation, Yale, 1954 (microfilm).

10. Fox, R.B., "The Pinatubo Negritos: Their Useful Plants and Material Culture," *The Philippine Journal of Science*, 81, 3–4, 1952, Manila.

11. Ackernecht, E.H., *Bulletin of the History of Medicine*, II, pp. 503–521, reprinted 1958 in *Reader in Comparative Religion: An Anthropological Approach*, William A. Lessa and Evon S. Vogt (eds.), Evanston, Illinois and White Plains, New York, 1942, pp. 343–353.

12. Grollig, Francis, and Haley, Harold B., *Medical Anthropology* (The Hague, Mouton, 1976).

13. Fox, "The Pinatubo Negritos."

14. Ibid, pp. 212–213.

15. Boster, James, "Agreement Between Biological Classification Systems Is Not Dependent on Cultural Transmission," *American Anthropologist*, 89, 4, December 1987, pp. 914–920.

16. For example:

1973: Berlin, Brent, "Folk Systematics in Relation to Biological Classification and Nomenclature," *Annual Review of Ecology and Systematics*, 4, pp. 259–271.

Also 1973: Berlin, Brent; Breedlove, Dennis; and Raven, Peter, "General Principles of Classification and Nomenclature in Folk Biology." *American Anthropologist*, 75, pp. 214–242.

1981: Berlin, Brent; Boster, James; and O'Neill, John P., "The Perceptual Bases of Ethnobiological Classification: Evidence from Aguaruna Jivaro Ornithology," *Journal of Ethnobiology*, 1, 1, pp. 95–108.

1986: Boster, James; Berlin, Brent; and O'Neill, John P., "The Correspondence of Jivaroan to Scientific Ornithology," *American Anthropologist*, 88, 3, pp. 569–583.

1967: Bulmer, Ralph,"Why Is the Cassowary Not a Bird? A Problem of Zoological Taxonomy Among the Karam of the New Guinea Highlands," *Man*, 2, pp. 1–25.

1966: Diamond, Jared M., "Classification System of Primitive People," *Science*, 151, pp. 1102–1104.

17. Boster 1987, p. 919.

18. Berlin, Brent, "The chicken and the egg-head revisited: Further Evidence for the Intellectualist bases of Ethnobiological Classification," in Darrel A. Posey and William Leslie Overal (eds.), *Ethnobiology: Implications and Applications. Proceedings of the First International Congress of Ethnobiology* (Belem, Brazil, Museo Emilio Goeldi, 1990), Vol. I, p. 19.

19. Ibid, p. 20.

20. Ibid, p. 231.

21. Ibid, pp. 19–35.

22. Cardwell, H., *From Watt to Clausius: The Rise of Thermodynamics in the Early Industrial Age* (Ames, Iowa State University Press, 1989).

23. Posey, Darrell, "Intellectual Property Rights and Just Compensation for Indigenous Knowledge," *Anthropology Today*, 6, 4, August 1990, pp. 13–16.

24. Rajasekaran, B.; Warren, Dennis M.; and Babu, S.C., "Indigenous Natural-Resource Management Systems for Sustainable Agricultural Development—A Global Perspective," *Journal of International Development*, 3, 4, July 1991, 387–401.

25. Posey, Darrell A.; Frechione, John; Eddins, John; Francelino Da Silva, Luiz; Myers, Debbie; Case, Diane; and MacBeath, Peter, "Ethnoecology as Applied Anthropology in Amazonian Development," *Human Organization*, 43, 2, Summer 1984, pp. 95–107.

26. Apichatvullop, Yaowalak; Compton, J. Lim, "Local Participation in Social Forestry," *Regional Development Dialogue*, 14, 1, Spring 1993, pp. 34–42.

27. Oba, Gufu, "Environmental Education for Sustainable Development among the Nomadic Peoples: The UNESCO-IPAL Experience in Northern Kenya," *Nomadic Peoples*, 30, 1992, pp. 53–73.

28. Diouf, Moustapha, "New Directions for Rural Development in Senegal: Reconsidering Past Experience in Agricultural Research and Extension Policy," *Crossroads*, 31, 1991, pp. 77–84.

29. Atlieri, Miguel A., "The Ecology and Management of Insect Pests in Traditional Agroecosystems," in *Ethnobiology: Implications and Applications*, Vol. 1, pp. 132–158.

30. Ibid.

31. Ward, Brian, "Reflections on Seven Years at the Asian Disaster Preparedness Center," *Disasters*, 17, 4, December 1993, pp. 357–363.

32. Juma, Celestous, *The Gene Hunters* (London, Zed Press, 1990).

33. Michael Balick, Director, New York Botanical Garden's Institute of Economic Botany, quoted in *Time*, September 23, 1991, p. 52.

34. Joice, Christopher, "Prospectors for Tropical Medicines," *New Scientist*, 19 October 1991, pp. 36–40.

35. Juma.

36. *African Diversity*, June 1990.

37. Joice, pp. 36–40.

38. Ibid.

39. "Take Two Roots, Call Me . . . How wild animals use nature's medicine chest," *Newsweek*, February 3, 1992, p. 53.

40. Reported in Morell, Virginia, "The Really Secret Life of Plants," *New York Times*, December 18, 1994.

SIX / MEDICINE

1. Bannerman, Robert H.; Burton, John; Ch'en, Wen-Chieh (eds.), *Traditional Medicine and Health Care Coverage: A Reader for Health Administrators and Practitioners*, 1983, World Health Organization, Geneva, p. 11.

2. Kurup, P.N.V., "Ayurveda," in Bannerman et al., p. 50–59.

3. Ibid.

4. Kurup, pp. 52–54.

5. Ibid.

6. Singh, R.H., and Sinha, R.N., "Ayurvedic concept of the psychosomatic basis of health and disease," *Indian Journal of the History of Science*, 1, 11, May 1976, pp. 75–79.

7. Prakasam, K. Swayam, "Treatment of anaemia with special reference to iron in ancient Indian medicine Ayurveda: a historical perspective," *Bulletin of Indian Institute of History of Medicine*, Vol. XXI, pp. 99–104; Dube, C.B., and Kansal, C.M., "Diseases due to deficiencies of vital principles in the body," *Indian Journal of History of Science*, 16, 1, May 1981, pp. 104–107.

8. Tiwari, C.M.; Tripathi, S.N.; and Upadhyay, B.N., "Concept of worm infestation and infectious diseases in Indian medicine," in Lal, Shyam Kishore, and Parkhe, Arun M. (eds.). *Chikitsa*, Vol. 1, Dharmatma Tatyajimaharaj Memorial Medical Relief Trust, Shivapuri-Akalkot, India, 1979, pp. 37–38. On Susruta's descriptions of the spread of infectious diseases, see Singh, L.M.; Thakal, K.K.; and Deshpande, P.J., "Susruta's contributions to the fundamentals of surgery," *Indian Journal of the History of Science*, 5, 1, 1970.

9. Dixit, S.K.; Bhardwaj, H.C.; Sharma, M.; Sharma, A.V.; "Pharmaceutical Process before 10th Century A.D." Paper presented at the Congress on Traditional Sciences and Technologies of India, Nov 28-Dec 3, 1993, IIT Bombay

10. Singh et al.

11. Deshpande, P.J.; Sharma, K.R.; and Prasad, G.C., "Contributions of Susruta to the Fundamentals of Orthopedic Surgery," *Indian Journal of the History of Science*, 5, 1, 1970.

12. Sharma, Pandit Shiv, "Ayurvedic Medicine Past and Present," *Progress in Drug Research*, Vol. XV (Basel & Stuttgart, 1971, reprinted by Dabur Pvt. Ltd., New Delhi, 1975), p. 192.

13. Deshpande et al.

14. Krishnamurty, K.H., *A Source Book of Indian Medicine: An Anthology* (Delhi, B.R. Publishing Corporation, 1991), p. 8, section 8.

15. Deshpande et al.

16. Singh et al.

17. Krishnamurty, p. 8, section 9.

18. Elphinston, M., in *History of India* (London, 1843), quoted in Sangwan Satpal "European Impressions of Science and Technology in India (1650–1850)," in G. Kuppuram and K. Kumudamani (eds.), *History of Science and Technology in India*, Vol. V, "Science and Technology" (Delhi, Sundeep Prakashan, 1990), p. 71.

19. Srikanta Murthy, K.R., "Professional Ethics in Ancient Indian Medicine," *Indian Journal of the History of Medicine*, 18, 1973, pp. 45–49.

20. Trawick, Margeret, "The Ayurvedic Physician as Scientist," *Social Science and Medicine*, 24, 1987, p. 12.

21. Sharma, Romesh; Chaturvedi, C.; and Tewari, P.V., "Advances in Ayurvedic Pediatrics," *The Journal of Research and Education in Indian Medicine* (Varanasi, India, 1988), p. v.

22. Mitra, Jyotir, "Methodology for experimental research in ancient India," *Indian Journal of the History of Science*, 5, 1, 1970, pp. 68–75.

23. Ibid.

24. Schukla, H.C., "Ideas of scientific measurement in basic principles of

Ayurveda with special reference to somametry," *Indian Journal of the History of Science,* 5, 2, 1970, pp. 371–378.

25. Ibid.

26. Kurup, pp. 52–54; Sen, S.N., "History of Science in Relation to Philosophy and Culture in Indian Civilization," *Occasional Paper 6, Project of History of Indian Science, Philosophy and Culture,* New Delhi, 1993, pp. 19–20.

27. Singh et al.

28. Ibid.

29. Kurup, pp. 52–54; Sen, pp. 19–20.

30. Sharma, p. 189.

31. Obeyesekere, Gananath, "Science, Experimentation, and Clinical Practice in Ayurveda," in Charles Leslie and Allen Young (eds.), *Paths to Asian Medical Knowledge* (Berkeley, University of California Press, 1992), pp. 160–176.

32. Silva, K.T., "Ayurveda, malaria and the indigenous herbal tradition in Sri Lanka," *Social Science and Medicine,* 33, 2, 1991, pp. 153–160.

33. Uragoda, C.G., *A History of Medicine in Sri Lanka,* Sri Lanka Medical Association, 1987.

34. Bose, D.M.; Sen, S.N.; and Subbarayappa, B.V., *A Concise History of Science in India* (New Delhi, Indian National Science Academy, 1971), p. 259.

35. Jaggi, O.P., *Indian System of Medicine,* Vol. 4 of *History of Science and Technology in India* (Delhi, Atma Ram & Sons, 1973), pp. 209–212.

36. Heyn, Birgit, *Ayurvedic Medicine: The Gentle Strength of Indian Healing* (Wellinborough, Northamptonshire, Thorsons Publishing Group, 1987), p. 101.

37. Jaggi, O.P. *Indian System of Medicine,* Vol. 4 of *History of Science and Technology in India* (Delhi, Atma Ram & Sons, 1973), p. 209–212.

38. Arrian, *The Voyages of Nearchus from Indus to Euphrates—an account of the first navigation attempted by the Europeans in the Indian Ocean,* edited by William Vincent, London, 1797, p. 229.

39. Bose et al., p. 258.

40. Filliozat, J., "Ayurveda and Foreign Contacts," *Indian Journal of History of Medicine,* 1, 1 (1956).

41. Jaggi, p. 208.

42. Chowdhury, Amiya Kumar Roy, *Man, Malady and Medicine—History of Indian Medicine* (Calcutta, Das gupta & Co. Ltd., 1988), p. 68.

43. Hamilton, W., "History of Medicine, Surgery and Anatomy—Healing Art of Hindustan," 1831, p. 43, quoted in Jaggi, p. 212.

44. Jaggi, p. 213.

45. Jolly, C., *Indian Medicine,* English translation by C.G. Kashikar, Poona, 1951, cited p. 208 of Jaggi.

46. Bose et al., p. 257.

47. Arrian, p. 229.

48. Jaggi, p. 209.

49. Arrian, p. 229.

50. Chowdhury, p. 113.

51. Siddiqui, M.Z., *Studies in Arabic and Persian Medical Literature* (Calcutta University, 1959).

52. Jaggi, pp. 214–216.

53. Singer, S., *A Short History of Science* (Oxford, 1943), pp. 148, 162.

54. Brockway, Lucille, *Science and Colonial Expansion* (New York, Academic Press, 1979).

55. Gaitonde, P.D., *Portuguese Pioneers in India: Spotlight on Medicine* (Bombay, Popular Prakashan Private Ltd., 1983), p. vii-viii.

56. Patterson, T.J.S., "The relationship of Indian and European practitioners of medicine from the sixteenth century," in Meulenbeld, G. Jan, and Wujastyk, Dominik (eds.), *Studies on Indian Medical History* (Groningen, Egbert Forsten, 1987), p. 120.

57. Gaitonde, p. xii.

58. Ibid.

59. Ibid., p. 110.

60. Patterson, p. 120.

61. Gaitonde, pp. 118–138.

62. Gaitonde, p. xii.

63. Ibid.

64. Ibid., p. 140.

65. Quoted in Gaitonde, p. 146.

66. Gaitonde, p. 109.

67. Ibid., pp. 146–147.

68. Zoysa, A. de, and Palitharatna, C.D., "Medical science in Sri Lanka," in Petitjean, P., *Science and Empires* (Kluwer, 1992), p. 113.

69. Boxer, C.R., *The Christian Century in Japan, 1549–1650* (University of California Press, 1951).

70. Sergio, Antonio, in *Breve Interpretaçao da Historia de Portugal* (Lisbon, 1972), quoted in Gaitonde, p. 152.

71. Roberts, R.S., "The Early history of the import of drugs into Britain," in F.N.L. Poynter, *The Evolution of Pharmacy in Britain* (London, Pitman Medical, 1965), p. 165.

72. Zoysa, and Palitharatna, p. 113.

73. Paragraph 25 of the biography of Hendrik Van Rheede by J. Heniger, published by A. A. Balkema, quoted in "*State of India's Health*," Indigenous Health Services, New Delhi, p. 138.

74. Quoted in *State of India's Health*, p. 138.

75. Ibid., p. 139.

76. Madras Group, "Indian Science and Technology in the Eighteenth Century," *Readings from the PPST Bulletin*, Madras, PPST Foundation, n.d., p. 30.

77. Manucci, Niccolao, *Storia Do Mogor or Mogul India (1653–1708)*, Vol. 1, translated with Introduction and Notes by William Irvine (London, John Murray, 1907), p. 301.

78. *State of India's Health*, p. 140.

79. Patterson, p. 127.

80. Krishnamurty, *A Source Book of Indian Medicine*, p. 8, section 27.

81. Ibid.

82. Ibid., p. 8, section 26.

83. Neuburger, Max, *History of Medicine,* translated by Ernest Playfair (London, Oxford University Press, 1919), quoted in Jaggi, p. 236.

84. Harrison, Mark, "Tropical medicine in nineteenth-century India," *British Journal of the History of Science*, 25, 1992, pp. 299–318.

85. Quoted in Edwardes, Michael, *British India 1772–1947* (London, Sidgwick, 1967), p. 125.

86. Ibid.

87. Chaudhury, Ranjit Roy, *Herbal Medicine for Human Health*, WHO, New Delhi, 1992, p. 5.

88. Rao, I. Sanjeeva, "Surgical treatment of Fistula-in-Ano: Historical evolution of a technique," *Bulletin of Institute of the History of Medicine*, Vol. iii, pp. 169–172.

89. Jain, S.K., *Medicinal Plants* (New Delhi, National Book Trust, 1968), p. 4.

90. Chittampalli, Pudma, and Mulcahy, F. David, "Honey and sugar in the treatment of wounds and ulcers in biomedicine and in the ayurveda," in Darrel A. Posey and William Leslie Overal (eds.), *Ethnobiology: Implications and Applications. Proceedings of the First International Congress of Ethnobiology*, Vol. 2 (Belem, Brazil, Museo Emilio Goeldi, 1990), pp. 155–164.

91. Labadie, R.P, and de Silva, K.T.D., "Centella asiatica (L.) Urban in Perspective: an evaluative account," in Meulenbeld, G. Jan and Wujastyk, Dominik (eds.), *Studies on Indian Medical History* (Gröningen, Egbert Forsten, 1987), p. 218.

92. Chowdhury, p. 141.

93. *Medical and Aromatic Plants Abstract* Vol. 6 No. 6 1984 p. 468; Vol. 8 No. 2 1986 p. 120; Vol. 8 No. 2 1986 p. 122; Vol. 12 No. 1 1989 p. 13; Vol. 13 No. 4 1991 p. 319; Vol. 11 No. 1 1989 p. 23; Vol. 13 No. 5 1991 p. 413; Vol. 15 No. 4 1993 p. 365; Vol. 10 No. 4 1988 p. 308; Vol. 7 No. 6 1985, p. 530; Vol. 11 No. 2 1989 p. 143; Vol. 6 No. 6 1984 p. 471; Vol. 7 No. 6 1985 p. 528; Vol. 7 No. 4 1985 p. 313; Vol. 9 No. 5 1987 p. 501; Vol. 6 No. 4 1984 p. 291; Vol. 13 No. 4 1991 p. 317; Vol. 9 No. 6 1987 p. 578.

94. Sharma, Romesh; Chaturvedi, C.; and Tewari, P.V. "Advances in Ayurvedic Pediatrics," *The Journal of Research and Education in Indian Medicine*, Varanasi, India, 1988, pp. 20–25.

95. Dupart, Thierry, *Contribution to the study of a folk medicine, the ayurveda: double-blind test between a holarrhema antidysenterica wall, And metronidazole in the treatment of acute intestinal amoebiasis* [*Essai en double aveugle entre un medicament à base d'écorces d'holarrhena antidysenterica wall, Et le metronidazole dans le traitement de l'amibiase intestinale aigue*]. MD thesis, Universite de Bourgogne (France), 1993, pp. 132.

96. *Medical and Aromatic Plants Abstract* Vol. 9 No. 4 1987 p. 402; Vol. 8 No. 2 1986 p. 122; Vol. 6 No. 6 1984 p. 471; Vol. 12 No. 2, 1990 p. 98; Vol. 9 No. 4 1987 p. 401; Vol. 11 No. 2 1989 p. 144; Vol. 10 No. 6 1988 p. 534; Vol. 10 No. 6 1988 p. 534; Vol. 11 No. 1 1989 p. 22; Vol. 8 No. 2 1986 p. 120; Vol. 12 No. 6 1990 p. 493; Vol. 9 No. 4 1987 p. 405; Vol. 6 No. 6 1984 p. 469; Vol. 6 No. 6 1984 p. 474; Vol. 10 No. 6 1988 p. 532.

97. Chaturvedi, G.N., *Clinical Studies on Kamla (jaundice) and Yakti Rogas (liver disorders) with Ayurvedic Drugs* (Central Council for Research in Ayurveda and Siddha, New Delhi, 1988).

98. Chaudhury, p. 5.

99. *Medical and Aromatic Plants Abstract* Vol. 6 No. 6 1984 p. 468; Vol. 6 No. 6 1984 p. 471; Vol. 9 No. 6 1987 p. 578; Vol. 7 No. 6 1985 p. 529; Vol. 7 No. 3 1985

p. 139; Vol. 8 No. 6 1986 p. 523; Vol. 6 No. 4 1984 p. 288; Vol. 9 No. 4 1987 p. 401; Vol. 7 No. 3 1985 p. 138; Vol. 8 No. 2 1986 p. 121; Vol. 13 No. 4 1991 p. 317; Vol. 11 No. 3 1989 p. 257; Vol. 11 No. 3 1989 p. 257; Vol. 7 No. 1 1985 p. 23; Vol. 9 No. 6 1987 p. 577; Vol. 6 No. 6 1984 p. 468; Vol. 7 No. 3 1985 p. 138; Vol. 10 No. 4 1988 p. 307; Vol. 7 No. 3 1985 p. 210; Vol. 6 No. 6 1984 p. 474; Vol. 11 No. 6 1989 p. 514; Vol. 9 No. 4 1987 p. 400.

100. Chaudhury, p. 5.

101. Dash, Bhagwan, and Basu, R.N., "Methods of Sterilization and Conception in Ancient and Medieval India," *Indian Journal of the History of Science*, 3, 1, 1969, pp. 9–24.

102. *Medical and Aromatic Plants Abstract* Vol. 6 No. 3 1984 p. 203; Vol. 15 No. 4 1993 p. 365; Vol. 7 No. 6 1985 p. 529; Vol. 7 No. 4 1985 p. 412; Vol. 13 No. 4 1991 p. 319; Vol. 7 No. 4 1985 p. 312; Vol. 7 No. 4 1985 p. 412; Vol. 7 No. 3 1985 p. 210; Vol. 6 No. 4 1984 p. 289; Vol. 7 No. 4 1985 p. 313; Vol. 8 No. 6 1986 p. 518; Vol. 10 No. 4 1988 p. 307; Vol. 10 No. 2 1988 p. 93; Vol. 6 No. 6 1984 p. 472; Vol. 10 No. 2 1988 p. 94; Vol. 11 No. 2 1989 p. 144; Vol. 7 No. 3 1985 p. 210.

103. Laxmipathy, A., "Anti-anxiety Effect of an Ayurvedic Compound Preparation—A Crossover Trial," Paper presented at the Congress on Traditional Sciences and Technologies of India, November 28–December 3, 1993, IIT Bombay.

104. Sharma et al., pp. 1–9.

105. Ayensu, Edward S., "Endangered plants used in traditional medicine," in Bannerman, Robert H.; Burton, John; and Ch'en, Wen-Chieh (eds.), *Traditional Medicine and Health Care Coverage: A Reader for Health Administrators and Practitioners* 1983 WHO, Geneva, p. 177.

106. Labadie and de Silva, p. 210.

107. Ayensu, p. 178.

108. *Medical and Aromatic Plants Abstract* Vol. 15 No. 4 1993 p. 363; Vol. 11 No. 3 1989 p. 254; Vol 6. No. 6 1984 p. 473; Vol. 6 No. 6 1984 p. 474; Vol. 7 No. 3 1985 p. 138; Vol. 9 No. 6 1987 p. 576; Vol. 13 No. 5 1991 p. 412; Vol. 10 No. 4 1988 p. 308; Vol. 9 No. 6 1987 p. 577; Vol. 11 No. 4 1989 p. 346; Vol. 13 No. 5 1991 p. 412; Vol. 9 No. 4 1987 p. 400; Vol. 6 No. 6 1984 p. 468; Vol. 9 No. 4 1987 p. 404; Vol. 8 No. 6 1986 p. 521; Vol. 7 No. 4 1985 p. 413; Vol. 10 No. 2 1988 p. 93; Vol. 13 No. 4 1991 p. 314; Vol. 12 No. 4 1990 p. 287; Vol. 13 No. 5 1991 p. 412.

109. Treleaven, J; Meller, S.; Farmer, P.; Birchall, D.; Goldman, J.; and Piller, G., "Arsenic and Ayurveda," *Leukemia-Lymphoma*, 10, 4–5, July 1993, pp. 343–345.

110. Karnik, C.R., and Jopat, P.D., "Researches towards finding of a new drug Tylophora asthmatica W.A.—A clinical investigation for Asthma and Bronchial infection," in Shyam Kishore Lal and Arun M. Parkhe (eds.), *Chikitsa*, Vol. 1, Dharmatma Tatyajimaharaj Memorial Medical Relief Trust, Shivapuri-Akalkot, India, 1979, pp. 64–68.

111. Salerno, John William, *Selective growth inhibition of human colon adenocarcinoma and malignant melanoma cell lines by sesame oil in vitro (ayurveda)*, Maharishi International University Ph.D., 1991, p. 98.

112. *Medical and Aromatic Plants Abstract* Vol. 9 No. 6 1987 p. 576; Vol. 11 No. 1

1989 p. 22; Vol. 11 No. 2 1989 p. 144; Vol. 9 No. 6 1987 p. 577; Vol. 12 No. 4 1990 p. 283; Vol. 9 No. 6 1987 p. 574.

113. Karnik and Jopat, pp. 64–68.

114. Weiner, Michael A., *Weiner's Herbal: Guide to herb medicine* (Mill Valley, CA, Quantum Books, 1991), p. 244.

115. Heyn, p. 32.

116. Dharam Vir, K.V.J., "Experimental and Clinical Study of the Tulasi (Ocimum sanctum Linn.) in Relation to Cholesterol and Arterial Hypertension," in *Bharat Nirman*, National Conference on Herbal Science, Constitution Club, New Delhi, 1989, pp. 53–64.

117. *Medical and Aromatic Plants Abstract* Vol. 12 No. 3 1990 p. 196; Vol. 10 No. 6 1988 p. 533; Vol. 11 No. 2 1989 p. 144; Vol. 8 No. 6 1986 p. 518; Vol. 6 No. 3 1984 p. 202; Vol. 6 No. 6 1984 p. 470; Vol. 6 No. 6 1984 p. 468; Vol. 7 No. 3 1985 p. 210; Vol. 12 No. 3 1990 p. 194 Vol. 8 No. 1 1986 p. 17; Vol. 13 No. 4 1991 p. 319; Vol. 9 No. 4 1987 p. 404; Vol. 9 No. 4 1987 p. 405; Vol. 9 No. 4 1987 p. 404; Vol. 6 No. 6 1984 p. 473; Vol. 7 No. 3 1985 p. 138; Vol. 10 No. 4 1988 p. 306; Vol. 10 No. 2 1988 p. 94; Vol. 7 No. 3 1985 p. 139.

118. *Medical and Aromatic Plants Abstract* Vol. 10 No. 6 1988 p. 533; Vol. 6 No. 5 1984 p. 374; Vol. 10 No. 4 1988 p. 308; Vol. 7 No. 1 1985 p. 24; Vol. 12 No. 2 1990 p. 97; Vol. 6 No. 6 1984 p. 471; Vol. 6 No. 5 1984 p. 378; Vol. 11 No. 2 1989 p. 142; Vol. 10 No. 2 1988 p. 91; Vol. 9 No. 4 1987 p. 400; Vol. 9 No. 5 1987 p. 501; Vol. 8 No. 6 1986 p. 518; Vol. 10 No. 5 1988 p. 429; Vol. 15 No. 4 1993 p. 364; Vol. 11 No. 2 1989 p. 142; Vol. 6 No. 5 1984 p. 378; Vol. 10 No. 4 1988 p. 309; Vol. 15 No. 4 1993 p. 365; Vol. 11 No. 2 1989 p. 142; Vol. 13 No. 4 1991 p. 312; Vol. 6 No. 6 1984 p. 474; Vol. 7 No. 1 1985 p. 25; Vol. 12 No. 4 1990 p. 283; Vol. 7 No. 4 1985 p. 412; Vol. 12 No. 4 1990 p. 285; Vol. 9 No. 4 1987 p. 403; Vol. 6 No. 6 1984 p. 470.

119. Shetty, B.R., and Laxmipathy, A., "Studies on the Rasayana Effect of an Ayurvedic Compound Drug in Apparently Normal Aged Persons," paper presented at the Congress on Traditional Sciences and Technologies of India, November 28–December 3, 1993, IIT Bombay.

120. *Medical and Aromatic Plants Abstract* Vol. 6 No. 3 1984 p. 203; Vol. 7 No. 3 1985 p. 139; Vol. 9 No. 6 1987 p. 577; Vol. 10 No. 3 1988 p. 212; Vol. 11 No. 2, 1989 p. 144; Vol. 7 No. 3 1985 p. 138; Vol. 10 No. 6, 1988 p. 531; Vol. 10 No. 2 1988 p. 93; Vol. 9 No. 6 1987 p. 576; Vol. 6 No. 6 1984 p. 470; Vol. 9 No. 4 1987 p. 403; Vol. 7 No. 4 1985 p. 313.

121. *Medical and Aromatic Plants Abstract* Vol. 6 No. 6 1984 p. 473; Vol. 8 No. 2 1986 p. 121; Vol. 7 No. 6 1985 p. 529; Vol. 6 No. 6 1984 p. 473; Vol. 9 No. 5 1987 p. 501; Vol. 8 No. 6 1986 p. 522; Vol. 12 No. 3, 1990 p. 195; Vol. 9 No. 6 1987 p. 574; Vol. 13 No. 4, 1991 p. 314; Vol. 6 No. 6 1984 p. 469.

122. Heyn, p. 32.

123. Chaudhury, p. 5.

124. Radford, Tim, "Trust me, I'm a witch doctor," *The Guardian*, London, Thursday, June 27, 1996, pp. 6–7.

125. Farnsworth, Norman R., "The NAPALERT database as an information source for application to traditional medicine," in Bannerman, Robert H.; Burton, John; and Ch'en, Wen-Chieh (eds.), *Traditional Medicine and Health Care*

Coverage: A Reader for Health Administrators and Practitioners, 1983, WHO, Geneva, p. 184.

126. Elisabetsky, Elaine, and de Moraes, Jaolo A.R., "Ethnopharmacology: A Technological Development Strategy," in *Ethnobiology: Implications and Applications,* Vol. 2, pp. 111–118.

127. Singh, V.K., and Khan, Abrar M., *Medicinal Plants and Folklore: A strategy towards the conquest of human ailments,* Today and Tomorrow Printers and Publishers, 1990, p. 4.

128. RAFI (Rural Advancement Foundation International), *The Role of Indigenous Knowledge in the Conservation and Development of Biodiversity,* 1992, p. 7.

129. Farnsworth, N.R., "Screening Plants for New Medicines," in Wilson, E.O. (ed.), *Biodiversity* (Washington, National Academy Press, 1988).

130. RAFI, p. 7.

131. Ibid., p. 2.

132. Gorinsky, Conrad, "Ethnobiology and Medicine," in *Ethnobiology: Implications and Applications,* Vol. 2, pp. 119–123.

133. Elisabetsky and de Moraes, pp. 111–118.

134. Weiner, p. 227.

135. Jain, S.K., *Medicinal Plants* (New Delhi, National Book Trust, 1968), p. 1.

136. Ibid., p. 2.

137. Ahmad, R.U., and P.C. Srivasta, "About the Utilization and Cultivation of Medicinal Flora of J&K State," in C.K. Atal and B.M. Kapur (eds.), *Cultivation and Utilization of Medicinal and Aromatic Plants,* RRL (CSIR), Jammu Tawi, 1977, pp. 150–153.

138. Krishnamurty, p. 8, section 9.

139. Chowdhury, p. 138.

140. Ayensu, p. 180.

141. Kurup, p. 55.

142. Farnsworth, "The NAPALERT database," p. 237.

143. Jain, p. vi.

144. Ibid., p. 1.

145. Farnsworth, "The NAPALERT database," p. 237.

146. Chaudhury, p. 10.

147. Ibid.

148. Sharma et al., p. v.

149. Ibid., pp. 25–125.

150. Ibid.

151. Singh and Khan, p. ii.

152. Ibid., p. 4.

153. Ibid.

154. Ibid., p. 59.

155. Ibid., p. 103.

156. Ibid., p. 163.

157. Akerle C.O., "Role of Traditional Medicine Programme in Health care delivery system update from W.H.O," keynote address, International Seminar on Unani Medicine, New Delhi, 1987.

158. Krishnamurty, p. 8, section 26.

159. Ibid.
160. McAlpine, Thorpe, and Warrier, Ltd., *EEC Market for Herbal Medicines 1990–1994*, London, 1990, p. 5.
161. Chopra, Deepak, *Perfect Health: The Complete Mind/Body Guide* (New York, Harmony Books, 1990).
162. Chaudhury, p. 4.
163. Chaudhury, p. 26.

SEVEN / MATHEMATICS

1. Zaslavasky, Claudia, *Africa Counts: Number and Pattern in African Culture* (New York, Lawrence Hill Books, 1973).
2. D'Ambrasio, Ubiratan, "The cultural dynamics of the encounter of two worlds after 1492 as seen in the development of scientific thought," *Science and the meeting of two worlds: Impact,* No. 167, UNESCO, Paris, Vol. 42, No. 3, 1992, pp. 205–215; D'Ambrasio, Ubiratan, *Etnomatematica. Arte ou Tecnica de Explicar e Conhecer* (Sao Paulo, Editora Atica, 1990).
3. Restivo, Sal Episteme, *Mathematics in Society and History* (Boston and London, Dordecht, 1992); Struik, Dirk J., "The Sociology of Mathematics Revisited: A Personal Note," *Science & Society,* Fall 1986, pp. 280–299.
4. Bose, D.M.; Sen, S.N.; and Subbarayappa, B.V., *A Concise History of Science in India* (New Delhi, Indian National Science Academy, 1971), p. 148.
5. Joseph, George Gheverghese, *The Crest of the Peacock: Non-European Roots of Mathematics* (London, Penguin, 1990), p. 234.
6. Srinivasiengar, C.N., *The History of Ancient Indian Mathematics* (Calcutta, The World Press Private, Ltd., 1967), pp. 3–5.
7. Bell, E.T., *The Development of Mathematics* (New York, McGraw Hill, 1945), pp. 54–63.
8. Singh, Navjyoti, "Insight into foundations of modern mathematics through a comparative study of foundations of mathematics in India, China and Greece," *PPST Bulletin,* May 1985, pp. 53–71.
9. Singh, Navjyoti, "Foundations of Logic in Ancient India: Linguistics and Mathematics," in Rahman, A. (ed.), *Science and Technology in Indian Culture: A Historical Perspective* (New Delhi, Nistads, 1984), pp. 79–106.
10. Singh, "Foundations of Logic in Ancient India," pp. 79–106.
11. Ibid.
12. Joseph, p. 217.
13. Singh, "Foundations of Logic in Ancient India," pp. 79–106.
14. Singh, "Insight into foundations of modern mathematics," pp. 53–71.
15. Ibid.
16. Srinivasiengar, pp. 12–15.
17. Singh, "Insight into foundations of modern mathematics," pp. 53–71.
18. Ibid.
19. Ibid.
20. Sangwan, Satpal, "European Impressions of Science and Technology in India (1650–1850)," in Kuppuram, G., and Kumudamani, K. (eds.), *History of Science and Technology in India,* Vol. V, "Science and Technology" (Delhi, Sundeep Prakashan, 1990), p. 65.

21. Robertson, W., *An Historical disquisition Concerning the knowledge which the Ancients had of India*, London, printed for A. Stratham and T. Condell, E. Balfour at Edinburgh, 1791, pp. 304, 308.

22. Srinivasiengar, p. 5.

23. Billard, Roger, "Aryabhata and Indian Astronomy," *Indian Journal of History of Science*, 12, 1977, pp. 207–224.

24. Jaggi, O.M., *History of Science and Technology in India*, Vol. 6, *Indian Astronomy and Mathematics* (Delhi, Atma Ram, 1986), p. 145.

25. Joseph, p. 22.

26. Abdi, Wazir Hasan, "Interaction of West Asian and Central Asian Science," in Kuppuram, G., and Kumudamani, K. (eds.), *History of Science and Technology in India*, Vol. IV, "Science" (Delhi, Sundeep Prakashan, 1990), p. 339.

27. Bose et al., p. 179–180.

28. Datta, Bibhutibhusan, and Singh, Avadhesh Narayan, *History of Hindu Mathematics, Part 1: Numeral Notation and Arithmetic* (Lahore, Motilal Banarsi Das, 1935), p. 146.

29. Bose et al., p. 179–180.

30. Smith D.E., *History of Mathematics*, Vol. 11 (Boston, Ginn and Co., 1958), pp. 115–116.

31. Bose et al., p. 180.

32. Quoted in Smith, D.E., p. 488.

33. Datta and Singh, p. 175.

34. Smith, D.E., p. 144–148.

35. *Bijaganita of Bhaskara II*, p. 99, quoted in Jaggi, *History of Science and Technology in India*, Vol. 6, *Indian Astronomy and Mathematics*, p. 197.

36. Jaggi, *Indian Astronomy and Mathematics*, p. 196.

37. Joseph, p. 272.

38. Ibid., p. 281.

39. Ibid., p. 18.

40. Bose et al., p. 134–135.

41. Bose et al., p. 180.

42. Dickson, Leonard Eugene, *History of Theory of Numbers*, Vol. 2 (New York, Chelsea Publishing Company, 1952, reprinted 1966), pp. 192–193.

43. Srinivasiengar, pp. 57–61.

44. Quoted in Bose et al., p. 167.

45. Singh, Parmanand, "The So-called Fibonacci Numbers in Ancient and Medieval India," *Historia Mathematica*, 12, 1985, pp. 229–244.

46. Datta, Bibhutibhusan, and Singh, Avadhesh Narayan, "Use of permutations and combinations in India," *Indian Journal of History of Science*, 27, 3, 1992.

47. Joseph, p. 275.

48. Quoted in Bose, et al., p. 169.

49. Joseph, pp. 276–279.

50. Srinivasiengar, p. 110.

51. Selenius, C., "Rationale of the Chakravala process of Jayadeva and Bhaskara 11," *Historia Mathematica*, 12, 229–44, p. 180.

52. Joseph, p. 242.

53. Ibid.

54. Ibid., p. 250.

55. Singh, Navjyoti, "Jain theory of measurement and theory of transfinite numbers," *Journal of Asiatic Society*, XXX, 1988.

56. Ibid.

57. Ibid.

58. Ibid., p. 78.

59. Singh, Navjyoti "Jain theory of measurement and theory of transfinite numbers."

60. Ibid.

61. Srinivasiengar, p. 24.

62. Joseph, p. 251.

63. Srinivasiengar, p. 81.

64. Bag, A.K., "Binomial Theorem in Ancient India," in Kuppuram and Kumudamani, "Science," p. 191.

65. Ibid., pp. 191–199.

66. Bose et al., p. 590.

67. Raina, Dhruv, and Singh, Navjyoti, "The Jaina Theory of Motion," *Arhat Vacana*, 1, September 1988, pp. 91–96.

68. Ibid.

69. Roy, Sourin, "Al-Biruni and Hindu Speculations on Gravitation," *Indian Journal of History of Science*, 10, 2, November 1975, p. 218.

70. Quoted in Roy, p. 220.

71. Al Biruni's Indica I, p. 276, quoted in Roy, p. 220.

72. Roy, p. 221.

73. Quoted in Roy, p. 218.

74. Quoted in Roy, p. 219.

75. Ibid.

76. Jaggi, *Indian Astronomy and Mathematics*, p. 227.

77. Sarasvati Amma, pp. 211–213.

78. *See* Boyer, Carl B., *The History of the Calculus and Its Conceptual Development* (New York, Dover, London; Constable, 1959), p. 108.

79. Sarasvati Amma, p. 217.

80. Joseph, p. 299.

81. Jain, L.C., "Divergent Sequences Locating Transfinite Sets," in *Trilokasara*, 12, 1, May 1977, p. 59.

82. Jain, L.C., "Set Theory in Jaina School of Mathematics," *Indian Journal of the History of Science*, 8, 1 and 2, 1973, pp. 1–24.

83. Jain, L.C., "Divergent Sequences Locating Transfinite Sets," in *Trilokasara*, 12, 1, May 1997, p. 59.

84. Srinivasiengar, pp. 67–69.

85. Parameswaran, S., "Kerala's Heritage in Mathematics," in Kuppuram and Kumudamani, Vol. 2, p. 177.

86. Ibid., pp. 183–184.

87. Bag, A.K., "Indian Literature on Mathematics During 1400–1800 AD," in Kuppuram and Kumudamani, *History of Science and Technology in India*, Vol. 2, p. 113.

88. Sarma, K.V., "Some Highlights of Astronomy and Mathematics in Medieval India," in Kuppuram and Kumudamani, *History of Science and Technology in India*, Vol. 2, p. 418.

89. Ibid.

90. Bag, A.K., "Madhava Sine and Cosine Series," in Kuppuram and Kumudamani, *History of Science and Technology in India*, Vol. 2, p. 318.

91. Joseph, p. 20.

92. Parameswaran, pp. 180–183.

93. Beg, "Madhava Sine and Cosine Series," p. 318.

94. Parameswaran, p. 185.

95. Sarasvati Amma, p. 108.

96. Parameswaran, p. 185.

97. Sarma, p. 418.

98. Ibid., p. 417.

99. Srinivasiengar, p. 154.

100. Sarma, p. 417.

101. Ramasubramanian, K.; Sirinivas, M.D.; and Sriram, M.S., "Heliocentric model of planetary motion in the Kerala school of Indian astronomy," paper presented at the Congress on Traditional Sciences and Technologies of India, November 28–December 3, 1993, IIT Bombay; Ramasubramanian, K.; Sirinivas, M.D.; and Sriram, M.S., "Modification of the earlier Indian planetary theory by the Kerala astronomers (c. 1500 AD) and the implied heliocentric picture of planetary motion," *Current Science*, 66, 10, 25 May 1994, p. 784.

102. Ibid.

103. Ibid.

104. Sarasvati Amma, pp. 159–166.

105. Ramasubramanian et al., "Modification of the earlier Indian planetary theory by the Kerala astronomers," p. 784.

106. Jhunjhunwala, Ashok, *Indian Mathematics: An Introduction* (New Delhi, Wiley Eastern, Limited, 1993), p. 1. Parthasarathi, Ranjani, and Jhunjhunwala, Ashok, "Modified straight division: A computer implementation of multiple-precision division," *Microprocessing and Microprogramming*, 41, 1995, pp. 193–209.

107. Jhunjhunwala, *Indian Mathematics*, pp. 2–6.

108. Jhunjhunwala, Ashok, "Recent Research in Indian Integral Algorithms," paper presented at the Congress on Traditional Sciences and Technologies of India, November 28–December 3, 1993, IIT Bombay.

109. Parthasarathi and Jhunjhunwala, pp. 193–209.

110. Jhunjhunwala, *Indian Mathematics*, p. 99.

111. Kline, Morris, *Mathematics in Western Culture* (New York, Oxford University Press, 1965), p. 240.

112. Krishnan, C.N.; Muthulaksmi, A.; and Srividya, V., "Modified Bhaskara Approximations for a Set of Common Functions," paper presented at the Congress on Traditional Sciences and Technologies of India, November 28–December 3, 1993, IIT Bombay.

113. Singh, "Foundations of Logic in Ancient India," pp. 79–106.

114. Sangal, Rajeev, "Recent Advances in Indian Grammatical Tradition," paper presented at the Congress on Traditional Sciences and Technologies of India, November 28–December 3, 1993, IIT Bombay. Bharati, Akshar; Chaitanya, Vineet; and Sangal, Rajeev, *Natural Language Processing: A Paninian Perspective* (New Delhi, Prentice-Hall of India Private, Limited, 1995).

115. Bharati, Akshar; Chaitanya, Vineet; and Sangal, Rajeev, *A Computational*

Grammar Based on Paninian Framework, Dept. of Computer Science, IIT Kanpur, 1993, p. 37.

116. Briggs, Rick, "Knowledge Representation in Sanskrit and Artificial Intelligence," *AI Magazine,* 6, 1, Spring 1985, pp. 32–39.

117. Srihari, S.N.; Rapaport, W.J.; and Kumar, D., *On Knowledge Representation Using Semantic Networks and Sanskrit,* Technical Report, Dept. of Computer Science, University at Buffalo, State University of New York, 1987.

118. Bharati et al., p. 65.

119. Ibid.

120. Ibid., p. iii.

121. Bharati, Akshar; Chaitanya, Vineet; and Sangal, Rajeev, "Paninian Theory Applied to English," *Computer Science and Informatics: Journal of the Computer Society of India,* 1996, forthcoming Bombay.

122. Ramanna, Raja, "Concept of discreetness, continuity and the Cantor continuum theory as related to the life-time and masses of elementary particles," *Current Science,* 65, 5, 25 September 1993.

123. Singh, "Insight into foundations of modern mathematics," pp. 60–65.

124. Kline, pp. 6–7.

125. Singh, "Insight into foundations of modern mathematics," pp. 60–65.

126. Sirinivas, M.D., "The Methodology of Indian Mathematics and Its Contemporary Relevance," *PPST Bulletin,* Serial No. 12, September 1987, p. 12–13.

127. Singh, "Insight into foundations of modern mathematics," pp. 60–65.

128. Ibid.

129. Sarma, p. 420–421.

130. Sirinivas, pp. 1–35.

131. Ibid., p. 12–13.

132. Jain, p. 70.

EIGHT / A SEARCH FOR NEW PSYCHOLOGIES

1. Huber, Jack, *Through an Eastern Window* (Boston, Houghton Mifflin Company, 1967), p. 1.

2. Huber.

3. Ibid., p. 108.

4. Ibid., pp. 106–117.

5. Peppetier, Kenneth R., *Mind as Healer: Mind as Slayer* (Delacorte Press, 1977), p. 26.

6. Miller, N.E., "Learning of Visceral and Glandular Responses," *Science,* 163, 1969, pp. 434–445.

7. Benson, Herbert, *The Relaxation Response* (New York, Avon Books, 1975), p. 79–80.

8. Anand, B.W.; China, G.S.; and Singh, B., "Some Aspects of Electroencephalographic Studies in Yogis," *Electroencephalography and Clinical Neurophysiology,* 13, 1961, pp. 452–456; "Studies on Shri Ramananda Yogi During His Stay in an Air-Tight Box," *Indian Journal of Medical Research,* 49, 1961, pp. 82–89; Kasamatsu, A., and Hirai, T., "An Electroencephalographic Study on the Zen Meditation (Zazen)," *Folia Psychiatrica et Neurologia Japonica,* 1966, pp. 315–336.

9. Corrick, James A., *The Human Brain: Mind Over Matter* (New York, Arco Publishing Inc., 1983), p. 165.

10. Facklam, Margery and Howard, *The Brain* (New York, Harcourt Brace Jovanovich Publishers, 1982), p. 44.

11. Benson, *The Relaxation Response,* p. 9–10.

12. Ibid., pp. 75–88.

13. Ornish, Dean, *Reversing Heart Disease,* pp. 140–42; Benson, *Relaxation Response,* pp. 75–88.

14. Ornish, pp. 140–142.

15. Peppetier, p. 191.

16. Benson, *The Relaxation Response,* pp. 75–88.

17. Ibid., p. 26–27.

18. Ibid., p. 110–112.

19. Ibid., p. 7.

20. Ibid., p. 168–169.

21. Benson, Herbert; Malhotra, M.S.; Goldman, Ralph F.; Jacobs, Gregg-D.; et al., "Three case reports of the metabolic and electroencephalographic changes during advanced Buddhist meditation techniques," *Behavioral Medicine,* 16, 2, Summer 1990, pp. 90–95.

22. Benson, H., "Body Temperature Changes During the Practice og gTummo yoga," *Nature,* 298, 1982, p. 402.

23. Benson, Herbert, *Your Maximum Mind* (New York, Random House, 1987), p. 22.

24. Wilson, Reid, *Don't Panic: Taking Control of Anxiety Attacks* (New York, Harper and Row, 1996).

25. Walsh, Roger, "Two Asian Psychologies and Their Implications for Western Psychotherapists," *American Journal of Psychotherapy,* XLII, 4, October 1988, pp. 546–548.

26. Ibid.

27. James, William, *Principles of Psychology,* 1910, reprinted 1950 Dover, New York, p. 424.

28. Varela, Francisco F.; Thompson, Evan; and Rosch, Eleanor, *The Embodied Mind: Cognitive Science and Human Experience* (Cambridge, MA, The MIT Press, 1993), pp. 31–33.

29. *Encyclopedia of Buddhism,* published by The Government of Ceylon, Colombo, 1961, Vol. 1, Fascicle 4, p. 559.

30. Walsh, Roger, "The Search for Synthesis: Transpersonal Psychology and the Meeting of East and West, Psychology and Religion, Personal and Transpersonal," *Journal of Humanistic Psychology,* 32, 1, Winter 1992, pp. 32–38.

31. Walsh, "The Search for Synthesis," pp. 32–38.

32. Ibid.

33. *Encyclopedia of Buddhism,* Vol. 1, Fascicle 1, p. 379.

34. Organ, Troy Wilson, *Philosophy and the Self: East and West* (Selingrove, Susquehanna University Press, 1987), p. 162.

35. Kalupahana, David, J., "The Buddhist conception of time and temporality," *Philosophy East and West,* XXIV, 2, April 1974, p. 186.

36. Bogart, Greg, "The Use of Meditation in Psychotherapy: A Review of the Literature," *American Journal of Psychotherapy,* XLV, 3, July 1991, pp. 383–412.

37. Ibid.

38. Ibid.

39. Goleman, Daniel, *The Meditative Mind* (New York, G.P. Putnam & Sons, 1988), p. 164.

40. Sudsuang, R.; Chentanez, V.; and Veluvan, K., "Effect of Buddhist meditation on serum cortisol and total protein levels, blood pressure, pulse rate, lung volume and reaction time," *Physiology-Behavior,* 50, 3, September 1991, pp. 543–548.

41. Peppetier, p. 250.

42. Ornish, pp. 140–142.

43. Ornstein, Robert, and Sobel, David, *The Healing Brain* (New York, Simon and Schuster, 1987), p. 155.

44. Cousins, Normam, *Head First: The Biology of Hope* (New York, E.P. Dutton, 1989), p, 236.

45. Ornish, pp. 140–142.

46. *U.S. News and World Report,* August 3, 1993.

47. Benson, *Your Maximum Mind,* pp. 138–144.

48. Ornish, pp. 140–142.

49. Peppetier, p. 250.

50. Beck, Aaron T., and Emery, Gary, *Anxiety Disorders and Phobias: A Cognitive Perspective* (New York, Basic Books Inc., 1985), pp. 210–231.

51. Moyers, Bill, *Healing and the Mind* (New York, Doubleday, 1993), pp. 71–86.

52. Warpeha, A., and Harris, J., "Combining traditional and nontraditional approaches to nutrition counseling," *Journal of the American Dietary Association,* 93, 7, July 1993, pp. 797–800.

53. Voigt, Harrison, "Enriching the sexual experience of couples: The Asian traditions and sexual counseling," *Journal of Sex and Marital Therapy,* 17, 3, Fall 1991, pp. 214–219.

54. Mikulas, William L.,"Buddhism and Behavior Modification," *Psychological Record,* 31, 3, Summer 1981, pp. 331–342.

55. De Silva, Padmal, "Buddhist psychology: A review of theory and practice," *Current Psychology Research and Reviews,* 9, 3, Fall 1990, pp. 236–254.

56. Ibid.

57. Ibid., pp. 661–678.

58. De Silva, Padmal, "Early Buddhist and modern behavioral strategies for the control of unwanted intrusive cognitions," *Psychological Record,* 35, 4, Fall 1985, pp. 437–443.

59. Moyers, p. xii.

60. Ibid.

61. Ibid., pp. 115–143.

62. Kabat-Zinn, Jon; Massion, Ann O.; Kristeller, Jean; Petersen, Linda Gay; Fletcher, Kenneth E.; Pbert, Lori; Lenderking, William R.; and Santorelli, Saki F., "Effectiveness of a Meditation-Based Stress Reduction Program in the Treatment of Anxiety Disorders," *American Journal of Psychiatry,* 149, 7, July 1992, pp. 936–943.

63. Beck, Aaron T., and Emery, Gary, *Anxiety Disorders and Phobias: A Cognitive Perspective* (New York, Basic Books Inc., 1985).

64. Ibid.

65. Ibid., pp. 232–251.

66. Wilson, Reid, *Don't Panic: Taking Control of Anxiety Attacks* (New York, Harper and Row, 1996), pp. 165–169.

67. Henley, Arthur, *Phobias: The Crippling Fears* (Secaucus, NJ, Lyle Stuart Inc., 1987).

68. Harrington, Geri, *The Asthma Self-Care Book: How to Take Control of Your Asthma* (New York, Harper Perennial, 1989), pp. 148–151.

69. Farber, E.M., and Nall, L., "Psoriasis: a stress-related disease," *Cutis*, 51, 5, May 1993, pp. 322–326.

70. Kaplan, K.H.; Goldenberg, D.L.; and Galvin, Nadeau M., "The impact of a meditation-based stress reduction program on fibromyalgia," *General Hospital Psychiatry*, 15, 5, September 1993, pp. 284–289.

71. Surwit, Richard, "Effects of Relaxation on Glucose Tolerance," *Diabetes Care*, 6, 1983.

72. Deepak, K.K.; Manchanda, S.K.; and Maheshwari, M.C., "Meditation improves clinicoelectroencephalographic measures in drug-resistant epileptics," *Biofeedback and Self-Regulation*, 19, 1, March 1994, pp. 25–40.

73. Ornish, pp. 142–146.

74. "Can transcendental meditation make you live long and prosper," *New Scientist*, 28 April 1990, p. 240.

75. Welch, Raquel, *Body and Mind*, 1989, HBO video.

76. Moyers, pp. 177–193.

77. Ornstein and Sobel, pp. 140–150.

78. Moyers, pp. 213–239.

79. Ornstein and Sobel, p. 29.

80. Goleman, Daniel, *The Meditative Mind* (New York, G.P. Putnam & Sons, 1988), p. 170.

81. Kiecolt, Glaser, "Psychosocial Enhancement of Immuneocompetence in a Geriatric Population," *Health Psychology*, 4, 1, 1985, pp. 25–41; Kiecolt, Glaser, "Modulation of Cellular Immunity in Medical Students," *Journal of Behavioural Medicine*, May 1986.

82. Ornish, pp. 140–142.

83. Ornstein and Sobel, p. 155.

84. Fromm, Gerhard H., "Neurophysiological speculations on Zen enlightenment," *Journal of Mind and Behavior*, 13, 2, Spring 1992, pp. 163–169.

85. Johnson, George, "Yes, There Is Such a Thing As Mind Over Matter," *The New York Times*, February 25, 1996, p. 3.

86. Goleman, Daniel, "Psychotherapy Found to Produce Changes in Brain Function Similar to Drugs," *The New York Times*, Thursday, February 15, 1996, B12.

87. Bogart, Greg, "The Use of Meditation in Psychotherapy: A Review of the Literature," *American Journal of Psychotherapy*, XLV, 3, July 1991, pp. 406–408.

88. Bogart, pp. 383–412.

89. Walsh, Roger, "The Search for Synthesis," p. 32.

90. Bogart, pp. 383–412.

91. Walsh, "The Search for Synthesis," p. 32.

92. Bogart, pp. 383–412.

93. Ibid.

94. Ibid.

95. Goleman, *The Meditative Mind*, p. 173.

96. Ornish, pp. 140–142.

97. Barth, George Francis, *Nova Yoga: The Yoga of the Imagination* (New York, Mason & Lipscomb Publishers, 1992).

98. Peppetier, pp. 197–200.

99. Goleman, *The Meditative Mind*, p. xxii.

100. Moyers, p. xii.

101. Walsh, "The Search for Synthesis," p. 32.

102. Goleman, *The Meditative Mind*, pp. 168–170.

103. "Yoga goes mainstream," *U.S. News & World Report*, May 16, 1994.

104. Moyers.

105. Wallis, Claudia, "Faith & Healing," *Time*, June 24, 1996.

106. "Yoga goes mainstream."

107. Maclaine, Shirley, *Going Within: A guide for inner transformation* (New York, Bantam Books, 1989).

108. Walsh, Roger, "Two Asian Psychologies and their Implications for Western Psychotherapists," *American Journal of Psychotherapy* Vol. XLII, No. 4, October 1988, pp. 543–545.

109. Goleman, Daniel, "Buddhist and Western psychology: Some commonalities and differences," *Journal of Transpersonal Psychology*, 13, 2, 1981, pp. 125–136.

110. Suzuki, D.T.; Fromm, Erich; and De Martino, Richard, *Zen Buddhism and Psychoanalysis* (New York, Grove Press Inc., 1960).

111. Goleman, "Buddhist and Western psychology," pp. 125–136.

112. Coward, Harold G., "Psychology and Karma," *Philosophy East and West*, XXXII, 1, January 1983, pp. 49–61.

113. Watts, Alan, *Psychotherapy East and West* (New York, Pantheon, 1961).

114. Noda, Shunsaku J., "Individualpsychologiische Gruppentherapie und Meditation/Adlerian Group Therapy and Meditation," *Zeitschrift für Individualpsycholgie*, 14, 2, 1989, pp. 121–128.

115. Kahn, Michael, "Vipassana meditation and the psychobiology of Wilhelm Reich," *Journal of Humanistic Psychology*, 25, 3, Summer 1985, pp. 117–128.

116. Rubin, Jeffrey B., "Meditation and psychoanalytic listening," *Psychoanalytic Review*, 72, 4, Winter 1985, pp. 599–613.

117. Epstein, Mark D., "On the neglect of evenly suspended attention," *Journal of Transpersonal Psychology*, 16, 2, 1984, pp. 193–205.

118. Engler, John, H., "Vicissitudes of the Self according to Psychoanalysis and Buddhism: A Spectrum Model of Object Relations Development," *Psychoanalysis and Contemporary Thought*, 6, 1, 1983, pp. 29–72.

119. Walsh, "Two Asian Psychologies," pp. 546–548.

120. Das, Ajit K., "Beyond self-actualization," *International Journal for the Advancement of Counselling*, 12, 1, January 1989, pp. 13–27.

121. Welwood, John, "Principles of inner work: Psychological and spiritual," *Journal of Transpersonal Psychology*, 16, 1, 1984, pp. 63–73.

122. Maslow, Abraham H., *Towards a Psychology of Being* (New York, Van Nostrad, 1968), p. 114.

123. Ibid.

124. Ibid., pp. 110, 119.

125. Ibid., p. 78.

126. Frankl, Victor E., "Self-transcendence as a human phenomenon," in Anthony J. Sutich and Miles A. Vich (eds.), *Readings in Humanistic Psychology* (New York, The Free Press, 1966), p. 115.

127. Takashima, Hiroshi, "Logotherapy and Buddhistic thought. Special: Viktor E. Frankl, 80 years," *International Forum for Logotherapy*, 8, 1, Spring/Summer 1985, pp. 54–56.

128 May, Rollo, *The Discovery of Being: Writings in Existential Psychology* (New York, W.W. Norton & Company, 1983), p. 58.

129. Allport, Gordon, *The Person in Psychology* (Boston, Beacon, 1968).

130. Goleman, *The Meditative Mind,* p. 147.

131. Kakar, Sudhir, "The Human Life Cycle: The Traditional Hindu View and Psychology of Erik Erikson," *Philosophy East and West*, XVIII, 3, July 1968, pp. 127–137.

132. Erikson, Erik, *Childhood and Society* (New York, Norton, 1963).

133. Goleman, *The Meditative Mind,* p. 147.

134. Becker, Susan K., and Forman, Bruce D., "Zen Buddhism and the psychotherapy of Milton Erickson: A transcendence of theory and self," *Psychology: A Journal of Human Behavior*, 26, 2–3, 1989, pp. 39–48.

135. Dewan, Mantosh J., and Gupta, Sanjay, "Congruence between Hindu philosophy and writings of Otto Rank," *Psychological Reports*, 70, 1, February 1992, pp. 127–130.

136. Russell, Elbert W., "Consciousness and the unconscious: Eastern meditative and Western psychotherapeutic approaches," *Journal of Transpersonal Psychology*, 18, 1, 1986, pp. 51–72.

137. Merkle, William Frank, *Bringing Empathic Awareness Beyond the Self: An Interdisciplinary Approach to the Healing Relationship*, Ph.D. diss., The Union Institute, 1993, pp. 270.

138. Nakamura, Hajime, "Buddhism," *Dictionary of the History of Ideas*, Philip B. Wiener (ed.) (New York, Scribner, 1973), p. 255.

139. Walsh, "The Search for Synthesis," pp. 19–45.

140. Walsh, "Two Asian Psychologies," pp. 546–548.

141. Clay, John, *R.D. Laing: A Divided Self* (London, Hodder, 1996).

142. Donaldson, Margaret, *Human Minds: An Exploration* (London, Penguin Books, 1992). See also Hoyland, John, "Modes of Mind," *New Scientist Supplement*, April 1993, pp. 18–19.

143. Donaldson, p. 227.

144. Langer, Ellen J., *Mindfulness* (Addison Wesley, 1989).

145. Langer, p. 31.

NINE / TRAVERSING FUTURE TECHNOLOGIES THROUGH SOME PAST CONCEPTS

1. *Bio Technology: Economic and Wider Impacts*, OECD, 1989, pp. 52–55.

2. Kaminuma, Tsuduchika, and Matsumoto, Gen (eds.), *Biocomputers: The Next Generation from Japan* (Chapman and Hall, 1991), p. 40.

3. "Denser, Faster, Cheaper: The Micro Chip in the 21st Century," *The New York Times*, Sunday, December 29, 1991, p. 5.

4. Calem, Robert E., "In Far More Gadgets, a Hidden Chip," *The New York Times*, January 2, 1994.

5. Port, Otis, "Creating Chips an Atom at a Time," *Business Week*, July 29, 1991, pp. 54–55; Clery, Daniel, "Memorable future for the lone electron," *New Scientist*, February 27, 1993; Abu-Mostafa, Yaser, and Psaltis, Demetri, "Optical Neural Computers," *Scientific American*, March 1987.

6. Kahn, Patricia, "Genome on the Production Line," *New Scientist*, 24 April 1993, pp. 32–36; James, Barry, "Sorting out the 'Library' of Genes," *International Herald Tribune*, December 23, 1993.

7. Watson, James D., "The Human Genome Project: Past, Present and Future," *Science*, 248, April 1990.

8. Berer, Marge, "The Perfection of Offspring," *New Scientist*, 124, Issue 1725, July 14, 1990, pp. 58–59.

9. "Chips unravel the genome," *New Scientist*, 6 May 1989, p. 36.

10. Hitomi, K., "Computer Aided Design, Manufacturing and Management," International Conference on Industrial Culture and Human Centered Systems, Tokyo Keizai University, 1990.

11. "Fetal transplants for Parkinson's disease," *Economist*, 6 January 1990.

12. Mahowald, Mary B.; Silver, Jerry; and Rakheson, Robert A., "The Ethical Options in Transplanting Fetal Tissue," *The Hastings Center Report*, 17, 1, February 1987, pp. 9–15.

13. Kahn, pp. 32–36; Nelsen, Rolf Haugaard, "Gene test may assess chance of conceiving a handicapped child," *New Scientist*, 18 July 1992, p. 19.

14. Associated Press Report, *New York Times*, March 2, 1993, "Lorenzo's Oil Malady: Biologists Find Gene."

15. Angier, Natalie, "Scientists Find Long-Sought Gene that Causes Lou Gehrig's Disease," *New York Times*, March 4, 1993.

16. Goonatilake, Susantha, *The Evolution of Information: Lineages in Gene, Culture and Artefact* (London, Pinter Publishers, 1991), pp. 131–135.

17. Wuketits, Franz M., *Evolutionary Epistemology and Its Implications for Human Kind* (Albany, SUNY Press, 1990), p. 53.

18. Nagel, Thomas, "What is it like to be a bat," *The Philosophical Review*, October 1974.

19. Latour, B., and Woolgar, S., *Laboratory Life: The Social Construction of Scientific Facts* (London, Sage, 1979); Barnes, Barry, *T. S. Khun and Social Science* (London and Basingstoke, Macmillan Press Ltd., 1982), p. 10.

20. Johnson-Laird, P.N., *The Computer and the Mind: An Introduction to Cognitive Science* (London, Fontana, 1988).

21. Mowrer, O.H., "Ego Psychology, Cybernetics and Learning Theory," in Buckley, Walter (ed.), *Modern Systems Research for the Behavioral Scientist* (Chicago, Aldyne Publishing, 1968), p. 339.

22. Laszlo, Erwin, *Introduction to Systems Philosophy* (New York, Harper, 1973), p. 293.

23. Macy, Joanna, *World as Lover, World as Self* (Berkeley, CA, Parallax Press, 1991), pp. 82–83.

24. Gruson, Lindsey, "When 'Mom' and 'Grandma' are One and the Same," *New York Times*, February 16, 1993.

25. Wheeler, David L., "Ethicist Urges Public debate on Medical Therapies that Could Cause Genetic Changes in Offspring," *Chronicle of Higher Education*, Vol. 37, Issue 24, February 27, 1991.

26. Wind, James P., "What Can Religion Offer Bioethics," *Hastings Center Report*, A Special Supplement, July/August 1990.

27. Scully, Thomas, and Scully, Celia, *Playing God: The New World of Medical Choices* (New York, Simon and Schuster, 1987).

28. Mable, Margot C. J., *Bioethics and the New Medical Technology* (New York, Atheneum, 1993).

29. Kinichiro, Kajikawa, "A New Field Emerges," *Hastings Center Report*, 19, 155, 4 July 1989; Callahan, Daniel, "Religion and the Secularization of Bioethics," *Hastings Center Report*, A Special Supplement, July/August 1990; Callahan, Daniel, and Campbell, Courtney S., "Theology, Religious Traditions, and Bioethics," *Hastings Center Report*, A Special Supplement, July/August 1990, 4; Campbell, Courtney S., "Religion and Moral Meaning in Bioethics," *Hastings Center Report*, A Special Supplement, July/August 1990, 4.

30. Representative of this literature are: Kabat-Zinn, Jon; Massion, Ann O.; Kristeller, Jean; Petersen, Linda Gay; Fletcher, Kenneth E.; Pbert, Lori; Lenderking, William R.; and Santorelli, Saki F., "Effectiveness of a Meditation-Based Stress Reduction Program in the Treatment of Anxiety Disorders," *American Journal of Psychiatry*, 149, 7, July 1992, pp. 936–943; Walsh, Roger, "Two Asian Psychologies and Their Implications for Western Psychotherapists," *American Journal of Psychotherapy*, XLII, 4, October 1988; Sweet, Michael J., and Johnson, Craig G., "Enhancing empathy: The interpersonal implications of a Buddhist meditation technique. Special Issue: Psychotherapy and religion," *Psychotherapy*, 27, 1, Spring 1990, pp. 19–29: De Silva, Padmal, "Buddhism and behavior modification," *Behavior Research and Therapy*, 22, 6, 1984, pp. 661–678; Goleman, Daniel, "Buddhist and Western psychology: Some commonalities and differences," *Journal of Transpersonal Psychology*, 13, 2, 1981, pp. 125–136; Donaldson, Margaret, *Human Minds: An Exploration* (London, Penguin Books, 1992), p. 227; Bogart, Greg, "The Use of Meditation in Psychotherapy: A Review of the Literature," *American Journal of Psychotherapy*, XLV, 3, July 1991, pp. 383–412.

31. "Anatta," *Encyclopedia of Buddhism*, Vol. 1, Published by The Government of Ceylon, 1961, Reprinted Colombo, 1984, p. 567.

32. "Anicca," *Encyclopedia of Buddhism*, Vol. 1, p. 569.

33. Rahula, Walpola, *What the Buddha Taught* (London, Gordon Fraser, 1978), p. 59.

34. Ibid., p. 65.

35. "Anatta," *Encyclopedia of Buddhism*, Vol. 1, p. 567.

36. Ibid., p. 569.

37. Kalupahana, D.J., and Tamura, Koyu, "Antarabhava," in Malalasekera, G.P. (ed.), *Encyclopedia of Buddhism*, Vol. 1, No. 1, January 1970, Ceylon Journal of Humanities, p. 441.

38. Jayatilleke K.N., *Survival and Karma in Buddhist Perspective* (Kandy, Buddhist Publication Society, 1980).

39. Ibid., p. 29.

40. Rahula, p. 65.

41. Feer, L. (ed.), *Samyutta Nikaya*, Vol. III (London, Pali Text Society, 1884), p. 57.

42. Rahula, p. 56.

43. Rahula.

44. Organ, p. 162.

45. Quoted in Gunaratne, V.F., *Buddhist Reflections on Death* (Kandy, Buddhist Publication Society, 1982), p. 12.

46. Kolm, Serge-Christopher, "The Buddhist theory of 'no-self,'" in Elster, Jon (ed.), *The Multiple Self* (Cambridge University Press, 1986), pp. 233–265.

47. Loy, David, "Avoiding the Void: The Lack of Self in Psychotherapy and Buddhism," *The Journal of Transpersonal Psychology*, 24, 2, 1992, pp. 151–179.

48. Page, Richard C., and Berkow, Daniel N., "Concepts of the Self: Western and Eastern Perspectives," *Journal of Multicultural Counseling and Development*, 19, April 1991, pp. 83–93.

49. Parfitt, Derek, *Reasons and Persons* (Oxford University Press, 1984).

50. Scheffler, Samuel, "Ergo: Less Ego," *The Times Literary Supplement*, 4 May 1984, No. 4231.

51. Parfitt, p. 212.

52. Ibid., p. 280.

53. Lewin, Shirley Robin, "The Person Vanishes," in *The Spectator*, 19 May 1984, Vol. 252, No. 8132.

54. Downie, R., and Telfer, V., *Respect for Persons* (London, George Allen and Unwin, 1969), p. 10.

55. Collins, Steven, "Buddhism in Recent British Philosophy and Theology," in Kalupahana, David J., and Weeraratne, W.G. (eds.), *Essays in Honour of N.A. Jayawickreme*, Jayawickreme Felicitation Volume Committee, Colombo, 1987.

56. Varela, Francisco F.; Thompson, Evan; and Rosch, Eleanor, *The Embodied Mind: Cognitive Science and Human Experience* (Cambridge, MA, The MIT Press, 1993), p. xviii.

57. Varela et al., p. xiv.

TEN / VIRTUAL REALITY

1. Biocca, Frank, "Communication Within Virtual Reality: Creating a Space for Research," in *Journal of Communication*, 42, 4, Autumn 1992, p. 14.

2. Bylinsky, Gene, "The Marvels of 'Virtual Reality,'" *Fortune*, June 3, 1991, pp. 138–143.

3. Gibson, William, *Neuromancer* (New York, Ace Science Fiction Books, 1984).

4. Balsamo, Anne, "The Virtual Body in Cyberspace," *Journal of Research in the Technology and Philosophy*, 12, Spring 1993, special issue on "Technology and Feminism," Joan Rothschild (ed.).

5. Goonatilake, Suran, "Virtually Certain of the Meaning of Life," *Computing*, 18 July 1991, p. 20.

6. Furness, Thomas A., III, "Fantastic Voyage," *Popular Mechanics*, December 1986, pp. 63–65.

7. Goonatilake, Suran, p. 20.

8. Krueger, Myron, *Artificial Reality 11* (Reading, MA, Addison-Wesley, 1991), p. 3.

9. Balsamo, pp. 119–140.

10. Welter, Therese R., "The Artificial Tourist: Virtual Reality Promises New Worlds for Industry," *Industry Week*, October 1, 1990, p. 66.

11. Goddard, Alison, "Virtual therapy reaches new heights," *New Scientist*, June 11, 1994, p. 6; "Space station orbits in virtual reality," *Machine Design*, 66, June 6, 1994, p. 16; Machlis, Sharon, "Virtual surgery: computers promise better training, techniques," *Design News*, 49[50], June 13, 1994, p. 44; Taylor, Russell M., and Chi, Vernon L., "Take a walk on the image with virtual-reality microscope display," *Laser Focus World*, 30, May 1994, p. 145–149; Whitehouse, Karen, "Embracing VR as a liberator," *IEEE Computer Graphics and Applications*, 14, May 1994, p. 90; Whitehouse, Karen "The museum of the future," *IEEE Computer Graphics and Applications*, 14, May 1994, p. 8–11; Gottschalk, Mark A., "Engineering enters the virtual world," *Design News*, 49[50], May 9, 1994, p. 23–24; Illman, Deborah L., "Researchers make progress in applying virtual reality to chemistry," *Chemical & Engineering News*, 72, March 21, 1994, pp. 22–25; O'Neill, Brian, "Putting virtual reality to work: Eye surgery simulator could help physicians learn and practice new techniques," *Simulation*, 61, December 1993, p. 417–418,

12. Helm, Michael, *The Metaphysics of Virtual Reality* (New York, Oxford University Press, 1993), p. vi.

13. Quoted in Rheingold, Howard, *Virtual Reality* (New York, Touchstone, 1991), p. 192.

14. Steuer, Jonathan, "Defining Virtual Reality: Dimensions Determining Telepresence," in *Journal of Communication*, 42, 4, Autumn 1992, pp. 75.

15. Shapiro, Michael A., and McDonald, Daniel G., "I'm Not a Real Doctor, but I Play One in Virtual Reality: Implications of Virtual Reality for Judgments about Reality," in *Journal of Communication*, 42, 4, Autumn 1992, p. 94.

16. Rheingold, p. 16.

17. Ibid., pp. 114–116.

18. Krueger, p. xii.

19. Ibid., p. 148.

20. Ibid., p. 127.

21. Helm, p. 30.

22. Ibid., p. 79.

23. Martin, Douglas, "Virtual Reality! Hallucination! Age of Aquarius! Leary's Back!" *The New York Times*, Sunday, May 20, 1990.

24. *Mondo 2000* #4, 1991, "The Carpal Tunnel of Love, Virtual Sex with Mike Saenz," interview with Jeff Milstead and Jude Milhon.

25. Goonatilake, Suran, p. 20.

26. Krueger, pp. 264–245.

27. Ibid., p. xi.

28. Rheingold, p. 186.

29. Ibid., p. 354.

30. Krueger, p. 201.

31. Helm, p. 24.

32. Stone, Allucquere Rosanne, "Will the Real Body Please Stand Up? Boundary Stories about Virtual Cultures," in *Cyberspace: First Space*, Benedikt, Michael (ed.) (Cambridge, MA, The MIT Press, 1991), pp. 81–119.

33. Helm, p. 130.

34. Krueger, p. xvi-xvii.

35. Helm, p. 155.

36. Ibid., p. 83.

37. Rheingold, p. 387.

38. Tart, Charles T., "Multiple personality, altered states and virtual reality: The world simulation process approach," *Dissociation Progress in the Dissociative Disorders*, 3, 4, December 1990, pp. 222–233.

39. Knox, David; Schacht, Caroline; and Turner, Jack, "Virtual Reality: A Proposal for Treating Test Anxiety in College Students," *College Student Journal*, 27, 3, September 1993, pp. 294–296.

40. Rheingold, p. 390.

41. Helm, p. 100.

42. Krueger, p. 245.

43. Thomas, David, "Old Rituals for New Space: Rites de Passage and William Gibson's Cultural Model of Cyberspace," in *Cyberspace: First Space*, pp. 31–49.

44. Stone, pp. 81–119.

45. Stenger, Nicole, "Mind is a Leaking Rainbow," in *Cyberspace: First Space*, pp. 50–58.

46. Helm, p. 117.

47. Shapiro and McDonald, p. 94.

48. Helm, p. 88–89.

49. Helm, p. 49.

50. Bishop, Donald H. (ed), *Indian Thought: An Introduction* (New York, John Wiley & Sons, 1975), pp. 145–147.

51. Barlingay, S.S., "Indian Epistemology and Logic," in Bishop, *Indian Thought*, pp. 148–171.

52. Quoted in Bishop, p. 3.

53. Joshi, J.N., "Metaphysics," in Bishop, *Indian Thought*, pp. 176–192.

54. Joshi, p. 179.

55. Ibid., p. 179.

56. Joshi, in Bishop.

57. Ibid., p. 179.

58. Ramakrishna Rao, K.S., "Jainism," in Bishop, *Indian Thought,* p. 94.

59. Joshi, p. 179.

60. Jacobson, Nolan Pliny, *The Heart of Buddhist Philosophy* (Southern Illinois University Press, Carbondale and Edwardsville, 1988), p. ix.

61. Bishop, Donald H., "Buddhism, in Bishop, *Indian Thought,* p. 130.

62. Joshi.

63. Ibid., p. 179.

64. Bishop, "Buddhism," p. 130.

65. Kola, pp. 233–265.

66. Joshi.

67. Ibid.

68. Ibid., p. 179.

69. Swami Nikhilananda, "The Three States (Avasthatraya)," *Philosophy East and West*, 11, 1, April 1952, pp. 66–75.

70. N. Mishra, "Samskaras in Yoga Philosophy and Western Psychology," *Philosophy East & West*, 2, 4, January, pp. 253–308.

71. Seelawimala, Madawala, and McKinly, Arnold, "*Sati* (Mindfulness) and

the structure of the mind in Early Buddhism," *The Pacific World*, Berkeley, No. 3, Fall 1987, pp. 3–15.

72. Larson, Gerald James, and Deutsch, Eliot (eds.), *Interpreting Across Boundaries: New Essays in Comparative Philosophy* (Princeton University Press, 1988).

73. Bishop, *Indian Thought*, pp. 364–383.

74. Barlingay, in Bishop, *Indian Thought*.

75. Ibid.

76. Ibid.

77. Ibid.

78. Ibid.

79. Jacobson, p. x.

80. *Encyclopedia of Buddhism*, Published by The Government of Ceylon, Colombo, 1966, Vol. 2, p. 60.

81. *Encyclopedia of Buddhism*, Vol. 1, Fascicle 1, p. 379.

82. Barlingay.

83. Ibid.

84. Cochran, Tracy, "Samsara Squared: Buddhism and Virtual Reality," *Tricycle: The Buddhist Review*, Fall 1992, pp. 77.

85. Ibid., pp. 76–81.

86. Ibid., p. 79.

87. Lifton, Robert Jay, *The Protean Self: Human Resilience in an Age of Fragmentation* (New York, Basic Books, 1993).

88. Loy, David, "Avoiding the Void: The Lack of Self in Psychotherapy and Buddhism," *The Journal of Transpersonal Psychology*, 24, 2, 1992, pp. 151–179; also Berger, Peter L.; Berger, Brigitte; and Kellner, Hanfred, *The Homeless Mind* (London, Penguin, 1973), pp. 74, 77.

89. Elster, Jon (ed.), *The Multiple Self* (Cambridge University Press, 1986).

90. Sliker, Gretchen, *Multiple Mind: Healing the Split in Psyche and World* (Boston and London, Shambala, 1992).

91. Organ, Troy Wilson, *Philosophy and the Self: East and West* (Selingrove, Susquehanna University Press, 1987), p. 162.

92. Rheingold, p. 191.

93. Helm, p. 86.

94. James, W., *The Varieties of Religious Experience* (New York, New American Library, 1902/1958), p. 298.

95. Walsh, Roger "Can Western Philosophers Understand Asian Philosophies? The Challenge and Opportunities of State-of-Consciousness Research," in Ogilvy, James (ed.), *Revisioning Philosophy* (Albany, SUNY Press, 1992), pp. 281–301.

96. Walsh, "Can Western Philosophers Understand Asian Philosophies?" pp. 281–301.

97. Ibid.

98. Walsh, Roger, "Two Asian Psychologies and Their Implicaitons for Western Psychotherapists," *American Journal of Psychotherapy*, XLII, 4, October 1988, pp. 546–548.

99. Walsh, "Can Western Philosophers Understand Asian Philosophies?" pp. 281–301.

100. Joshi.

101. Krueger, p. 3.

102. Ogilvy, *Revisioning Philosophy*, p. x.

ELEVEN / DIGGING DEEPER

1. Frank, Phillip, "Einstein's Philosophy of Science," in Coley, Noel G., and Hall, Vance M.D., *Darwin to Einstein: Primary Sources on Science and Belief* (Longmanns London and Open University, 1980), p. 148.

2. Quoted in Elman, Benjamin A., "Nietzsche and Buddhism," *Journal of the History of Ideas*, Oct.-Dec. 1983, p. 671.

3. Helm, Michael, *The Metaphysics of Virtual Reality* (New York, Oxford University Press, 1993), p. 130.

4. Urwick, B.J., *The Message of Plato* (London, Allen & Unwin, 1920).

5. Vissilis, Vitsaxis, *Plato and Upanishads* (New Delhi, Arnold-Heinemans, 1977).

6. Ibid.

7. Scharfstein, Ben-Ami, *Philosophy East/Philosophy West: A Critical Comparison of Indian, Chinese, Islamic and European Philosophy* (Oxford, Basil Blackwell, 1978), p. 30

8. Lewis, Leta Jane, "Fichte and Samkara," *Philosophy East and West*, XII, 4, January 1963, pp. 301–311.

9. McEvilly, Wayne, "Kant, Heidegger and the Upanishads," *Philosophy East and West*, XII, 4, January 1963, pp. 311–319.

10. Hoffman, Yoel, "The Possibility of Knowledge: Kant and Nagarjuna," in Ben-Ami Scharfstein et al. (eds.), *Philosophy East/Philosophy West*, p. 27

11. Ibid., p. 287.

12. Price, H.H., "The Present Relations between Eastern and Western Philosophy," *The Hiberts Journal*, L111, April 1955, p. 229.

13. Nakamura, Hajime, "Buddhism," *Dictionary of the History of Ideas*, Phillip B. Wiener (ed.) (New York, Charles Scribner, 1973), p. 255

14. Lovejoy, Arthur O., "Schopenhauer as an Evolutionist," in Glass, Bentley; Temkin, Owsoi; and Straus, William L. (eds.), *Forerunners of Darwin 1745–1859* (Baltimore, John Hopkins Press, 1968), pp. 415–438.

15. Faber, M.D., "Back to a Crossroad: Nietzsche, Freud and the East," *New Ideas in Psychology*, 16, 1, 1988, pp. 25–45.

16. Ibid., p. 4.

17. Ibid., p. 7.

18. Ibid., p. 15.

19. Hanna, Fred J., "The transpersonal consequences of Husserl's phenomenological method," *Humanistic Psychologist*, 21, 1, Spring 1993, pp. 41–57.

20. Smith, H., Foreword, to *The Three Pillars of Zen*, Kapleau, P. (ed.) (Garden City, NY, Doubleday, 1965), p. xii.

21. Ibid., p. 16.

22. Riepe, Dale, "The Indian Influence in American Philosophy: Emerson to Moore," *Philosophy East and West*, XVII, 14, January-October 1967, pp. 124–137.

23. Inada, Kenneth K., and Jacobson, Nolan B. (eds.), *Buddhism and American Thinkers* (Albany, SUNY Press, 1984), p. vii.

24. Ma, Alexander Kuang Shin, *On the Buddhist logical approach to semantics. Part one: The causal theory of perception*, Vols. I and II, Ph.D. diss., Georgetown University, 1981, p. 548.

25. Glasenapp, Helmuth, "Indian and Western Metaphysics," *Philosophy East and West*, III, 3, October 1953, p. 230.

26. Ibid., p. 231

27. Ibid.

28. Mohanty, J. N., "Consciousness and Knowledge in Indian Philosophy," *Philosophy East and West*, XXIX, 1, January 1979, pp. 3–11.

29. Rose, Mary Carman, "Investigative interrelatedness between the study of the human mind and present day philosophy," *Philosophy East and West*, XXIX, 2, April 1979, pp. 189–200.

30. Post, E.L., "Introduction to a General Theory of Elementary Propositions," *American Journal of Mathematics*, 43, 1921, pp. 163–185.

31. Chi, Richard S.Y., "Buddhist Logic and Western Thought," in Inada and Jacobson, p. 111.

32. Ibid., p. 115.

33. Rao, K.S. Ramakrishna, "Jainism," in Bishop, *Indian Thought*, p. 95.

34. Hoffman, F.J., "Rationality in Early Buddhist Four Fold Logic," *Journal of Indian Philosophy*, 10, 1982; Staal, J.F., *Exploring Mysticism* (Penguin, 1976); Staal, J.F., "Making Sense of the Buddhist tetra lemma," *Philosophy East and West: Essays in Honour of T.M.P. Mahadevalen*, H.D. Lewis (ed.), 1976; Murti, T.R.V., *Central Philosophy of Buddhism* (London, 1955).

35. *Dighanikayal*, pp. 22–23, quoted in the excellent overview article on *Chatuskoti* by R.D. Gunaratne in the *Buddhist Encyclopaedia*, Vol. V, p. 256, 1990.

36. Hoffman, F.J., 1982, p. 333.

37. Gunaratne, R.D. "Logical forms of Chatuskoti: A New Solution," *Philosophy East and West*, 36, 3, 1986.

38. *Encyclopedia of Buddhism*, Published by The Government of Ceylon, Colombo, 1979, Vol. V, p. 258.

39. *Encyclopedia of Buddhism*, Vol. V, p. 259.

40. Oppenheimer, J. Robert, *Science and the Common Understanding* (London, Oxford University Press, 1954).

41. Schneider, Ivo, "The Contributions of the Sceptic Philosophers Arcesilas and Carneades to the Development of an Inductive Logic Compared with Jaina-Logic," *Indian Journal of the History of Science*, 12, 2, November 1977, p. 274.

42. Quoted in Schneider, p. 177.

43. Schneider, p. 178.

44. Ibid., p. 179.

45. Kosko, Bart, and Isaka, Satoru, "Fuzzy Logic," *Scientific American*, July 1993, pp. 76–81.

46. McNeill, Daniel, and Freiberger, Paul, *Fuzzy Logic* (New York, Simon & Schuster, 1993), p. 319.

47. Goldberg, David E., *Genetic Algorithms in Search, Optimization and Machine Learning* (Reading, MA, Addison-Wesley, 1988).

48. Kneale, William and Martha, *The Development of Logic* (London, Oxford University Press, 1971), p. 1.

49. Quine, Willard V.O., *Elementary Logic* (New York, Harper Torch Books, 1965), p. 1.

50. Ellington-Wough, Ter, "Algebraic and Geometric Logic," *Philosophy East and West*, XXIV, 1, January 1974, pp. 23–41.

51. Ibid., p. 25.

52. Ibid., p. 26.

53. Ibid.

54. Ibid., p. 29.

55. Ibid., p. 39.

56. Ma, p. 548.

57. Bohr, Niels, in "The Bohr-Einstein dialogue," in French and Kennedy (eds.), *Niels Bohr: A Centenary Volume* (Harvard University Press, 1985).

58. Eells, Ellery, and Skyrms, Brian (eds.), *Probability and Conditionals: Belief Revision and Rational Decision* (Cambridge University Press, 1992); Eells, Ellery, *Probabilistic causality* (Cambridge University Press, 1991).

59. Kalupahana, David, *Causality, the Central Philosophy of Buddhism* (The University Press of Hawaii, 1975), pp. 124–125.

60. Jain, L.C., "System Theory in Jaina School of Mathematics," *Indian Journal of the History of Science*, 14, 1, May 1979, pp. 31–61.

61. "Positive disintegration—an interview with Joanna Macy," *Tricycle: The Buddhist Review*, Spring 1993, pp. 68–73.

62. Macy, Joanna, *World as Lover, World as Self* (Berkeley, CA, Parallax Press, 1991), p. 59.

63. Ibid., p. 66.

64. *Samyutta Nikaya* 111.57.

65. Mowrer, O.H., "Ego Psychology, Cybernetics and Learning Theory," in Buckley, Walter, *Modern Systems Research for the Behavioral Scientist* (Chicago, Aldine Publishing, 1968), p. 339.

66. Quoted in Gunaratne, V.F., *Buddhist Reflections on Death* (Kandy, Buddhist Publication Society, 1982), p. 12.

67. Macy, pp. 67–72.

68. Ibid., pp. 82–83.

69. Laszlo, Erwin, *Introduction to Systems Philosophy* (New York, Harper, 1973), p. 293.

70. Macy, p. 83.

71. Laszlo, p. 170.

72. *Samyutta Nikaya* 11.62.

73. *Anguttara Nikaya*, quoted in Oldenburg, H., *Buddha: His Life, His Doctrine, His Order* (Delhi, Indological Book House, 1971), p. 243.

74. Macy, p. 92.

75. Goonatilake, Susantha, *Evolution of Information: Lineages in Gene, Culture and Artefact* (London, Pinter Publishers, 1991).

76. Goudge, A. Thomas, "Evolutionism," *Dictionary of the History of Ideas*, Philip P. Weinir (ed.), Vol. 2, pp. 124–188 (Charles Scribner's Sons, New York, 1973), p. 175.

77. Goudge, pp. 174–189.

78. Dasgupta, Surendranath, *Natural Science of the Ancient Hindus*, Indian Council of Philosophical Research, Motilal Banarsidass, Delhi, 1987, p. 55.

79. Seal, B.N., *The Positive Sciences of the Ancient Hindus* (London, 1915), pp. 7–8.

80. A representative sample of some of the issues being faced in this discourse is seen, for example, in several of the papers in "The International Seminar on Evolutionary Systems," Vienna, March 1995, organized by the Konrad Lorenz Institute, among others.

81. For a detailed treatment of these lineages see Goonatilake, *Evolution of Information*.

82. Schrödinger, E., *What is Life?* (London, Cambridge University Press, 1943), p. 93.

83. Nair, Ranjit, "The Science of Schrödinger and the Vision of the Vedanta," paper presented at "Erwin Schrödinger Symposium," Sorbonne, Paris, 1992, and "Indo-Austrian Workshop," University of Delhi, 1994, and forthcoming as a chapter in D. Tiemersma (ed.), *Between Science and Culture*, Kluwer 1997.

84. von Glasenapp, Helmuth, *Immortality and Salvation in Indian Religions,* translated from the German by E.F.J. Payne (Calcutta, Susil Gupta India Private Ltd., 1963), p. iv.

85. Ibid., p. 17.

86. Ibid.

87. Chapple, Christopher, *Karma and Creativity* (Albany, State University of New York Press, 1986), p. 3.

88. Chapple, p. xi.

89. Humphreys, p. 37.

90. Harvey, Peter, *An Introduction to Buddhism: Teachings, History and Practice* (Cambridge University Press, 1990).

91. Chapple, p. 5.

92. Humphreys, p. 31.

93. Ghosh, Shyam, *Hindu Concept of Life and Death* (New Delhi, Munshiram Manoharlal Publishers, 1968), p. 218.

94. Ibid., p. 209.

95. Humphreys, p. 52.

96. Ghosh, p. 218.

97. Humphreys, p. 52.

98. "Cula-Kammavibhanga Sutta," in Nanamoli Thera, *The Buddha's Words on Kamma* (Kandy, Buddhist Publication Society, 1975).

99. Nyanatilake, Mahathera, *Karma Rebirth* (Kandy, Buddhist Publication Society, 1982), p. 5.

100. *Sutta Nipata* 4. 230–231.

101. Humphreys, p. 49.

102. Varela, Francisco J., Maturana, H. R., and Uribe, R., "Autopoiesis: The Organization of Living Systems, Its Characterization and a Model," *Bio-Systems*, 5, 4, 1974, pp. 187–196.

103. Dawkins, Richard, *The Selfish Gene* (Oxford University Press, 1989), dust jacket, and pp. 19–20.

104. *Bhagavad-Gita*, chapter 11, verses 13 and 22, quoted in *The Case for Rebirth* (Kandy, Buddhist Publication Society, 1973), p. 6.

105. Quoted in Gunaratne, *Buddhist Reflections on Death*, p. 41.

106. *Encyclopedia of Buddhism*, Vol. 2, p. 119.

107. Jayatilleke K.N., *Survival and Karma in Buddhist Perspective* (Kandy, Buddhist Publication Society, 1980).

108. Varela, et al., "Autopoiesis," pp. 187–196; Varela, Francisco J.; Thompson, Evan; and Rosch, Eleanor, *The Embodied Mind: Cognitive Science and Human Experience* (MIT, 1993), 110.

109. Varela et al., *Embodied Mind*, pp. 121–123.

110. Varela et al., "Autopoiesis," pp. 187–196.

111. Morin, E., "Self and Autos," in *Autopoiesis—A theory of Living Organization*, M. Zeleny (ed.) (New York, North Holland, 1981), p. 133.

112. Inada, Kenneth K., "Time and temporality—A Buddhist Approach," *Philosophy East and West*, XXIV, 2, April 1974, pp. 171–179, esp. p. 171.

113. Whitehead, Alfred North, *Process and Reality* (New York, Macmillan, 1929).

114. Neville, Robert C., "Buddhism and Process Philosophy," in Inada and Jacobson.

115. Quoted in Jacobson, 1988, p. 23.

116. A representative sample of some of the issues being faced in this discourse is seen, for example, in several of the papers in "The International Seminar on Evolutionary Systems," Vienna, March 1995, organized by the Konrad Lorenz Institute, among others.

117. For useful hints see, for example, Swearer, Donald K., "Control and Freedom: The Structure of Buddhist Meditation in the Pali Suttas," *Philosophy East and West*, XXIII, 4, October 1973, pp. 435–457.

118. Varela et al., *Embodied Mind*, p. xviii.

119. Varela et al., *Embodied Mind*, p. 21.

120. Ibid., p. 13.

121. Ibid., p. xiv.

122. Ibid., p. 21–22.

123. Ibid., p. 23.

124. Ibid., p. 34–51.

125. Ibid., pp. 51–81.

126. Ibid., pp. 105–130.

127. Ibid., pp. 131–145.

128. Ibid., pp. 148–181.

129. Quoted in Varela et al., *Embodied Mind*, p. 198.

130. Varela et al., *Embodied Mind*, p. 199.

131. Ibid., p. 210.

132. Ibid., p. 212.

133. Ibid., pp. 213–214.

134. Ibid., p. 254.

135. McLaughlin, Andrew, "Images and Ethics of Nature," *Environmental Ethics*, 7, 4, Winter 1985, pp. 293–320.

136. Olds, Linda E., "Integrating Ontological Metaphors: Hierarchy and Interrelatedness," *Soundings*, 72, 2–3, Summer-Fall 1992, pp. 403–420.

TWELVE / TOWARD A NEW MILLENNIUM

1. Brown, Richard, "Social Theory as Metaphor: On the Logic of Discovery for the Sciences of Conduct," *Theory and Society*, 1976, 3, 2, Summer, pp. 169–197.

2. Patterson, Gordon, "Paradigms, Puzzles and Root Metaphors: George Christoph Lichtenberg and the Exact Sciences," *Journal of Mind and Behavior*, 1982, 3, 3–4, Summer-Autumn, pp. 275–287.

3. Reck, Andrew J., and Hare, Peter H., "Pepper and Recent Metaphilosophy," *Journal of Mind and Behavior*, 1982, 3, 3–4, Summer-Autumn, pp. 207–216.

4. Peters, Ted, "Metaphor and the Horizon of the Unsaid," *Philosophy and Phenomenological Research*, 1978, 38, 3, March, 355–363.

5. Hallberg, Margareta, "Science and Feminism, Vetenskop och feminism," *Sociologisk-Forskning*, 1990, 27, 3, pp. 27–34.

6. Jansen, Sue Curry, "The Ghost in the Machine: Artificial Intelligence and Gendered Thought Patterns," *Resources for Feminist Research/Documentation sur La Recherche Feministe*, 1988, 17, 4 December, 4–7.

7. Gould, Stephen Jay, "The Politics of Evolution," *Psycho History–Review*, 1983, 11, 2–3, Spring 15–35.

8. Agassi, Joseph, "Anthropomorphism in Science," *The Dictionary of the History of Ideas*, Philip P. Wiener (ed.) (Charles Scribner's Sons, New York, 1973), Vol. 1, pp. 87–91.

9. Haddne, Richard W., "Social Relations and the Content of Early Modern Science," *The British Journal of Sociology*, 1988, 39, 2, June, 255–280.

10. Manier, Edward, "'External factors' and 'Ideology' in the Earliest Drafts of Darwin's Theory," *Social Studies of Science*, 1987, 17, 4 November, pp. 581–609.

11. Gruber, Howard E., "History and Creative Work: From the Most Ordinary to the Most Exalted," *Journal of the History of the Behavioral Sciences*, 1983, 19, 1 January, 4–14.

12. Manier, pp. 581–609.

13. Berman, Bruce, "The Computer Metaphor Bureaucratizing the Mind," *Science as Culture*, 1989, 7, pp. 7–42.

14. PSYCHON Year Book, Gravenhage 1986.

15. Van-Besien, Fred, "Metaphors in Scientific Language," *Communication and Cognition*, 1989, 22, 1, pp. 5–22.

16. Young, Robert M., "Association of Ideas," *Dictionary of the History of Ideas*, Wiener, Philip P. (ed.) (Charles Scribner's Sons, New York, 1973), Vol. 1, pp. 111–118.

17. Ibid., p. 113.

18. Ibid., p. 113.

19. Ibid., p. 114.

20. Ibid., p. 114.

21. Ibid., p. 115.

22. Ibid., pp. 116–117.

23. Judge, Anthony N., "Metaphors as Transdisplinary Vehicles of the Future," paper, *Conference on Science and Tradition: Transdisciplinary Perspectives on the Way to the 21st Century*, Union des Ingenieurs et des Techniciens utilisant la Langue Française, with UNESCO, Paris, December 1992.

24. Ibid.

25. Buttiner, Anne, "Musing on Helicon; root metaphors and geography," *Geoqrafiska Annaler* 64B 1982, pp. 89–96.

26. Nudler, Oscar, quoted in Judge 1992.

27. Klein, Julie Thompson, *Interdisciplinarity: History, Theory and Practice* (Wayne State University, Detroit), p. 93.

28. Ibid.

29. Goonatilake, Susantha, *Aborted Discovery: Science and Creativity in the Third World* (London, Zed, 1984), pp. 157–168.

30. Judge 1992.

31. Glaser, Barney G. and Strauss, Anselm S., *The Discovery of Grounded Theory: Strategies for Qualitative Research* (Aldine Publishing Co., Chicago, 1967).

32. Bag, A.K., "Mathematical and Astronomical Heritage of India," *Occasional Paper 14*, Project of History of Indian Science, Philosophy, and Culture, New Delhi, 1993, p. 2.

33. Goonatilake, Susantha, *Crippled Minds: An Exploration into Colonial Culture* (New Delhi, Vikas, 1982).

34. Kanigel, Robert, *The Man Who Knew Infinity: A Life of the Genius Ramanujan* (New York, Charles Scribner's Sons, 1991).

35. Nandy, Ashis, *Alternative Sciences* (New Delhi, Allied Publishers Pvt. Ltd., 1979).

36. Datta, Bibhutibhusan and Singh, Awadhesh Narayan, "Use of permutations and combinations in India," *Indian Journal of History of Science*, 27, 3, 1992.

37. Dharmapal, Bharatiya Chitta Manas and Kala Centre for Policy Studies, Madras, 1993, p. 24.

38. Sarma, K.V., "Some Highlights of Astronomy and Mathematics in Medieval India," in Kuppuram G. and Kumudamani K., *History of Science and Technology in India* (Delhi, Sundeep Prakashan, 1990), Vol, 2, p. 418.

39. Ramasubramanian, K.; Sirinivas, M.D.; and Sriram, M.S., "Heliocentric model of planetary motion in the Kerala school of Indian astronomy," paper, Congress on Traditional Sciences and Technologies of India, November 28–December 3, 1993, IIT Bombay.

40. Pingree, D. Jyothisastra, *Astral and Mathematical Literature* (Wiesbaden, Suhrkamp, 1981), p. 118.

41. Sirinivas, M.D., "The Methodology of Indian Mathematics and Its Contemporary Relevance," *PPST Bulletin*, Serial No. 12, September 1987, p. 1.

42. Dharmapal, pp. 27–28.

43. Varela et al., 1993, p. 22.

44. Hess, David J., *Science and Technology in a Multicultural World: The Cultural Politics of Facts and Artifacts* (New York, Columbia University Press, 1995), New York.

Index

SUSANTHA GOONATILAKE, Fellow of the World Academy of Sciences, is trained in both engineering and sociology. He is Senior Consultant on Science and Technology for the United Nations and a member of the faculty of the New School for Social Research, New York, and the Vidyartha Center for Science and Society. He has taught at universities and research institutes in Asia, Europe, and America. His numerous publications include *Aborted Discovery: Science and Creativity in the Third World, Crippled Minds: An Exploration into Colonial Culture,* and *The Evolution of Information: Lineages in Gene, Culture and Artefact.*

DATE DUE
